*Experimental Methods in the Physical Sciences*

VOLUME 29B

ATOMIC, MOLECULAR, AND OPTICAL PHYSICS:
ATOMS AND MOLECULES

# EXPERIMENTAL METHODS IN THE PHYSICAL SCIENCES

Robert Celotta and Thomas Lucatorto, *Editors in Chief*

*Founding Editors*

L. MARTON
C. MARTON

Volume 29B

# Atomic, Molecular, and Optical Physics: Atoms and Molecules

*Edited by*

F. B. Dunning
*Department of Physics*
*Rice University, Houston, Texas*

*and*

Randall G. Hulet
*Department of Physics*
*Rice University, Houston, Texas*

## ACADEMIC PRESS

San Diego   New York   Boston   London   Sydney   Tokyo   Toronto

This book is printed on acid-free paper. ∞

Copyright © 1996 by ACADEMIC PRESS, INC.

Academic Press, Inc.
A Division of Harcourt Brace & Company
525 B Street, Suite 1900, San Diego, California 92101-4495

*United Kingdom Edition published by*
Academic Press Limited
24-28 Oval Road, London NW1 7DX

International Standard Serial Number: 1079-4042

International Standard Book Number: 0-12-475976-9

PRINTED IN THE UNITED STATES OF AMERICA
96  97  98  99  00  01  BC  9  8  7  6  5  4  3  2  1

# CONTENTS

## 12. Excited Level Lifetime Measurements
J. E. LAWLER and T. R. O'BRIAN

## 13. Doppler-Free Spectroscopy
JAMES C. BERGQUIST

# CONTRIBUTORS

Numbers in parentheses indicate the pages on which the authors' contributions begin.

ANNE M. ANDREWS (273), *National Institute of Standards and Technology, Molecular Physics Division, Gaithersburg, Maryland 20899*

JAMES C. BERGQUIST (255), *National Institute of Standards and Technology, Boulder, Colorado 80303*

J. BLAND-HAWTHORN (363),[1] *Department of Space Physics and Astronomy, Rice University, Houston, Texas 77251*

CURTIS C. BRADLEY (129), *Department of Physics and Rice Quantum Institute, Rice University, Houston, Texas 77251*

J. M. BROWN (85), *Department of Physical Chemistry, Oxford OX1 3QZ, England*

OLIVIER H. CARNAL (341), *Holtronic Technologies S.A., Marin, Switzerland*

G. CECIL (363), *Department of Physics and Astronomy, University of North Carolina, Chapel Hill, North Carolina 27599*

A. CHUTJIAN (49), *Jet Propulsion Laboratory, California Institute of Technology, Pasadena, California 91109*

K. M. EVENSON (85), *National Institute of Standards and Technology, Boulder, Colorado 80303*

T. F. GALLAGHER (115, 325), *Department of Physics, University of Virginia, Charlottesville, Virginia 22503*

TIMOTHY J. GAY (95), *Behlen Laboratory of Physics, University of Nebraska, Lincoln, Nebraska 68588*

H. HOTOP (191), *Fachbereich Physik, Universität Kaiserslautern, D-67653 Kaiserslautern, Germany*

RANDALL G. HULET (129), *Department of Physics and Rice Quantum Institute, Rice University, Houston, Texas 77251*

G. SAMUEL HURST (171), *Institute of Resonance Ionization Spectroscopy, The University of Tennessee, Knoxville, Tennessee 37932*

[1]Current address: Anglo-Australian Observatory, P.O. Box 296, Epping, NSW 2121.

CARTER KITTRELL (393), *Department of Chemistry and Rice Quantum Institute Rice University, Houston, Texas 77251*

J. E. LAWLER (217), *Department of Physics, University of Wisconsin—Madison, Madison, Wisconsin 53706*

JABEZ J. MCCLELLAND (145), *Electron Physics Group, National Institute of Standards and Technology, Gaithersburg, Maryland 20899*

JÜRGEN MLYNEK (341), *Fakultät für Physik, Universität Konstanz, D-7750 Konstanz, Germany*

MICHAEL D. MORSE (21), *Department of Chemistry, University of Utah, Salt Lake City, Utah 84112*

T. R. O'BRIAN (217), *Radiometric Physics Division, National Institute of Standards and Technology, Gaithersburg, Maryland 20899*

O. J. ORIENT (49), *Jet Propulsion Laboratory, California Institute of Technology, Pasadena, California 91109*

JAMES E. PARKS (171), *Institute of Resonance Ionization Spectroscopy, The University of Tennessee, Knoxville, Tennessee 37932*

NORMAN F. RAMSEY (1), *Lyman Physics Laboratory, Harvard University, Cambridge, Massachusetts 02138*

R. D. SUENRAM (273), *National Institute of Standards and Technology, Molecular Physics Division, Gaithersburg, Maryland 20899*

C. R. VIDAL (67), *Max Planck Institute für Extraterrestriche Physik, D-8046 Garching bei München, Germany*

LINDA YOUNG (301), *Argonne National Laboratory, Argonne, Illinois 60439, and Joint Institute for Laboratory Astrophysics, University of Colorado, Boulder, Colorado 80309*

# PREFACE

Since the publication in 1967 of "Atomic Sources and Detectors," Volumes 4A and 4B of this series, the field of atomic, molecular, and optical physics has seen exciting and explosive growth. Much of this expansion has been tied to the development of new sources, such as the laser, which have revolutionized many aspects of science, technology, and everyday life. This growth can be seen in the dramatic difference in content between the present volumes and the 1967 volumes. Not all techniques have changed however, and for those such as conventional electron sources, the earlier volumes still provide a useful resource to the research community. By carefully selecting the topics for the present volumes, Barry Dunning and Randy Hulet have provided us with a coherent description of the methods by which atomic, molecular, and optical physics is practiced today. We congratulate them on the completion of an important contribution to the scientific literature.

Beginning with Volume 29A, the series is known as *Experimental Methods in the Physical Sciences* instead of *Methods of Experimental Physics*. The change recognizes the increasing multidisciplinary nature of science and technology. It permits us, for example, to extend the series into interesting areas of applied physics and technology. In that case, we hope such a volume can serve as an important resource to someone embarking on a program of applied research by clearly outlining the experimental methodology employed. We expect that such a volume would appear to researchers in industry, as well as scientists who have traditionally pursued more academic problems but wish to extend their research program into an applied area. We welcome the challenge of providing an important and useful series of volumes for all of those involved in today's broad research spectrum.

Robert J. Celotta<br>
Thomas B. Lucatorto

# VOLUMES IN SERIES

# EXPERIMENTAL METHODS IN THE PHYSICAL SCIENCES

(formerly Methods of Experimental Physics)

*Editors-in-Chief*

*Robert Celotta and Thomas Lucatorto*

# 1. THERMAL BEAM SOURCES

## Norman F. Ramsey

Lyman Physics Laboratory, Harvard University
Cambridge, Massachusetts

## 1.1 Introduction

The different kinds of sources for beams of neutral atoms and molecules include thermal sources for slow beams, jet sources for supersonic beams, and fast beam sources formed by neutralization of ion beams. Since the last two are discussed in later chapters, this chapter concentrates on the many varieties of thermal sources. In discussions of thermal sources, the words atomic and molecular are often used interchangeably because the basic design principles are similar for atomic and molecular sources.

The earliest atomic beam sources were simple containers with narrow apertures, which were either rectangular or circular. Gases that are noncondensable at ordinary temperatures pass at the correct source pressure to the source through tubes. Typically the source pressures are a few torr (millimeters of Hg or 133.32 Pa) for apertures whose smallest linear dimension are a few hundredths of a millimeter. Such sources are often cryogenically cooled to obtain slower molecules or lower rotational states. For the study of atoms, such as hydrogen which normally occur in polyatomic form, it is often necessary to have an electric discharge in the source to dissociate the molecules [1]. If elevated temperatures are required to obtain the desired vapor pressures, the sources are usually heated ovens with suitable exit apertures.

It is often necessary to diminish the emission of unused source material to conserve the material, to provide long uninterrupted operating times, or to diminish pumping or contamination problems in the source region. The on-axis flux with atomic beam thermal sources depends primarily on the source vapor pressure. To decrease the loss of source material from wrongly directed source molecules, different techniques are used. These are called dark wall, bright wall, or recirculating, depending on whether the beam intercepted by the wall is not reemitted, is reemitted, or is recirculated [2–6].

All thermal sources depend on principles of kinetic theory summarized in the next section.

1

EXPERIMENTAL METHODS IN THE PHYSICAL SCIENCES
Vol. 29B

## 1.2 Theoretical Principles

### 1.2.1 Effusion from Thin-Walled Apertures

The sources of molecules in many molecular-beam experiments consist of small chambers which contain the molecules in a gas or vapor at a few torr pressure and which have small circular apertures or narrow slits about 0.02 mm wide and 1 cm high. Except where otherwise indicated, the equations in this section apply to apertures of any shape provided the width, $w$, is taken as the smallest cross-sectional dimension of the aperture. The width of the slit and the pressure are such that there is molecular effusion [1, 7] as contrasted to hydrodynamic flow. Under such conditions, the number, $dQ$, of molecules which will emerge per second from the source slit traveling in solid angle $d\omega$ at angle $\theta$ relative to a normal to the plane containing the slit jaws is, by elementary kinetic theory arguments [1, 7],

$$dQ = (d\omega/4\pi)n\bar{v}\cos\theta A_s, \tag{1.1}$$

where $n$ is the number of molecules per unit volume, $\bar{v}$ is the mean molecular velocity inside the source, and $A_s$ is the area of the source slit. For an ideal gas of pressure $p$ and absolute temperature $T$,

$$p = nkT. \tag{1.2}$$

The total number $Q$ of molecules that should emerge from the source in all directions can be found by integrating Eq. (1.1) over the $2\pi$ solid angle of the forward direction. In this case, and with $d\omega$ taken as $2\pi\sin\theta d\theta$ so that the integration goes from $\theta$ equals 0 to $\frac{1}{2}\pi$, one immediately obtains from the integration that

$$Q = \tfrac{1}{4}n\bar{v}A_s. \tag{1.3}$$

Two assumptions are inherent in Eqs. (1.1) and (1.3). One of these is that every molecule which strikes the aperture passes through it and does not have its direction changed. This assumption is valid only if the thickness of the slit jaws is as discussed later in this section. The other assumption is that the spatial and velocity distributions of the molecules inside the source are not affected by the effusion of the molecules. The strict requirement for this condition is that

$$w \ll \lambda_{Ms}, \tag{1.4}$$

where $w$ is the slit width and $\lambda_{Ms}$ the mean free collision path inside the source. By the usual kinetic theory demonstration,

$$\lambda_{Ms} = \frac{1}{n\sigma\sqrt{2}}, \tag{1.5}$$

where $\sigma$ is the molecular collision cross section. For air at room temperature, $\lambda_{Ms} = 300$ meters at $10^{-6}$ torr. For the small collision angles that are often significant with well-defined molecular beams, the effective mean free paths are usually considerably smaller than the above.

In actual practice it is found that for most purposes, effusive sources are effective when

$$w \sim \lambda_{Ms}, \qquad (1.6)$$

If Eq. (1.4) or (1.6) is not satisfied, a partial hydrodynamic flow results with the creation of a jet instead of free molecular flow. This may create turbulence and a widened beam, or it may provide a cooled jet beam with the advantages discussed later in the chapter on supersonic beams. It is significant that the restrictions in Eqs. (1.4) or (1.6) depend only on the width and not on the height of the slit or with circular apertures on the radius of the aperture. Consequently, approximately the same source pressure can be used with a slit whose width equals the radius of a circular aperture. On the other hand, if the slit is a high one, it will have a greater area and produce a correspondingly greater beam intensity.

## 1.2.2 Effusion from Long Channels

The cosine law of molecular effusion that is implied by Eq. (1.1) was established by the pioneer work of Knudsen [1]. However, if the thin-walled aperture assumed above is replaced by a canal-like aperture of appreciable length, the molecules which enter the canal at a considerable angle will strike the canal wall and have a smaller chance of escaping. In this way the angular distribution of the emergent beam is changed considerably [1].

One consequence of the change is that the total amount of gas which emerges from the source is diminished, but that which emerges in the direction of the canal is undiminished, provided the pressure is sufficiently low for collisions inside the canal to be negligible—that is, provided $\lambda_{Ms} \geq \ell$, where $\ell$ is the canal length. This improvement of beam intensity per amount of source material consumed is of great value in many molecular beam experiments such as those with radioactive isotopes. The effectiveness of the canal in reducing the total number of emerging molecules is expressed in terms of a factor $1/\kappa$, which is such that Eq. (1.3) with a canal-like aperture is replaced by

$$Q = (1/\kappa)\tfrac{1}{4}n\bar{v}A_s. \qquad (1.7)$$

Claussing and others [1] have calculated the values of $1/\kappa$ for apertures of a number of different shapes making the bright wall assumption mentioned in Section 1.1 Their results for slits of various shapes are as follows, if $w$ is the aperture width, $h$ the height, $r$ the radius, and $\ell$ the canal length:

(a) Any shape aperture of very short length or with $\ell = 0$:

$$1/\kappa = 1. \tag{1.8}$$

(b) Long circular cylindrical tube with $\ell \gg r$:

$$1/\kappa = \frac{8}{3}\frac{r}{\ell}. \tag{1.9}$$

(c) Long rectangular slit with $h \gg \ell$, $\ell \gg w$:

$$1/\kappa = \frac{w}{\ell}\ln\frac{\ell}{w}. \tag{1.10}$$

(d) If the aperture is of a shape different from any of the above, a poorer approximation to $1/\kappa$ can often be obtained for large $\ell$ from the expression of Knudsen [1]:

$$1/\kappa = (16/3) \Big/ \left(A_s \int_0^\ell \frac{o}{A^2}\,d\ell\right), \tag{1.11}$$

where $o$ is the periphery and $A$ the area of a normal cross section at position $\ell$ along the length of the canal. $A_s$ is present just to cancel that in Eq. (1.7).
(e) If the cross section of the canal remains unaltered along its length, Eq. (1.11) reduces to

$$1/\kappa = (16/3)A/\ell o. \tag{1.12}$$

It should be noted that Eq. (1.9) is a special case of Eq. (1.12).

Although the preceding calculations apply to single channels, they can also be applied to a large number of parallel channels, provided that they do not significantly interfere with each other. Successful sources have been made from arrays of small-diameter tubes and from multichanneled glass. All of the foregoing channel results are based on the bright wall assumption. With the same geometry, but with dark walls or with recirculating ovens, the effective $\kappa$ factor can be much larger.

## 1.2.3 Molecular Beam Intensities

If the pressure in the molecular-beam apparatus is sufficiently low that no appreciable amount of the beam is scattered out, and if no collimating slit or similar obstruction intercepts the beam on its way to the detector, the theoretical beam intensity can readily be calculated from the expressions of the preceding section.

Let $A_d$ be the area of the detector, $\ell_0$ be the length of the apparatus from source to detector, and $I$ be the beam intensity or number of molecules which strike the detector per second. Then from Eq. (1.1),

$$I = \frac{1}{4\pi} \frac{A_d}{\ell_0^2} n\bar{v}A_s.$$

$$(1.13)$$

If $p'$ is the source pressure in torr, $M$ the molecular weight (not the weight of the molecule), $T$ the absolute temperature in kelvins $A_s$ and $A_d$ in square centimeters, and $\ell_0$ in centimeters, the preceding equation can be re-expressed as

$$I = 1.118 \times 10^{22} \frac{p'A_sA_d}{\ell_0^2\sqrt{(MT)}} \text{ molecules sec}^{-1}.$$

$$(1.14)$$

The foregoing expression for the beam intensity assumes that there is no attenuation of the beam by scattering. If there is such attenuation in any chamber through which the beam passes, such as the separating chamber, the resultant beam intensity can be obtained by multiplying Eq. (1.14) by the attenuation factors appropriate to each such chamber. If $\lambda_{kv}$ is the mean free path for a beam molecule of velocity $v$ in passing through the $k$th chamber whose length is $\ell_k$, the attenuation factor for that chamber and that molecular velocity is $\exp(-\ell_k/\lambda_{kv})$. With the very long narrow beams used in molecular-beam experiments, collisions which deflect a molecule only slightly are sufficient to eliminate the molecule from the beam. As a result, the appropriate value for $\lambda_{kv}$ in molecular-beam experiments is often less than the mean free path appropriate to other, less sensitive experiments.

## 1.2.4 Molecular Beam Shapes

If a collimator slit is in position, the apparent beam intensity measured at a specific location is no longer given by Eq. (1.14), but rather depends on what portion of the beam is intercepted by the collimator. In general, the measured beam intensity depends on the widths of the source, collimator, and detector slits, and on their relative positions. The beam shape can most easily be calculated by first considering the dependence of the beam intensity on detector position in the limit when the detector is infinitely narrow, i.e., the beam intensity that reaches the detector but is not integrated over a detector slit of appreciable width. In an atomic beam, the transverse dimensions are small compared to the beam length, so the angles can be treated as small in this calculation. The beam shape can easily be determined with the aid of Figure 1 to be that given in Figure 2,

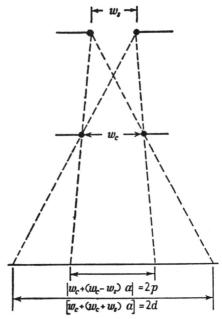

FIG. 1. Relation of source and collimator widths to beam shape. In the penumbra region the intensity varies linearly with displacement, because the amount of exposed source varies in this way. Consequently, the beam is of trapezoidal shape [1].

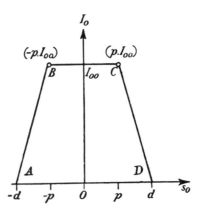

FIG. 2. Beam shape with detector of negligible width. The beam intensity $I_0$ is plotted as a function of the transverse displacement $s_0$ from the center of the beam [1].

where

$$p = \tfrac{1}{2}|w_c + (w_c - w_s)a|, \tag{1.15}$$

$$d = \tfrac{1}{2}[w_c + (w_c + w_s)a], \tag{1.16}$$

$$a \equiv \ell_{cd}\ell_{sc} = \ell_{sd}/\ell_{sc} - 1 = r - 1, \tag{1.17}$$

$$r = \ell_{sd}/\ell_{sc}, \tag{1.18}$$

and where $\ell_{cd}$ is the distance from the collimator to detector, $\ell_{sc}$ the distance from source to collimator, $\ell_{sd}$ the distance from source to detector, $w_s$ the width of the source slit, $w_c$ the width of the collimator slit, $p$ the half-width of the top of the trapezoidal beam shape, and $d$ the half-width of the bottom of the trapezoidal beam shape. The absolute value signs are included to make the result valid in the case when the collimator slit is so narrow that there is no position at the detector location for which the source is completely unobscured. An important special case is that for which $w_s = w_c = w$ and $a = 1$; the preceding equations then give

$$p = \tfrac{1}{2}w; \qquad d = \tfrac{3}{2}w. \tag{1.19}$$

The trapezoidal character of the beam is apparent from the fact that full intensity is received at a detector position for which the source is completely unobscured. On the other hand, if the source is partially obscured, the amount of obscuration varies linearly with position at the detector. From this linear characteristic, it is apparent that the beam intensity per unit detector width $I_0(s_0)$ is given as a function of the observation position $s_0$ at the detector position by the following:

(a) In region AB, where $-d < s_0 < -p$,

$$I_0(s_0) = I_{00} \frac{d + s_0}{d - p}. \tag{1.20}$$

(b) In region BC, where $-p < s_0 < -p$,

$$I_0(s_0) = I_{00}. \tag{1.21}$$

(c) In region CD, where $p < s_0 < d$,

$$I_0(s_0) = I_{00} \frac{d - s_0}{d - p}. \tag{1.22}$$

If the collimator and source widths are such that in the region BC the source is completely unobscured, $I_{00}$ is the beam intensity per unit detector width that would be observed in the absence of a collimator, as in Eq. (1.13). If, on the other hand, the source is partially obscured in this region, $I_{00}$ is reduced in proportion to the amount of obscuration.

If the beam is measured with a detector of appreciable width, for any specific location of the detector the observed total beam intensity is the integral of all of Figure 2 that is included within the detector slit width. Consequently, the beam intensity observed as a detector is moved across the beam is similar to Figure 2, but the sharp edges are rounded off by the effect of the detector slit width.

### 1.2.5 Velocity Distribution in a Volume of Gas

In a volume of gas, such as inside the source, the velocity distribution of the molecules is in accordance with the Maxwell distribution law [1]. If $dN$ is the number of molecules in the velocity interval $v$ to $v + dv$ and $N$ is the total number of molecules, this law gives

$$\frac{dN}{dv} = Nf(v) = \frac{4N}{\sqrt{\pi}} \frac{1}{\alpha^3} v^2 \exp(-v^2/\alpha^2), \tag{1.23}$$

where

$$\alpha = \sqrt{(2kT/m)}, \tag{1.24}$$

and where $m$ is the mass of the molecule. From (1.24) it follows immediately that $d^2N/dv^2 = 0$ when $v = \alpha$. Therefore, $\alpha$ may be given the simple physical interpretation of the most probably molecular velocity inside the source.

### 1.2.6 Velocity Distribution in Molecular Beams

On superficial consideration, one might expect that Eq. (1.23) would also be applicable to molecules in the molecular beam. However, such is not the case, as was first observed experimentally by O. Stern and later justified theoretically by A. Einstein. In particular, the probability of a molecule emerging from the source slit is proportional to the molecular velocity, as is apparent either physically or mathematically from the explicit velocity dependence in Eq. (1.1). The velocity dependence in the beam is then proportional to Eq. (1.23) multiplied by $v$. The constant of proportionality can then be obtained by normalizing to the full beam intensity $I_0$. Then, if $I(v)dv$ is the beam intensity from molecules between velocities $v$ and $v + dv$,

$$I(v) = \frac{2I_0}{\alpha^4} v^3 \exp(-v^2/\alpha^2). \tag{1.25}$$

By a simple integration (by parts and with the new variable $y = v^2/\alpha^2$), one can confirm that the integral of $I(v)$ over all velocities is $I_0$.

### 1.2.7 Characteristic Velocities

With the different velocity distributions for molecules in a volume and for those in a beam as implied by Eqs. (1.23) and (1.25), the various kinds of most probable and average velocities differ in two cases. Thus, from Eq. (1.25) the most probable velocity in the beam can be found, that is, the velocity for which $dI(v)/dv = 0$. The result is that the most probable velocity in the beam, $v_{pB}$, is $\sqrt{3/2}\alpha$, instead of $\alpha$ as found for molecules in the source. Other characteristic velocities can similarly be found. Let $v_p$ indicate a most probable velocity, $\bar{v}$ an average velocity, $V$ a root mean square velocity, and the subscripts V and B velocities inside a volume or in a beam. Then the following relations apply:

(a) Inside a volume of gas:

Most probable velocity:

$$v_{pV} = \alpha = \sqrt{(2kT/m)}. \tag{1.26}$$

Average or mean velocity:

$$\bar{v}_V = \int_0^\infty vf(v)dv = (2/\sqrt{\pi})\alpha = 1.13\alpha. \tag{1.27}$$

Root mean square velocity:

$$V_V = \left[\int_0^\infty v^2f(v)dv\right]^{1/2} = \sqrt{(3/2)}\alpha = 1.22\alpha. \tag{1.28}$$

(b) In a molecular beam:

Most probable velocity:

$$v_{pB} = \sqrt{(3/2)}\alpha = 1.22\alpha. \tag{1.29}$$

Average velocity:

$$\bar{v}_B = (3/4)\sqrt{\pi}\alpha = 1.33\alpha. \tag{1.30}$$

Root mean square velocity:

$$V_B = \sqrt{2}\alpha = 1.42\alpha. \tag{1.31}$$

Median velocity:

$$v_{mdB} = 1.30\alpha. \tag{1.32}$$

The preceding characteristic velocities can be related to the temperature of the molecules with the aid of Eq. (1.24).

## 1.3 Sources

### 1.3.1 Sources for Noncondensable Gases

For noncondensable gases such as $H_2$, the source problem is so simple as to be almost trivial. A tube carrying gas to a small volume with an adjustable slit opening for the beam to emerge is all that is required. Ordinarily the source is attached to a liquid nitrogen trap so that it can be cooled. Because of the thermal contractions on cooling, the source should be made of a single material (stainless steel is often used) and should not be soft-soldered. The trap walls should be kept thin to diminish liquid air evaporation by thermal conduction along the walls. A typical noncondensable gas source is shown at the bottom of the liquid nitrogen trap in Figure 3. It is often desirable to provide some directivity to the beam to diminish the total gas flow. This can be done with bright wall channels, as discussed later in greater detail.

With $H_2$ an increase in precision could presumably be obtained if a liquid $H_2$ reservoir were used to cool the source because of the lower velocities of the emerging gas. However, so far no such source has been used. A more convenient alternative has been cooling of the source with a flow of cold helium from a liquid helium reservoir as described by Code and Ramsey [9]. A small heater in the liquid helium storage reservoir is controlled by the source temperature to adjust the flow of cold gaseous helium.

In some cases it is desirable to change from a noncondensable gas source to a heated-oven source without loss of vacuum. Figure 3 illustrates a mechanism which permits such changes [1].

One complication with noncondensable gases that is not present with low vapor pressure substances is the need for much higher pumping speeds in the source chamber. With low vapor pressure substances, the source pump is effectively supplemented by the condensation of the source material. In typical experiments with molecular hydrogen [1], a source-chamber diffusion pump with a rated speed of 1420 liters per second was used.

There is an optimal distance between the source slit and the first fore slit of the separating chamber. If this distance is too great, excessive amounts of scattering will occur in the source chamber, and if the distance is too small, the separating chamber pressure will be too great. In reference [1] it is shown that the optimum separation should be such that, if the subscript 1 indicates the source chamber and 2 the separating chamber, the quantity

$$(\ell_1/\lambda_1) - (\ell_2/\lambda_2)$$

is a minimum. The criteria become more complicated when one includes effects of molecular reflection on the separating slits and of appreciable slit thickness.

## 1.3.2 Heated Ovens

For substances whose vapor pressures are too low at room temperature, heated ovens must be used. Various designs of heated ovens have been used, depending on the material to be vaporized, the length of the run desired, etc.

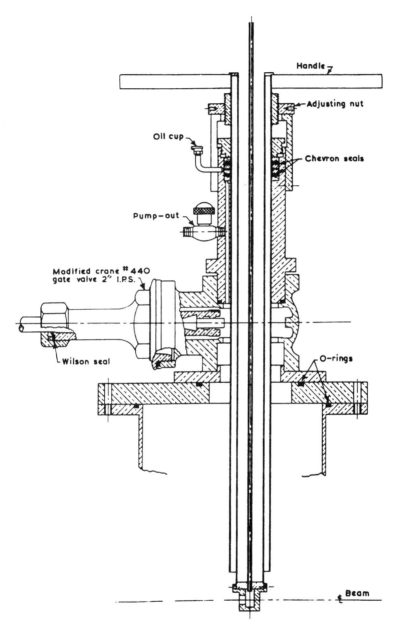

Fig. 3. Source for noncondensable gases at the bottom of a liquid nitrogen trap. The source mounting is designed so the noncondensable source can be removed and replaced by a heated oven without breaking the system vacuum [1]. The distance from the handle to the beam line is 0.5 m.

The material of choice for oven construction is determined by certain criteria: It must not melt at the required temperature, it must not react or alloy with the material being studied, and it must inhibit excessive creep to the slit jaws. For most molecular beam substances, iron, nickel, or monel metal is suitable for ovens, with iron being cheaper and more easily machined, while nickel is somewhat more permanent and more easily reuseable. A gold-plated iron oven with gold-plated slits often provides a more stable beam than a plain iron one. Melted hydroxides react vigorously with iron, but Millman and Kusch [1] have found that silver ovens are suitable with hydroxides. Indium and gallium alloy with iron, but indium beams have been produced effectively in a pure molybdenum oven at 1000°C [1]. A satisfactory beam of gallium cannot be produced in this way, but Renzetti [1] has shown that graphite ovens are suitable, as are graphite inserts into molybdenum ovens, with channels, wells, and slit jaws of graphite. A graphite insert has also been found to improve the stability of indium beams, although molybdenum slit jaws have been used in such cases because they are then less easily broken. An $Al_2O_3$ crucible inside a graphite oven, with the crucible and oven separated by a tantalum foil to prevent the reduction of the crucible, has been found by Lew [11] to be especially suitable for atomic aluminum. For praseodymium, Lew [12] has used a $ThO_2$ crucible inside a molybdenum oven, as this diminishes the chances of the slits being clogged with molten metal.

A typical oven design by Kusch and others is shown in Figure 4 [1, 13]. It should be noted that the heating elements are closer to the slit jaws than the loading well, so the slit is at a higher temperature than the rest of the oven, an effect that is enhanced by the cut under the slit. This temperature gradient

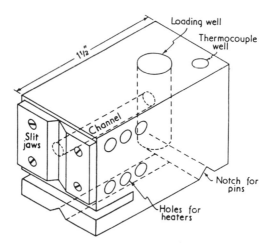

FIG. 4. Schematic view of a typical molecular beam oven [1].

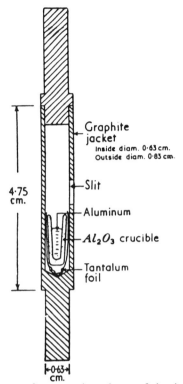

FIG. 5. An oven for generating a beam of aluminum atoms.

diminishes clogging of the slits. The hole above the well is ordinarily sealed with a tightly fitting conical plug. The slit jaws are usually held in place by four screws made of the same material as the oven and jaws. The slit jaws and the mating oven surface are lapped to prevent leakage. The channel to the slit is of a diameter slightly greater than the desired beam height. The oven is usually mounted on three tapered tungsten pins which fit into the V-shaped notches shown in Figure 5. The thermocouple for measuring the oven temperature, often chromel alumel, is inserted into the hole indicated. Sometimes the thermocouple is insulated from the oven with ceramic tubing. The heater inserted into each hole of the oven typically consists of 50 cm of 0.010 in. wolfram (tungsten) wire coiled with a hand drill about a mandril of 0.075 in. diameter and placed within a Stupakoff ceramic tube of about 3.5 mm outer diameter. Quartz tubing is sometimes used as an insulator, but it is more expensive and more fragile. The number of heaters varies from two to 20, depending on the desired temperature. Ordinarily two adjacent coils are wound from a single piece of tungsten wire. Connections between the coils are often made mechanically with nickel connectors with set screws. Alternatively, and with greater reliability, the inter-connections are made by spot-welding the tungsten wires to intermediate nickel wires (tungsten does not weld to tungsten as effectively as to nickel). Tantalum heater wires are now usually used, because they are less likely to break and can be used repeatedly. Alternating current, controlled with a variac and stabilized with a saturable reactor in the primary, is ordinarily used for heating. A one-to-one insulating transformer is often used to diminish the seriousness of a short of the heater wires to ground.

A quite different oven design was used by Lew [11] for atomic Al, as shown in Figure 5. An aluminum oxide crucible is placed in a thin-walled graphite jacket. Tantalum foil is wrapped around the crucible to prevent its contact with the graphite. The aluminum wire is placed in the crucible, and heating is by passing alternating current directly through the graphite jacket. With a water-cooled radiation shield surrounding the oven, about 800 watts of power is necessary to raise the temperature in the vicinity of the slit to 1670 K. Above 1700 K, the aluminum oxide begins to decompose. A charge of 0.2 gram lasts about six hours. Sometimes these ovens are heated by electron bombardment.

When atomic beams are deflected by nuclear magnetic moments, the deflections are so small that narrow apertures must be used and arrangements must be made to hold the aperture to a precise position. However, for many experiments, such as the electric deflection of polar molecules, much wider beams are used and the sources are simpler. For example, Hilborn et al. [14] use a simple source oven made from a stainless steel tube 20 cm long and 1 cm in diameter, with a wall thickness of 0.02 cm. The tube has a 0.045 cm hole which acts as the oven aperture. The tube is heated by passing a 50 amp AC current through it.

When alkali beams of Na, K, etc., are produced, molecules such as $Na_2$ are usually present in the beam at the same time to amounts of about 0.5%. This is an advantage when such molecules are to be studied, but can cause confusion when their presence is not recognized. The molecular fraction usually diminishes at higher oven temperatures. With some molecules, such as NaCl and NaF, polymerization occurs and dimers or higher polymers, such as $(NaCl)_2$, are in the beam [1].

Many molecular-beam source materials such as K, Rb, and Cs react with water and air with explosive violence, so care must be exercised in handling the materials and in opening used ovens that may not be completely empty. Some materials such as Na can be loaded safely by cleaning them under a liquid—for example, petroleum ether—and quickly inserting them in the source before the liquid has evaporated. Such a procedure, however, is not adequate for Rb and Cs, which are usually distilled in vacuum into a small glass capsule which is broken when the oven plug is inserted into an oven flushed with helium, nitrogen, or other inert gas. An alternative procedure is to generate the reactive vapor in the oven [11]. For example, if a mixture of Na or Ca metal is inserted in an oven along with CsCl and heated to about 200°C, a steady beam of Cs atoms emerges [11]. Also, a Na beam may be produced from a mixture of NaCl and fresh Ca chips.

Table I lists many of the substances from which satisfactory beams have been formed, along with the required oven temperatures [1]. In cases where the oven temperature is measured with a chromel–alumel thermocouple, the temperature is given in terms of the EMF with the cold junction at room temperature. An EMF of 1 mV corresponds to a temperature difference of about 25°C; exact calibrations are given in the handbooks. These temperatures are only approximate and vary with the experiments. Ovens should only slowly be brought to their final temperature.

### 1.3.3 Heated Filament Sources

As an alternative to an oven, the surface of a heated filament is sometimes used as the source for a molecular-beam experiment when the source material requires a very high temperature [1]. However, owing to the limited amount of material that can be stored, filaments are only rarely used.

TABLE I.    Oven Temperatures[a]

| Substance | Temperature (millivolts) | Substance | Temperature (millivolts) |
|---|---|---|---|
| Li | 26–29 | CsCl | 25 |
| Na | 14–16 | LiI | 22–25 |
| K | 10–11 | RbI | 26 |
| Rb | 9 | Li, Na, and K meta | 35–38 |
| Cs | 7–8 | and tetraborates | |
| In | 36–45 | NaFBeF2 | 33 |
| Ga | 40–45 | KFBeF$_2$ | 33 |
| LiF | 37 | NaClAlCl$_3$ | 13 |
| NaF | 35 | KClAlCl$_3$ | 13 |
| KCl | 35 | NaOH | 25–28 |
| KF | 32 | KOH | 25–28 |
| RbF | 28 | InCl$_3$ | 10 |
| CsF | 27 | Al | 1670 K |
| LiCl | 28 | Pe | 2000 K |

[a] This table lists the oven temperatures that are suitable for the formation of molecular beams with the listed substances [1]. Since the oven temperatures are ordinarily measured in millivolts of EMF with a chromel–alumel thermocouple whose cold junction is at room temperature, the oven temperature is listed in such units. One millivolt corresponds to a temperature difference of approximately 25°C; exact calibrations are given in handbooks. In a few cases temperatures are listed directly. Temperatures are brightness temperatures as determined with an optical pyrometer at 0.65 μm.

### 1.3.4 Laser Desorption Sources

When only a short pulse of thermal atoms is desired, a pulsed laser desorption source is used [6, 15]. For this, a pulsed heating laser bombards a target containing the sample material to temperatures up to 2500 K in times of the order of 10 ns. The energy of the desorption pulse may be controlled. Above some threshold, ions are produced, but below the threshold, neutral atoms are desorbed. The beam of desorbed atoms emerges from the target with essentially a Maxwell distribution and speeds typically of 300 m/s.

### 1.3.5 Sources for Dissociated Atoms

Often atoms such as H, D, or Cl need to be studied and must be dissociated from their ordinary form $H_2$, $D_2$ or $Cl_2$.

One means of dissociating $H_2$ molecules is the use of a heated tungsten oven; from such a source, some of the hydrogen emerges in atomic form [1]. Although this method was used in the Lamb–Retherford experiment [16], it is now seldom used.

In early atomic-beam experiments, the atoms were produced in a Wood's discharge tube [1], but foreign matter, such as pieces of metal, dust, wax, and sharp edges of glass, reduced the yield of atomic hydrogen. Even the electrodes of the discharge tube had to be kept a long distance from the region in which the atomic hydrogen was desired.

A better alternative source for atomic hydrogen, the halogens, and other atoms is an electrodeless radio-frequency discharge. Figure 6 shows a dissociation source [17] for an atomic hydrogen maser. The dissociation takes place in a

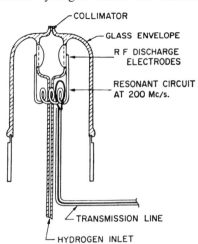

Fig. 6. Atomic hydrogen source [17].

spherical Pyrex bulb about 6 cm in diameter with a 0.5 mm source aperture and is generated by a 200 MHz 10 watt discharge. A high-intensity discharge increases the atom production, which is monitored by the intense deep red glow of the discharge. The discharge operating pressure is about 0.3 torr. The bulb is forced-air cooled, and the wall of the source chamber is water-cooled. Normally the tube must age for a few hours before the discharge acquires its characteristic red color. Molecular hydrogen is introduced to the bulb by a conventional gas handling system or through a heated palladium leak.

## 1.3.6 Sources to Conserve the Emitted Source Substance

It is often important to design the source to reduce the amount of unwanted material released. With noncondensable gases, this is often necessary to lower the speed requirements for the source pump. With heated ovens, it is often required to conserve source material, to extend the source life, and to diminish problems from the deposition of source materials in other parts of the apparatus. Long-lived sources are particularly important for atomic clocks, especially in satellites in the Global Positioning System (GPS).

The on-axis flux of atoms from the thermal sources depends primarily on the vapor pressure in the source. To decrease the loss of source material from wrongly directed molecules, various techniques are used. These techniques are often divided into three classes, referred to as (1) dark wall, (2) bright wall, and (3) recirculating, depending on whether the beam intercepted by a boundary wall is not reemitted, is reemitted, or is recirculated.

Dark wall collimators should involve simple geometric shadowing—that is, the collimators should just cut off the source emissions in useless directions. Although satisfactory for short running periods, they can introduce undesirable characteristics. For example, a dark wall oven for cesium [1] can use a carbon collimator to absorb every striking atom, but the carbon soon saturates and the cesium deposited in the walls either reevaporates, affecting the beam shape, or sticks, changing the size or shape of the collimator.

In bright wall sources, the directionality for the beam is provided by long channels through which the beam passes [1]. In some bright wall sources a single channel is used. The effectiveness of the channel is measured by the $\kappa$ factor of Eqs. (1.7)–(1.12). Typical dimensions for a bright wall $H_2$ source with a single aperture are 0.015 mm wide, 3 mm high, and 0.25 mm thick [1]. Zacharias and others [18] have successfully used stacks of hypodermic needles, channels of crimped foil, or channeled glass [17, 18] in the aperture to provide high directivity, as shown in Figure 7. Although the channel walls do not absorb the atoms striking them, they do diminish the number of wrongly directed atoms, since many of them bounce back to the source.

FIG. 7. Source with multiple channels.

Although bright wall sources avoid the saturation problems of the dark wall sources, they provide less effective collimation because of the atoms that emerge from the walls [2]. Neither the bright wall nor the dark wall sources utilize the wrongly directed atoms. This disadvantage can be overcome by recirculating ovens [3–5], using either the capillary action of a steel mesh [4] or porous wicking walls [5] as in a heat pipe, to return excess material to the source.

A simple wicking wall source invented by Drullinger [5] for a beam of circular cross-section is shown in Figure 8. The outer casing material is nonporous, but the inner substrate A consists of a porous wicking material, such as sintered tungsten. The substrate A is nearly saturated with the source material for the beam. A variety of source materials can be used, including cesium, alkali metals, and suitable organic compounds. Heating coils or resistive self-heating are used to maintain the oven temperature slightly above the melting point of the source material, with the source end B being slightly hotter than the rest of the oven. Capillary wetting action then develops a thin liquid layer of source material over the entire surface of the bore. The portion of the vapor which passes from the point where it evaporates to the output without striking the liquid layer will become part of the beam, but any of the material which strikes the liquid layer will condense and be drawn back into the substrate by capillary action. The same capillary action serves to distribute the source material throughout the substrate, so the saturated porous substrate also serves as a reservoir for the materials.

The substrate for such an oven may be composed of any suitable porous material, provided that the source material wets, but does not chemically react with, the substrate. Substrates have been formed of various sintered metals, including tungsten, molybdenum, and stainless steel. Porous nickel, copper, and alumina silicates should also be suitable. A water beam source has been constructed using cloth gauze as a substrate. The porosity of the substrate can also be varied within different parts of the oven to control potential sources of undesirable emissions. In forming the substrate, it is essential that the surface of the substrate remain porous. For example, boring a hole into a block of substrate

FIG. 8. Wicking recirculation source [5, 18].

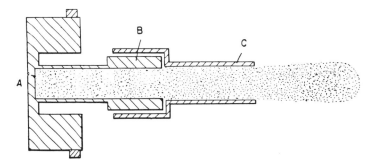

FIG. 9. Wicking recirculation source combined with bright wall collimator [18].

may close the pores, so the bore must be chemically etched to reopen them. Suitable prebored substrates are available commercially from Spectra-Mat, Inc., of Watsonville, California.

Drullinger [5] has successfully combined a wicking recirculating source of large storage capacity with a bright-wall collimator as in Figure 9. The oven substrate is enlarged at the source end A to provide a greater reservoir capacity. The wicking walls B are followed by a bright wall collimator C. The temperature gradients are such that region A is hotter than B, while the temperature of region C is intermediate. Such a source for a cesium beam atomic clock has been designed to operate about 10 years without the need for refilling with cesium.

# References

1. Ramsey, N. F., ed. (1956, 1985). *Molecular Beams*. Oxford University Press, Oxford, U.K.
2. Giordmaine, J. A., and Wang, T. C. (1960). *J. Appl. Phys.* **31,** 463.
3. Swenumson, R. D., and Even, U. (1981). *Rev. Sc. Instr.* **52,** 559.
4. Lambropoulos, M., and Moody, S. E. (1977). *Rev. Sc. Instr.* **48,** 131.
5. Drullinger, R. E., United States Patent Number 4,789,779 (Dec. 6, 1988).
6. Crawford, J. E. (1992). *Int. Conf. Ser.* **128,** 7 (edited by C. M. Miller and J. E. Parks). Inst. of Physics Publ., Bristol and Philadelphia.
7. Kennard, E. H. (1938). *Kinetic Theory of Gases*. McGraw-Hill Book Company, Inc., New York.
8. Zacharias, J. R. (1954). *Phys. Rev.* **94,** 751.
9. Code, R. F., and Ramsey, N. F. (1971). *Rev. Sci. Instr.* **42,** 896.
10. Kusch, P. (1950). *Phys. Rev.* **78,** 615.
11. Lew, H. (1949). *Phys. Rev.* **76,** 1086.
12. Lew, H. (1953). *Phys. Rev.* **91,** 619 and private communication.
13. Miller, R. C., and Kusch, P. (1955). *Phys. Rev.* **99,** 1314.
14. Hilborn, R. C., Gallagher, T. F., and Ramsey, N. F. (1972). *Journ. Chem. Phys.* **56,** 85.

15. Chu, S., Holberg, I., Bjorkholm, J., Cable, A., and Ashikin, A. (1985). *Phys. Rev. Lett.* **55**, 48.
16. Lamb, W. E., and Retherford, R. C. (1947). *Phys. Rev.* **72**, 241.
17. Kleppner, D., Berg, H. C., Crampton, S. B., Ramsey, N. F., Vessot, R. F. C., Peters, H. E., and Vanier, J. (1965). *Phys. Rev.* **138A,** 972.
18. Zacharias, J. R., Yates, J. G., and Haun, R. D. (1956). *Proc. Inst. Radio Eng.* **43**(3), 364. Fine collimators useful over a wide pressure range are supplied by Permionics, Inc., of Southbridge, Massachusetts.

# 2. SUPERSONIC BEAM SOURCES

## Michael D. Morse

Department of Chemistry, University of Utah
Salt Lake City, Utah

## 2.1 Introduction

Within the past 20 years, supersonic molecular beams have been put to ever-increasing use as a tool of atomic and molecular physics. In particular, supersonic expansions provide a means of preparing a molecular beam with a well-defined kinetic energy, which is particularly useful for crossed–beam and beam–surface scattering experiments. In addition, the nearly complete conversion of enthalpy to directed mass flow leads to the production of molecular beams with extremely low internal temperatures, allowing weakly bound van der Waals complexes to be prepared and spectroscopically characterized. The low internal temperatures achieved in supersonic expansions also lead to a dramatic reduction in spectral congestion and have permitted interpretable spectra to be recorded for many species which would otherwise be impossible to investigate. Clever techniques have allowed radicals, refractory species, and ions to be investigated as well. Most recently, the dramatic cooling achievable in a supersonic expansion has permitted the van der Waals molecule $^4He_2$ to be produced and detected [1], a remarkable achievement given that the ground state is predicted to be only bound by 0.0006 to 0.0011 $cm^{-1}$ (0.8 to 1.6 mK) [2].

Here we review the salient features of supersonic expansions and their use in chemical physics.

## 2.2 Effusive Beams

The expansion of a high-pressure gas from a reservoir into vacuum can occur via two physically distinct limiting cases, depending on the relationship between the mean free path of the gas molecules in the reservoir, $\lambda_0$, and the diameter (or width), $D$, of the expansion orifice (or slit). As these characteristic dimensions approach the limit $\lambda_0 \gg D$, the number of collisions suffered by a molecule as it leaves the reservoir approaches zero, and an effusive beam is generated with a velocity distribution of

$$P(v)dv = \frac{m^2}{2(RT_0)^2} v^3 \exp\left[ - \frac{mv^2}{2RT_0}\right] dv, \qquad (2.1)$$

21

EXPERIMENTAL METHODS IN THE PHYSICAL SCIENCES
Vol. 29B

where $m$ is the molar mass of the molecule, $R$ the gas constant, and $T_0$ the temperature of the reservoir [3]. This distribution is determined by $T_0$ and $m$, but is otherwise independent of details such as the pressure in the reservoir and the presence of other species effusing from the expansion orifice. Because no collisions occur during the expansion process, effusive beams provide a means of probing the conditions inside the reservoir at an elevated temperature and have been usefully combined with mass spectrometric methods to evaluate equilibrium constants and provide thermochemical data for gas-phase reactions at elevated temperatures. This is the basis for the field of Knudsen effusion mass spectrometry [4]. Effusive beams are discussed more thoroughly in the preceding chapter, and the reader interested in these topics is referred there for more details [3].

## 2.3 Supersonic Expansions from Circular Nozzles (Axisymmetric Jets)

### 2.3.1 The Idealized Continuum Model

In the limit of $D \gg \lambda_0$, molecules escaping from the reservoir suffer many collisions during the expansion process, which is then governed by gas dynamic equations similar to those used in the design of aircraft [5]. The theory in its most basic form was first worked out in 1951 [6], and experimental proof of the fundamental principles was provided shortly thereafter [7, 8]. At high source pressures, the effects of gas viscosity and heat transfer may be neglected, and the gas flow may be treated as an adiabatic, isentropic expansion. The adiabatic assumption leads to the conservation of the sum of the enthalpy and the kinetic energy of directed mass flow as the gas expands into vacuum [5], so that

$$H(x) + \tfrac{1}{2}mu(x)^2 = \text{constant}, \tag{2.2}$$

where $H(x)$ is the molar enthalpy of the gas at position $x$ from the point of expansion and $u(x)$ is the average flow velocity at this position. This equation limits the ultimate flow velocity, $u$, to the value

$$u_{\max} = \sqrt{\frac{2H(T_0)}{m}}, \tag{2.3}$$

where $H(T_0)$ is the molar enthalpy of the gas at the temperature of the source reservoir, $T_0$. For calorically perfect gases which have a constant-pressure molar heat capacity, $C_p$, that is independent of temperature, $H(T)$ is given as $C_p T$, and the ultimate flow velocity becomes

$$u_{\max} = \sqrt{\frac{2C_p T_0}{m}}. \tag{2.4}$$

For an ideal monatomic gas, $C_p = \frac{5}{2}R$, and Eq. (2.4) predicts terminal velocities of $1.77 \times 10^5$, $7.86 \times 10^4$, $5.59 \times 10^4$, $3.86 \times 10^4$, and $3.08 \times 10^4$ cm/s for He, Ne, Ar, Kr, and Xe, respectively, expanding from a reservoir at 300 K. In practice, velocities closely approaching these values are readily achieved, particularly for He, Ne, and Ar, which are much less subject to condensation and clustering than their heavier analogues. In general, the assumption of an isentropic expansion is reasonably well validated by measurements of local temperatures at varying distances downstream from the point of expansion [10].

For a calorically perfect ideal gas [defined by the relationships $dH = C_p dT$ (with $C_p$ independent of temperature) and the ideal gas law, $p = \rho kT$], adiabatic isentropic expansion from a reservoir at conditions $(p_0, T_0, \rho_0)$ to final conditions $(p_1, T_1, \rho_1)$ then gives

$$\frac{T_1}{T_0} = \left(\frac{p_1}{p_0}\right)^{(\gamma-1)/\gamma}; \quad \frac{\rho_1}{\rho_0} = \left(\frac{p_1}{p_0}\right)^{1/\gamma}; \quad \text{and} \quad \frac{\rho_1}{\rho_0} = \left(\frac{T_1}{T_0}\right)^{1/(\gamma-1)}, \quad (2.5)$$

where $\rho$ is the density of the gas and $\gamma$ is the heat capacity ratio $C_p/C_v$, equal to 1.6666 (5/3) for monatomic gases, 1.4 (7/5) for diatomic gases ignoring vibration, and 1.2857 (9/7) for diatomic gases at high temperatures where vibration contributes $Nk$ to the heat capacity. These equations imply that an expansion carried out with a large pressure ratio will lead to a large reduction in temperature.

The propagation of sound in a gas is also an example of an adiabatic, nearly isentropic process [5], and the speed of sound, $a$, in an ideal gas is given as

$$a(T) = \sqrt{\frac{\gamma RT}{m}}. \quad (2.6)$$

The Mach number, $M(x)$, is then defined as the ratio of the average flow velocity of the expanding gas at position $x$, $u(x)$, to the local speed of sound of the gas, $a(x)$, i.e., $M(x) = u(x)/a(x)$. The substantial cooling associated with supersonic expansion reduces the local speed of sound of the expanding gas tremendously, while the average flow velocity, $u(x)$, increases as random motion is converted into directed mass flow (but never exceeding $u_{max}$, as discussed earlier). The expansion is said to become supersonic when the ratio $u(x)/a(x)$ (or the Mach number, $M(x)$), increases beyond unity. Experimentally, it is not difficult to achieve Mach numbers of 20 or greater in a supersonic expansion.

Combining Eqs. (2.2), (2.5), and (2.6), it is possible to express the temperature, pressure, and density ratios in terms of the Mach number, $M(x)$, as

$$\frac{T(x)}{T_0} = W^{-1}; \quad \frac{p(x)}{p_0} = W^{-\gamma/(\gamma-1)}; \quad \frac{\rho(x)}{\rho_0} = W^{-1/(\gamma-1)}, \quad (2.7)$$

where

$$W \equiv 1 + \frac{\gamma - 1}{2} M(x)^2.$$

Finally, the functional dependence of the Mach number on downstream distance, $x$, has been calculated using the method of characteristics [11], and an accurate fitting formula has been developed [12], providing

$$M(x) = A \left( \frac{x - x_0}{D} \right)^{\gamma - 1} - \frac{1}{2} \frac{[(\gamma + 1)/(\gamma - 1)]}{A [(x - x_0)/D]^{\gamma - 1}}, \qquad (2.8)$$

where $D$ is the orifice diameter, and $x_0/D$ and $A$ depend on $\gamma$ as given in Table I. The translational temperature predicted from Eqs. (2.7) and (2.8) is displayed in Figure 1 for the expansion of an ideal gas, as a function of the distance from the point of expansion, measured in units of the nozzle diameter, $x/D$.

Using Eq. (2.8) in combination with Eq. (2.7), one obtains asymptotic expressions for the temperature, pressure, and density ratios at large distances $x/D$ as

$$\frac{T(x)}{T_0} = B \left( \frac{x - x_0}{D} \right)^{-2(\gamma - 1)};$$

$$\frac{p(x)}{p_0} = C \left( \frac{x - x_0}{D} \right)^{-2\gamma}; \qquad (2.9)$$

$$\frac{\rho(x)}{\rho_0} = F \left( \frac{x - x_0}{D} \right)^{-2},$$

where $B$, $C$, and $F$ are constants determined from $A$ and $\gamma$. Note that $\rho(x)/\rho_0$ displays the inverse-square law decrease expected in the limiting region where the molecules expand along straight but diverging streamlines.

TABLE I.   Values of Parameters Used in Eqs. (2.8) and (2.13)

| $\gamma$ | $x_0/D$ | $A$ | $G$ |
|---|---|---|---|
| 1.666 (5/3) | 0.075 | 3.26 | 2.03 |
| 1.40  (7/5) | 0.40 | 3.65 | 2.48 |
| 1.30 | 0.70 | 3.90 | 2.80 |
| 1.286 (9/7) | 0.85 | 3.96 | 2.85 |
| 1.20 | 1.00 | 4.29 | 3.38 |
| 1.10 | 1.60 | 5.25 | 4.75 |
| 1.05 | 1.80 | 6.44 | 7.04 |

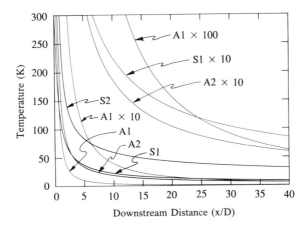

FIG. 1. Translational temperatures achieved in idealized continuum flow expansions. Results for axisymmetric jets with $\gamma = 1.666$ (5/3) (monatomic gases), labeled by A1, and for axisymmetric jets with $\gamma = 1.400$ (7/5) (diatomic gases with frozen vibrations), labeled by A2, are given as functions of the downstream distance, $x/D$. The corresponding results for slit nozzle expansions are labeled S1 and S2, respectively. For the A1, A2, and S1 examples, the function is plotted with the ordinate expanded by a factor of 10 or 100, indicated by A1 × 10, etc.

The distribution of molecular velocities, $P(v)$, can also be derived and is given as

$$P(v) \propto v^3 \exp\left[ -\frac{m(v - u(x))^2}{2RT(x)} \right], \qquad (2.10)$$

where $T(x)$ and $u(x)$ represent the local temperature and average flow velocity in the expansion, with $T(x)$ calculated from Eqs. (2.7) and (2.8). The average flow velocity is calculated through Eq. (2.2), which provides

$$u(x) = \sqrt{2[H(T_0) - H(T(x))]/m}, \qquad (2.11)$$

where $H$ is again the molar enthalpy as a function of temperature. Normalized velocity distributions are displayed in Figure 2 for a Maxwell–Boltzmann distribution at 300 K, along with the results of Eq. (2.1) for an effusive beam with source temperature 300 K and the results of Eq. (2.1) for a supersonically expanded beam reaching Mach numbers of 2, 5, 10, 20, and 30 from a 300 K source. In all cases the gas is assumed to be $^4$He; for other species the velocity distribution scales according to $m^{-1/2}$.

## 2.3.2 Departures from the Idealized Continuum Model

The idealized continuum model assumes that collisions occur with sufficient frequency for equilibrium to be maintained throughout the expansion process. At

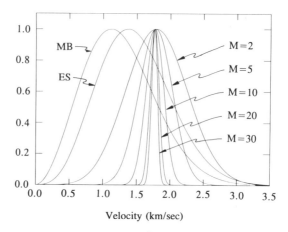

FIG. 2. Velocity distributions obtained for $^4$He expansions from a reservoir at 300 K, normalized to a peak value of unity. The Maxwell–Boltzmann distribution of atomic speeds in the reservoir is labeled MB, the distribution of velocities along the centerline of an effusive source is labeled ES, and the velocity distributions along the centerline of an expansion attaining Mach numbers of 2, 5, 10, 20, and 30 are indicated by the designations M = 2, M = 5, etc.

some point in the expansion, however, the collision frequency drops to such a low level that a particular degree of freedom may fall out of equilibrium. This is the beginning of the transition from continuum flow to free molecular flow, which ultimately leads to gas phase molecules isolated in a (nearly) collision-free environment. Generally, rotational relaxation occurs more readily than vibrational relaxation, and translational energy exchange occurs even more easily than rotational relaxation. For these reasons, it is typically the vibrational degrees of freedom which fall out of equilibrium first, followed by the rotational degrees of freedom. Finally, collisions become so infrequent that translational equilibration ceases, and the continuum model becomes invalid.

Knowing the distribution of velocities, $P(v)$, and the density, $\rho$, at various positions downstream, and making some simple assumptions concerning the effectiveness of collisional energy transfer, it can be shown that a real supersonic expansion will reach a terminal Mach number, $M_T$, beyond which no further cooling occurs [10]. $M_T$ has been shown to depend on a constant depending only on $\gamma$ (given by $G$, and listed in Table I), $\lambda_0/D$, and a collisional effectiveness parameter, $\varepsilon$, which denotes the maximum fractional change in the mean random velocity per collision, giving [10, 13]

$$M_T = G \left( \frac{\lambda_0}{D\varepsilon} \right)^{-(\gamma-1)/\gamma} . \qquad (2.12)$$

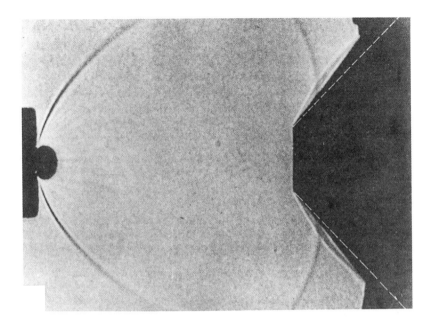

FIG. 3. Shadowgram of a freely expanding jet of $N_2$ emerging from a reservoir held at a pressure of 100 atm into a vacuum chamber maintained at 50 torr. The barrel shock is clearly visible, as is the terminal shock wave (Mach shock wave). A skimmer is used to transmit the central portion of the expansion into a second chamber, and an attached shock wave is visible on the outside of the skimmer. Reprinted, with permission, from reference [16].

Assuming a single component gas in the source reservoir, this may be written as

$$M_T = G(\sqrt{2}\sigma\rho_0 D\varepsilon)^{(\gamma-1)/\gamma}, \qquad (2.13)$$

where $\sigma$ is the collision cross section and $\rho_0$ is the number density of gas molecules in the source reservoir. Experiments using argon have demonstrated that $M_T$ is given by $M_T = 1.17\,(\sqrt{2}\sigma\rho_0 D)^{0.4}$, where the exponent 0.4 follows from Eq. (2.13) with $\gamma = 5/3$. $M_T$ is related to $\rho_0 D$ (or at fixed source temperature, to $p_0 D$), so that if a greater degree of cooling is desired, the product $p_0 D$ must be increased.

Kinetic models [14] and Monte Carlo simulations [15] show that when the collision frequency becomes sufficiently small, the velocity of a given molecule in the expansion stops changing, simply because it encounters no further collisions. Beyond this point the distribution of molecular velocity components along the expansion axis ceases to evolve. The distribution of molecular velocities perpendicular to the expansion axis continues to narrow, however, because molecules with large perpendicular components of velocity are driven off the axis by their own momentum. Thus, it becomes possible to define two translational temperatures, $T$ and $T_\perp$. The parallel temperature component is related to the distribution of molecular speeds along the axis and stops evolving when the terminal Mach number is reached. The perpendicular temperature component continues to drop, however, purely as a result of geometrical effects.

### 2.3.3 Interaction with Background Gases

In any real apparatus the idealized continuum model will also break down when the expanding gas becomes sufficiently rarefied that its density approaches that of the background gases. At this point collisions between the background gases and the expanding gas cause the molecular velocity distribution to be randomized, so that the expansion becomes irreversible, leading to a serious increase in entropy. If the background gas density is high, the expansion travels through a shock zone in which the molecules are rapidly decelerated; if the background gas density is low, a molecule in the jet experiences a series of individual scattering events, and the randomization of the velocity distribution occurs over an extended region of space. The thickness of the shock zone thus depends on the density of the background gas.

For free expansion into vacuum, two types of shock zones develop. A barrel shock forms around the centerline of the expansion, resembling a paraboloid of revolution centered on the jet axis, opening from the expansion orifice. This ends at a second shock zone, called the Mach disk, which forms a nearly flat terminal shock wave perpendicular to the centerline of the beam. For high background gas densities these structures are well defined and are easily observed through Schlieren photography, as is shown in Figure 3 [16]. Within the volume defined by these limiting shock zones, the description of the jet offered earlier is quite adequate. This volume may then be further subdivided into a region where there are sufficient collisions to maintain equilibrium so that the continuum model is valid, followed by a region farther downstream where nearly collisionless flow occurs. This latter region has been termed "the zone of silence," in recognition of the fact that the speed of sound in the expanding gas [as given in Eq. (2.6)] has dropped to nearly zero, essentially eliminating the possibility of acoustic waves in the expanding gas flow. All of this comes to a crashing halt, however, when the expanding molecules reach the Mach disk and their velocities are randomized by collisions.

The location of the Mach disk depends only on the ratio between the source and background pressures, $p_0$ and $p_B$ (and is independent of $\gamma$), although the thickness of the shock zone depends on the absolute magnitude of $p_B$ [17]. If the location of the Mach disk is denoted by $x_M$, its position is correctly given by the simple formula

$$x_M = 0.67 D \sqrt{\frac{p_0}{p_B}}, \tag{2.14}$$

for values of the pressure ratio $p_0/p_B$ in the range from 15 to 17,000 [12].

## 2.3.4  Seeded Beams

By adding a small mole fraction of a second species to the gas in the reservoir, one may create a seeded molecular beam. In the limit of high source density (high source pressure, $p_0$), the idealized continuum model of gas flow works well, and the behavior of the supersonic expansion is the same as that for a pure gas with molecular weight and heat capacity taken as the weighted average of the corresponding properties of the gases making up the mixture [9]. Under these high source pressure conditions, both component gases reach the same flow velocity and temperature at the same point in the expansion. For a gas mixture consisting of a low concentration of a heavy seed gas in a light carrier gas, the limiting flow velocity given by Eq. (2.3) is dominated by the more abundant, lighter gas, and it becomes possible to accelerate the seed gas to high kinetic energies. Furthermore, by virtue of the narrowing of the velocity distribution, these high kinetic energies are very well defined. As an example, by using a heated nozzle and a mixture of 0.1% xenon in hydrogen, xenon atoms have been accelerated to a kinetic energy of 30.2 ± 1.3 eV [18]. The ability to accelerate neutral species to such high, precisely defined energies has been quite useful in chemical dynamics investigations [19].

At low to intermediate source densities, the heavier gas molecules tend to lag behind the lighter gas molecules. In the expansion process, the heavier molecules are repetitively bombarded by the lighter gas molecules and are slowly accelerated to the full beam velocity. If the total source pressure is too low, however, less than complete acceleration is achieved, leading to "velocity slip" [9]. The phenomenon of velocity slip has been extensively investigated, both experimentally [20] and theoretically [21]. A simple model of seeded supersonic beams has been recently presented which provides useful expressions for terminal temperatures, velocity slip, enrichment factor, and beam intensity, and the validity of the model is documented by comparison to the properties of $I_2$ seeded in rare gas molecular beams [22].

A related effect observed in low- to intermediate-density seeded beams is a tendency of the heavy species to be concentrated on the jet axis. Multiple collisions with the lighter carrier gas during expansion tend to accelerate the component of velocity along the jet axis, but are less effective in modifying the perpendicular component of velocity. As a result, the initial trajectory of the heavy particle is bent toward the jet axis, leading to an increase in concentration of the heavy species along the axis [23]. A theoretical analysis of the phenomenon has been proposed [24] and confirmed by experiment [25].

This concentration effect can be further enhanced by employing an annular nozzle to provide a sheath flow circumscribing the flow from a central orifice [26]. By seeding the carrier gas flowing from the central orifice with the molecule of interest, and providing a larger flow of pure carrier gas through the annular nozzle, gas dynamic focusing will drive the seed gas toward the centerline of the expansion. This has been shown to provide a factor of 30 increase in seed gas concentration on the centerline of the expansion [26]. This method is well adapted to interfacing capillary column gas chromatography to jet expansions, since the heavy analyte molecules are more easily detected if they are concentrated along the jet axis [27].

The primary advantages of seeded beams are that they allow the seed gas to be accelerated to the beam velocity of the carrier gas, and also allow the internal degrees of freedom of the seed gas to be cooled substantially, approaching the translational temperature achieved in the expansion. This latter aspect has been particularly useful for spectroscopic investigations, which are further considered later.

### 2.3.5 Skimmed Beams

Although many experiments such as those based on laser-induced fluorescence may conveniently be conducted in the free jet zone of silence, it is often advantageous to skim the free supersonic jet. This allows the supersonic beam to be collimated and transmitted into a second chamber which is operated at much lower pressure, a tremendous advantage for beam–surface or crossed-beam chemical dynamics experiments and for spectroscopic studies employing photo-ionization mass spectrometry for detection.

Two methods have been employed to skim the supersonic free jet and allow the transmission of a molecular beam into a second chamber. In the simplest [28], a low background pressure ($10^{-3}$ to $10^{-4}$ torr) is maintained in the source chamber, preventing the formation of serious shock waves in front of the skimmer, which is typically conical in shape with a full angle of 25 to 30°. The skimmer is placed some distance from the back wall of the first chamber, so that there is little interference from gas molecules scattering off of the wall. The lack of significant shock waves allows the molecular beam to be passed into the second chamber without serious attenuation.

In the alternative method [29, 30], a relatively high background pressure ($10^{-2}$ to 1 torr) is maintained in the source chamber, leading to a very pronounced barrel shock and Mach disk. A precisely manufactured skimmer is inserted into the free jet zone of silence, and the central core of the supersonic expansion is transmitted into the second chamber. This leads to a well-defined system of shock waves attached to the skimmer. With careful skimmer design, however, no shock zone develops on the leading edge of the skimmer (which would impede the flow into the second chamber), and no shock waves are attached to the interior portion of the skimmer (which would disrupt the supersonic flow in the second chamber). The optimal skimmer design has been determined empirically [30] and has a full included internal angle of 45° and a full included external angle of 55°. The influence of skimmer length, diameter, and distance from the expansion orifice has been discussed [30]. Although the problem of skimmer interference in the high-pressure source has been solved, such sources seem to require extreme care for proper operation [26].

## 2.4 Clustering

From the very earliest investigations [31], clusters have been readily formed in supersonic expansions, and this has led to tremendous advances in our understanding of weakly bound cluster species. Clusters generated in supersonic molecular beams have now been investigated by microwave [32], infrared [33], and visible/ultraviolet spectroscopy [34]; they have even been structurally studied through electron diffraction methods [35]. Ionic clusters have also been studied using the techniques of supersonic expansions [36]. The production of clusters in a supersonic expansion also affords one of the cleanest methods of studying the initial steps in nucleation of a supersaturated vapor, chiefly because collisional processes in the jet are well understood [37].

Unlike translational cooling, which results from binary collisions, condensation of monomers to form clusters requires three-body collisions with the third body carrying off the energy. The total number of binary collisions occurring during the expansion is proportional to $p_0 D$ [38], and for this reason the terminal Mach number (a measure of the translational cooling) is a function of $p_0 D$, as given in Eq. (2.13). The total number of three-body collisions occurring during the expansion is proportional to $p_0^2 D$, while the mass throughput through the nozzle is proportional to $p_0 D^2$ [38]. Thus, the production of clusters will be maximized while maintaining a given gas throughput by increasing the reservoir pressure, $p_0$, and decreasing the nozzle diameter, $D$, to maintain a constant value of $p_0 D^2$.

A combined theoretical and experimental study of the influence of nozzle shape on cluster formation has been presented in which a sharp-edged orifice, a slightly converging "ideal" orifice, and a straight, capillary tube orifice were examined (see Figure 4) [39]. Although all three were calculated to give similar flow fields at large distances from the nozzle, the three differed significantly in their flow fields near the nozzle. As gas expands through an orifice, the gas flow

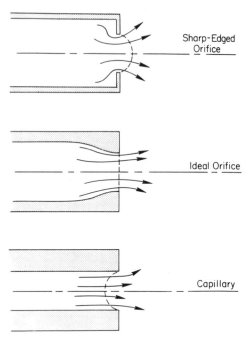

Fɪɢ. 4. Qualitative depiction of the streamlines and sonic surfaces for three types of nozzles shown in cross section. Reprinted with permission from H. R. Murphy and D. R. Miller, *J. Phys. Chem.* **88,** 4474 (1984). Copyright 1984, American Chemical Society.

velocity increases. The locus of points at which the flow velocity equals the local speed of sound is termed the "sonic surface," and it was in the shape and location of the sonic surface that the three nozzle types differed most significantly. The sharp-edged orifice led to a convex sonic surface located outside of the nozzle, the slightly converging "ideal" orifice led to a planar sonic surface at the end of the nozzle, and the straight, capillary nozzle led to a concave sonic surface located a short distance inside the capillary tube. The capillary nozzle was found to generate argon dimers in roughly twice the concentration of the other two nozzles, presumably because gas passes through the sonic surface earlier in the expansion in this design, thereby allowing greater cooling early in the expansions, before the gas density drops significantly.

Clustering can also be enhanced by confining the expansion after it passes through the sonic surface by the addition of a diverging section downstream of the nozzle throat [37]. This limits the expansion to a smaller angular range, thereby maintaining a high density during the initial portion of the expansion, which is most critical for the growth of clusters. A disadvantage is that a boundary layer develops on the inside of the diverging section, which tends to limit the terminal Mach number and ultimate temperature achieved in the expansion [40].

Clustering of a condensible gas also proceeds more readily if it is diluted in an inert, noncondensible gas such as helium or neon [37]. Presumably, the helium or neon acts to thermalize the growing clusters, so that the sequential addition of monomers is not seriously hampered by the heat of condensation which is released in the process. Finally, the growth of clusters is considerably aided by the use of a slit nozzle geometry, for which the gas density drops linearly, rather than quadratically, with distance from the source. Three-body collisions are therefore much more frequent than in the axisymmetic jet, greatly enhancing the formation of clusters or van der Waals complexes [40].

## 2.5 Supersonic Expansions from Slit Nozzles (Plane Jets)

### 2.5.1 The Idealized Continuum Model

The idealized continuum model applied to axisymmetric jets in Section 1.3.1 can also be applied to the ideal slit expansion. The expansion orifice is taken to be a slit opening of width $D$ and of length $L$. If $L \gg D$ and one is concerned about conditions at distances $x \ll L$ downstream, it is sufficient to consider a slit of infinite length. A supersonic slit expansion then ensues if $D \gg \lambda_0$, where $\lambda_0$ is again the mean free path in the reservoir.

In an idealized continuum slit supersonic expansion, it is still assumed that the gas flows without heat transfer to the walls and without viscosity, and an adiabatic, isentropic expansion is obtained. Thus, Eqs. (2.2)–(2.7) still obtain for gas expansion from a slit nozzle. The critical distinction between expansion from a circular orifice and from a slit nozzle is that in the latter case expansion occurs only in one transverse dimension.

The functional dependence of the Mach number achieved in a slit supersonic expansion as a function of the distance downstream, $x$, is given by [41]

$$M(x) = A_s \left( \frac{x - x_0}{D} \right)^{(\gamma-1)/2} - B_s \left( \frac{x - x_0}{D} \right)^{-(\gamma-1)/2}, \qquad (2.15)$$

where $A_s$, $B_s$, and $x_0/D$ are parameters depending on the ratio of heat capacities, $\gamma$. Specific values are provided in Table II. Using Eq. (2.15) in combination with

TABLE II.   Values of Parameters Used in Eq. (2.15)

| $\gamma$ | $x_0/D$ | $A_s$ | $B_s$ |
|---|---|---|---|
| 1.66̄ (5/3) | −0.218 | 3.06 | 1.21 |
| 1.40  (7/5) | 0.810 | 3.46 | 1.36 |
| 1.286 (9/7) | 0.205 | 3.83 | 1.67 |

Eq. (2.7), asymptotic expressions for the temperature, pressure, and density ratios at a downstream distance $x$ are obtained (subject to the constraint $D \ll x \ll L$):

$$\frac{T(x)}{T_0} = C_s \left(\frac{x - x_0}{D}\right)^{-(\gamma - 1)};$$

$$\frac{p(x)}{p_0} = F_s \left(\frac{x - x_0}{D}\right)^{-\gamma}; \qquad (2.16)$$

$$\frac{\rho(x)}{\rho_0} = G_s \left(\frac{x - x_0}{D}\right)^{-1},$$

where $C_s$, $F_s$, and $G_s$ are constants determined from $A_s$ and $\gamma$. Because in a slit nozzle gas expands in only one transverse dimension, $\rho(x)$ decreases as $x^{-1}$, not $x^{-2}$ as is characteristic of an axisymmetric expansion. For this reason, the exponents of $[(x - x_0)/D]$ in Eqs. (2.16) are decreased by a factor of 2 as compared to Eqs. (2.9). The slower decrease in molecular density and temperature in a slit nozzle expansion provides much higher binary and three-body collision rates, which are of great utility in the production of clusters. The two- and three-body collision rates for hard sphere argon atoms emerging from a reservoir at $p_0 = 100$ torr, $T_0 = 300$ K through a circular orifice 1 mm in diameter vs. a slit nozzle 1 mm in width have been calculated [40] and are shown in Figure 5. Collisions persist to far greater downstream distances in a slit nozzle expansion. This also suggests that slit nozzle expansions should be more effective in cooling recalcitrant degrees of freedom [41].

Slit nozzle expansions offer important advantages in spectroscopic applications. By using a probe laser beam directed parallel to the slit axis, it is possible to obtain the long path length needed for direct absorption measurements [42]. Further, because the decrease in density with distance is not so rapid as in the axisymmetric jet, one gains in absorption strength because of increased density as well as path length. Finally, because the gas streamlines diverge only in the direction perpendicular to the slit axis, spectroscopic interrogation of the sample using radiation propagating parallel to the slit reduces the Doppler broadening substantially [43]. This has been particularly useful in pulsed slit-jet expansion infrared absorption studies of hydrogen bonded species such as $N_2 \cdot HF$ [44].

## 2.5.2 Transition to Axisymmetric Flow

Even if the idealized continuum model remains valid, the characteristics of any real slit expansion will depart from those given in Eqs. (2.15) and (2.16) at distances $x \approx L$, where there will be a gradual transition from a one-dimensional expansion to the two-dimensional expansion that characterizes an axisymmetric

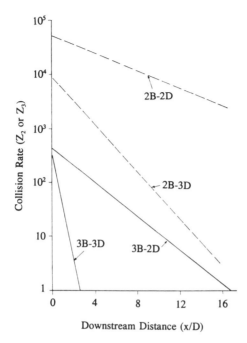

FIG. 5. Dependence of collision frequency on axial distance in an argon free jet expansion for source conditions of 100 torr and 300 K, and a nozzle diameter (or width, in the case of a slit expansion) of 1.0 mm. The two-body collision rates are indicated by dashed lines and the label "2B," three-body collision rates by solid lines and the label "3B." An axisymmetric jet is indicated by "3D," a wedge-shaped jet from an infinitely long slit nozzle by "2D." Reprinted, with permission, from reference [40].

jet. No complete treatment of the transition from plane to axisymmetric flow has yet been reported for the idealized continuum model. For large aspect ratios ($L/D$), the transition from plane to axisymmetric flow is expected quite far downstream, where the low collision rate may well make the assumptions of the idealized continuum model invalid. Such nonequilibrium effects have been investigated both theoretically and experimentally [45]. Unlike axisymmetric jets, where two translational temperatures ($T_\parallel$ and $T_\perp$) develop because of the loss of translational equilibration, slit nozzle expansions develop three distinct temperatures, associated with radial motion away from the source ($T_r$), motion along the long axis of the slit ($T_z$), and motion along the short axis of the slit ($T_\theta$). At large distances these translational temperatures are such that $T_r > T_\theta > T_z$ [45].

### 2.5.3 Structure of Shock Waves in a Slit Nozzle Expansion

As discussed in Section 1.3.3, the shock wave structure of an axisymmetric jet is axisymmetric, consisting of a barrel shock and the Mach disk. Because of the reduced symmetry of the slit nozzle expansion, the shock structure associated with a slit expansion is considerably more complex. In the limit of an infinitely long slit nozzle, however, the shock structure again retains the symmetry of the source. It consists of a barrel shock opening from the slit and a roughly planar Mach shock zone perpendicular to the expansion axis. The barrel shock is similar to the locus of points swept out by translating a parabola along the slit axis. The position, $x_M$, of the Mach shock in the limit of an infinitely long slit is given by

$$x_M = 1.23 \, D \left( \frac{p_0}{p_B} \right)^{0.775} \tag{2.17}$$

for a slit expansion, where $p_0$ is the pressure in the reservoir, and $p_B$ is the background pressure in the expansion chamber [46].

When a finite slit expansion ($L/D < \infty$) is employed (as is necessary in any real experiment), the shock structure is more complex, with lateral shock waves attached to the ends of the slit. These have been examined experimentally for high aspect ratio ($L/D = 450$ to $960$) slits [46]. As the barrel shock opens up, the lateral shock waves close in, so that the aspect ratio of the shock structure initially mimics the ratio $L/D$ of the slit, but decreases downstream. In certain $p_0/p_B$ and $L/D$ regimes, the expansion is still terminated by a Mach shock zone, but in other cases the expansion is terminated when the lateral shock waves approach one another and overlap before the Mach shock is encountered. For cases where the expansion is terminated by such overlap the downstream location, $x_L$, of the intersection (for aspect ratios $200 < L/D < 1000$ and pressure ratios $p_0/p_B < 5000$) is described by [46]

$$x_L = 40 \, D \left( \frac{p_0}{p_B} \right)^{0.34}. \tag{2.18}$$

It is also observed that the location of the Mach shock for a source of finite aspect ratio departs significantly from that given in Eq. (2.17) [46].

## 2.6 Implementation and Applications

### 2.6.1 Pulsed Sources

As demonstrated by Eq. (2.13), achieving low temperatures (or high Mach numbers) in an axisymmetric jet requires large values of the pressure–diameter product, $p_0 D$. Since the mass throughput of the nozzle is proportional to $p_0 D^2$, pumping considerations will ultimately determine the cooling attainable for given nozzle diameter $D$. Similar considerations also apply to slit nozzles.

A practical solution is to employ pulsed supersonic expansions, thereby reducing the duty cycle (and the pumping requirements) considerably. Pulsed expansions are particularly useful for applications involving pulsed lasers.

A question that arises in the implementation of a pulsed nozzle design is how long the nozzle must remain open for a supersonic expansion to fully develop. This duration will ultimately limit the useful duty cycle of the pulsed valve. A useful rule of thumb is that the pulsed valve must be fully open for a time $\Delta t \geq 4D/a_0$ for a supersonic expansion to fully develop, where $a_0$ is the speed of sound in the reservoir, given by Eq. (2.6) [47]. Here "fully open" implies that the gas flow rate is limited by the throat diameter of the nozzle, instead of any constrictions around the valve mechanism. Indeed, if the gas flow is limited by constrictions associated with the valve mechanism, the system will never reach the fully developed supersonic flow expected for the nozzle diameter, even if it is operated continuously. For the rare gases helium, neon, and argon expanding through a 1 mm orifice, this inequality requires the nozzle be fully open for 4, 9, and 12 µs, respectively, for the full development of supersonic flow. Presently no mechanical design exists for a pulsed valve that can fully open and close in such a short time period, so it may be assumed that full supersonic flow is achieved in a pulsed valve, provided it is opened sufficiently that flow is limited by the nozzle throat diameter.

Numerous pulsed valve designs have been described, and some are even commercially available. An early design used a pulsed electromagnet to repel a diamagnetic diaphragm, thereby opening a seal and allowing gas to flow for a period of 350 µs [48]. Another approach now rarely employed involved the use of a sliding piston [49] or other sliding seal [50] (typically driven by a solenoid) to open a nozzle for a period of 1 to 3 ms. An alternative is to discharge a capacitor through a "hairpin" consisting of a fixed conductor and a spring, with the result that the induced magnetic fields cause the conductors to repel one another, opening a valve and allowing gas to flow for a period of 350 µs [51, 52]. This basic design, which is commercially available [53] and is illustrated in Figure 6, can produce pulses as short as 10 µs in duration [54]. This represents the shortest gas pulse ever created in the laboratory.

Piezoelectric crystals can also be used to open and close a pulsed valve, producing short pulses (100 µs to 10 ms in duration), with repetition rates to 750 Hz [55]. Such valves are now commercially available for researchers using pulsed supersonic beams [56].

Another approach in common use employs a solenoid to lift a poppet from a sealing surface, allowing gas to flow. After the current pulse to the solenoid is terminated, a spring closes the valve. One design has been developed for use to temperatures as high as 670 K [57], while others have used modified commercial fuel injector valves [58], one of which has produced pulses 600 µs in duration, with a prominent inital peak having a FWHM of 150 µs [59]. This modified solenoid-spring design is also commercially available [60].

A variant of this approach uses one solenoid coil to open the valve and a second to close it [61–63]. At least one such device has provided pulses as short

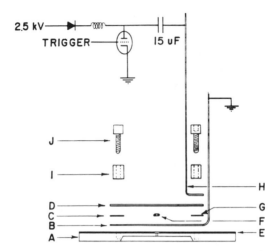

FIG. 6. An example of the "hairpin" pulsed valve, employing a stainless steel baseplate (A), a copper return conductor (B), a copper spacer (C), a copper-plate spring steel strip (D), an insulating coating of epoxy (E), a Viton O-ring (F), a Teflon spacer (G), a copper high-voltage input lead (H), plastic clamp bars (I), and a clamping screws (J). The nozzle is driven by discharging the capacitor charged to 25 kV through the elements H, D, C, and B, causing magnetic repulsion between the elements labeled B and D. This lifts element D off of the O-ring, allowing gas to flow. Reprinted, with permission, from reference [52].

as 10 μs in duration [64]. Details of such a valve are displayed in Figure 7. In most applications a spring is used to hold the valve closed when the solenoid currents are turned off.

Finally, two designs for pulsed slit expansions have been reported. One design uses two tight-fitting concentric cylinders with slits cut into them, which are rotated relative to one another. A gas pulse is emitted only during the short time that the slits are aligned providing a pulse duration of 150 μs [65]. Another design (shown in Figure 8) uses a solenoid to accelerate a plunger, which acts as a hammer, striking a rod connected to an elastomer seal over a slit orifice [66]. The impact causes the valve to open quickly, and a leaf spring is then used to close the valve, producing reproducible pulses 150–600 μs in duration.

With all of these designs now available, pulsed supersonic beam sources have become quite commonplace. Indeed, the maturity of pulsed supersonic expansion technology has enabled the focus of research to be shifted from technology development to applications.

### 2.6.2 Spectroscopic Applications

Spectroscopic applications of supersonic beams have become increasingly important since the pioneering experiments of Smalley, Levy, and Wharton first

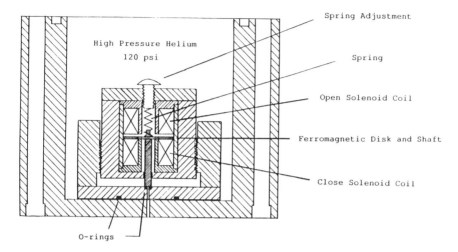

Spring Adjustment

High Pressure Helium
120 psi

Spring

Open Solenoid Coil

Ferromagnetic Disk and Shaft

Close Solenoid Coil

O-rings

FIG. 7. An example of a double solenoid pulsed valve, in which a current is supplied to the *open* solenoid coil, which pulls the ferromagnetic disk and shaft up, allowing gas to flow through the small O-ring. A current through the *close* solenoid coil then pulls the ferromagnetic disk and shaft down, resealing the O-ring. A spring holds the valve closed when the solenoid currents are turned off. Reprinted, with permission, from reference [63].

established the advantages of the technique [67]. Since then, several reviews have been published concerning the application of supersonic jet spectroscopy to stable molecules, complexes, radicals, and ions [33, 34, 38, 68, 69]. The field of supersonic jet spectroscopy has become so vast that it is impossible to review it in a short article such as this. For this reason only the most promising techniques for the generation and study of jet-cooled clusters, radicals, and ions are considered, omitting stable species entirely.

The supersonic nozzle, particularly when operated in a pulsed mode, can be easily combined with other techniques for the generation and jet-cooling of reactive species, some of which are quite fragile. The earliest investigations of jet-cooled radicals involved species such as NaNe and NaAr, generated using a sodium oven as the reservoir in a supersonic expansion employing neon or argon [70]. Other radicals, such as $CH_2$ [71], CN [72], SH and $S_2$ [73], HCO [74], and allyl [75], were subsequently produced and cooled by laser photolysis in the high-pressure zone of a supersonic expansion, a method which is certain to remain productive well into the future.

Other methods for generating jet-cooled radicals and ions employ discharge techniques [76]. A particularly successful method for obtaining spectra with a high degree of rotational cooling uses a corona discharge near the exit orifice of a supersonic expansion, as schematically illustrated in Figure 9 [77, 78]. This

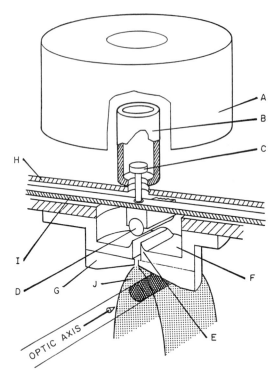

FIG. 8. Cutaway view of a slit pulsed valve. In operation, a current pulse through solenoid (A) accelerates plunger (B) against rod (C), which is connected to the seal assembly. The elastomer seal (D) is lifted from nozzle holder (E), permitting gas in plenum (F) to flow through interchangeable nozzles (G). The seal assembly continues upward until it hits stop (H). A leaf spring (I) returns the seal to the resting position, closing the valve. The resulting wedge-shaped expansion (J) provides a long path length, high-density sample at low (5–15 K) temperature. Reprinted, with permission, from reference [66].

method has been shown to be capable of producing fragile polyatomic radicals such as methyl nitrene ($CH_3N$:), which is produced in an excited electronic state. A continuous corona discharge source with a slit geometry has also been developed, providing reduced Doppler widths and narrowing the rotational linewidths [79]. A variant of the corona-excited discharge technology has also been applied to pulsed systems [80].

Another significant advance in the use of supersonic expansions to study radicals employs a flash pyrolysis source just prior to expansion into vacuum [81, 82]. In the initial application of this method, flash pyrolysis of $t$-butyl nitrite led to the production of jet-cooled methyl radicals with a rotational temperature of 40 K [83]. The technique has been used to record spectra of jet-cooled

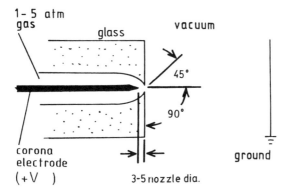

FIG. 9. Schematic diagram of a corona discharge nozzle, reprinted with permission from reference [78]. Important points are (1) the free expansion into a 90° half-angle on the low-pressure side, (2) the 45° half-angle of taper on the high-pressure side, and (3) the location of the corona electrode 3–5 nozzle diameters from the nozzle throat. The nozzle must be an electrical insulator, such as glass. Electrical continuity is formed by ionic conduction in the vacuum to the vacuum chamber.

propargyl [84], allyl [85], vinyl [86], $CCl_2$ [87], $SiCl_2$ [88], and $GeCl_2$ [89] radicals, along with formyl cyanide (HCOCN) [90] and cyclobutadiene [91]. A schematic drawing of a flash pyrolysis nozzle is shown in Figure 10. Again, it is clear that this technique will be widely used in the future.

Laser ablation of a target in the throat of a supersonic expansion has also become a common technique for generating jet-cooled metal and semiconductor clusters [92]. Several nozzle designs have been described, employing metal rods [62], disks [93], or wires [94]. In combination with laser-induced fluorescence or resonant two-photon ionization spectroscopy, this has become a powerful method for obtaining information on metal-based radicals such as metal dimers and trimers [95], metal diatomics MX [96], and larger clusters. This source has also been used for detailed studies of metal cluster reactivity [97] and bond strengths [98], and for production of carbon clusters for spectroscopic study [99]. Indeed, the soccer-ball structure of the famous $C_{60}$ molecule, buckminster-fullerene, was first proposed to explain the anomalous intensity of the mass 720 peak in the mass spectrum generated by laser vaporization of graphite in a pulsed supersonic jet [100]. More recently, metal–carbon structures with the chemical formula of $M_8C_{12}$ have been found to be particularly stable, and a dodecahedral structure has been proposed [101]. It is certain that the laser-vaporization supersonic expansion method will be a productive technique for the study of reactive chemical species for many years to come. A particular advantage of this method is that cations and anions are produced in the laser-induced plasma, and these are jet-cooled in the subsequent expansion into vacuum as well [102]. The

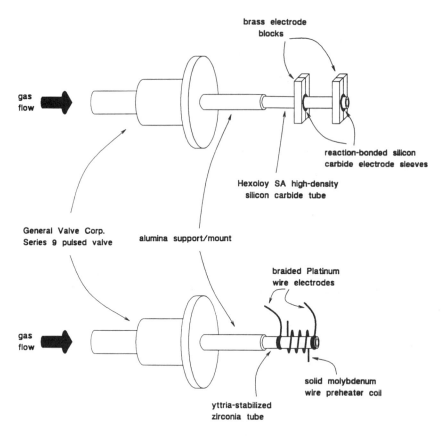

FIG. 10. Two pulsed pyrolysis nozzles used for the flash pyrolytic production of hydro-carbon radicals at temperatures up to 1800°C. The heated sections at the end of the extension tube are 1–2 cm long, corresponding to contact times with the hot zone of ≈10 μs for radical precursors seeded in helium carrier gas. Reprinted, with permission, from reference [82].

production of jet-cooled ions can be enhanced by directing excimer laser light into the nozzle source just prior to expansion [103].

In another interesting development, a collimated, 1 keV electron beam is used to ionize clusters produced in a supersonic expansion by crossing the supersonic cluster beam close to the expansion orifice [104]. Both positively and negatively charged clusters were generated, indicating that some clusters successfully captured the slow electrons which were detached by electron bombardment. In another experiment, an ultraviolet light source has been used to eject low-energy photoelectrons from a metal surface close to the expansion orifice of a supersonic expansion, allowing clusters to nucleate on the low energy electrons to produce stable cluster anions of $(H_2O)_n^-$, $(NH_3)_n^-$, etc. [105]. Supersonic beams have also been crossed near the expansion orifice with the output beams of effusive ovens, allowing the high-temperature species effusing from the oven to be picked up and cooled in the supersonic expansion [106]. By using an expansion of water vapor in helium and a sodium oven, for example, it has been possible to produce the surprising adducts $Na(H_2O)_n$ [107]. With these examples of clever hybrids of supersonic expansions and other techniques, it seems our only limitation in the future uses of supersonic beams lies in our own imaginations.

## Acknowledgments

I gratefully acknowledge support of my on-going research program, which uses supersonic beams extensively, by the National Science Foundation under Grant Number CHE-9215193 and by the Petroleum Research Fund, administered by the American Chemical Society.

## References

1. Luo, F., McBane, G. C., Kim, G., Giese, C. F., and Gentry, W. R. (1993). *J. Chem. Phys.* **98**, 3564; Schollkopf, W., and Toennies, J. P. (1994). *Science* **266**, 1345.
2. Liu, B., and McLean, A. D. (1989). *J. Chem. Phys.* **91**, 2348; Vos, R. J., van Lenthe, J. H., and van Duijneveldt, F. B. (1990). *J. Chem. Phys.* **93**, 643; Tawa, G. J., Whitlock, P. A., Moscowitz, J. W., and Schmidt, K. W. (1991). *Int. J. Supercomput. Appl.* **5**, 57.
3. Ramsey, N. F. (1993). "Thermal Beam Sources," in *Methods of Experimental Physics: Atomic, Molecular, and Optical Physics*, Volume II, R. Hulet and F. B. Dunning (eds.). Academic Press, New York; Ramsey, N. F. (1990). *Molecular Beams*, Oxford University Press, Oxford, U.K.
4. See, for example, Gingerich, K. A. (1980). *Faraday Symp. Chem. Soc.* **14**, 109, and references therein.
5. See, for example, Liepmann, H. W., and Roshko, A. (1957). *Elements of Gas-dynamics*, John Wiley & Sons, Inc., New York; or Zucrow, M. J., and Hoffman, J. D. (1976). *Gas Dynamics*, Vol. I, John Wiley & Sons, Inc., New York.
6. Kantrowitz, A., and Grey, J. (1951). *Rev. Sci. Instr.* **22**, 328.
7. Kistiakowsky, G. B., and Slichter, W. P. (1951). *Rev. Sci. Instr.* **22**, 333.
8. Becker, E. W., and Bier, K. (1954). *Z. Naturforsch.* **9A**, 975; Becker, E. W., and Henkes, W. (1956). *Z. Phys.* **146**, 320.
9. Anderson, J. B. (1974). "Molecular Beams from Nozzle Sources," in *Molecular Beams and Low Density Gasdynamics*, P. P. Wegener (ed.). Marcel Dekker, Inc., New York.
10. Anderson, J. B., and Fenn, J. B. (1965). *Phys. Fluids* **8**, 780.
11. Owen, P. L., and Thornhill, C. K. (1948). *Aeronaut. Res. Council, Great Britain, R & M*, 2616.
12. Ashkenas, H., and Sherman, F. S. (1966). In *Rarefied Gas Dynamics*, 4th Symposium, Vol. II, J. H. de Leeuw (ed.), p. 84. Academic Press, New York.
13. McClelland, G. M., Saenger, K. L., Valentini, J. J., and Herschbach, D. R. (1979). *J. Phys. Chem.* **83**, 947.
14. Hamel, B. B., and Willis, D. R. (1966). *Phys. Fluids* **9**, 829.

15. Bird, G. A. (1970). *AIAA J.* **8,** 1998.
16. Bier, K. and Hagena, O. (1966). In *Rarefied Gas Dynamics*, 4th Symposium, Vol. II, J. H. de Leeuw (ed.), p. 260. Academic Press, New York.
17. Bier, K., and Schmidt, B. (1961). *Z. Angew. Phys.* **13,** 34.
18. Campargue, R., Lebehot, A., Lemmonier, J. C., and Marette, D. (1980). *Rarefied Gas Dynamics*, 12th Symposium, S. S. Fisher (ed.). *Progress in Astronautics and Aeronautics*, **74,** Part 2, 823.
19. Parks, E. K., and Wexler, S. (1984). *J. Phys. Chem.* **88,** 4492; Russell, J. A., Hershberger, J. F., McAndrew, J. J., Cross, R. J., and Saunders, M. (1984). *J. Chem. Phys.* **88,** 4494.
20. Hagena, O., and Henkes, W. (1960). *Z. Naturforsch.* **15A,** 851; Abuaf, N., Anderson, J. B., Andres, R. P., Fenn, J. B., and Marsden, D. G. (1967). *Science* **155,** 997.
21. Anderson, J. B. (1967). *Entropie* **18,** 33.
22. DePaul, S., Pullman, D., and Friedrich, B. (1993). *J. Phys. Chem.* **97,** 2167.
23. Anderson, J. B. (1967). *Am. Inst. Chem. Eng.* **13,** 1188.
24. Sherman, F. S. (1965). *Phys. Fluids* **8,** 773.
25. Rothe, D. E. (1966). *Phys. Fluids* **9,** 1643.
26. Hayes, J. M. (1987). *Chem. Rev.* **87,** 745.
27. Stiller, S., and Johnston, M. V. (1987). *Anal. Chem.* **59,** 567.
28. Fenn, J. B., and Deckers, J. (1963). In *Rarefied Gas Dynamics*, 3rd Symposium, Vol. I, J. A. Laurmann (ed.), p. 497. Academic Press, New York; Fenn, J. B., and Anderson, J. B. (1966). *Rarefied Gas Dynamics*, 4th Symposium, Vol. II, J. H. de Leeuw (ed.). Academic Press, New York; Anderson, J. B., Andres, R. P., Fenn, J. B., and Maise, G. (1966). In *Rarefied Gas Dynamics*, 4th Symposium, Vol. II, J. H. de Leeuw (ed.). Academic Press, New York.
29. Campargue, R. (1964). *Rev. Sci. Instrum.* **35,** 111; Campargue, R. (1966). In *Rarefied Gas Dynamics*, 4th Symposium, Vol. II, J. H. de Leeuw (ed.). Academic Press, New York.
30. Campargue, R. (1984). *J. Phys. Chem.* **88,** 4466.
31. Becker, E. W., Bier, K., and Henkes, W. (1956). *Z. Physik* **146,** 333.
32. Legon, A. C. (1983). *Ann. Rev. Phys. Chem.* **34,** 275; Legon, A. C., and Millen, D. J. (1986). *Chem. Rev.* **86,** 635; also see *J. Chem. Phys.* **78,** 3483–3552 (Flygare memorial issue).
33. Nesbitt, D. J. (1988). *Chem. Rev.* **88,** 843.
34. Levy, D. H. (1980). *Ann. Rev. Phys. Chem.* **31,** 197; Levy, D. H. (1981). *Adv. Chem. Phys.* **47,** 323; Heaven, M. C. (1992). *Ann. Rev. Phys. Chem.* **43,** 283.
35. Bartell, L. S., and French, R. J. (1990). *Rev. Sci. Instrum.* **60,** 6468; Bartell, L. S., and French, R. J. (1989). *Z. Phys. D* **12,** 7; Bartell, L. S. (1986). *Chem. Rev.* **86,** 491.
36. Castleman, A. W., Jr., and Keesee, R. G. (1986). *Chem. Rev.* **83,** 589.
37. Hagena, O. F. (1974). "Cluster Beams from Nozzle Sources," in *Molecular Beams and Low Density Gasdynamics*, P. P. Wegener (ed.). Marcel Dekker, Inc. New York; Hagena, O. F. (1981). *Surf. Sci.* **106,** 101.
38. Smalley, R. E., Wharton, L., and Levy, D. H. (1977). *Accts. Chem. Res.* **10,** 139.
39. Murphy, H. R., and Miller, D. R. (1984). *J. Phys. Chem.* **88,** 4474.
40. Ryali, S. B., and Fenn, J. B. (1984). *Ber. Bunsenges. Phys. Chem.* **88,** 245.
41. Sulkes, M., Jouvet, C., and Rice, S. A. (1982). *Chem. Phys. Lett.* **87,** 515.
42. Amirav, A., Even, U., and Jortner, J. (1981). *Chem. Phys. Lett.* **83,** 1.

43. Veeken, K., and Reuss, J. (1985). *Appl. Phys. B* **38**, 117.
44. Lovejoy, C. M., and Nesbitt, D. J. (1987). *J. Chem. Phys.* **86**, 3151.
45. Beylich, A. E. (1981). In *Rarefied Gas Dynamics*, 12th Symposium, Vol. II, *Progress in Astronautics and Aeronautics* **74**, Sam S. Fisher (ed.), p. 710.
46. Dupeyrat, G. (1981). In *Rarefied Gas Dynamics*, 12th Symposium, Vol. II, *Progress in Astronautics and Aeronautics* **74**, Sam S. Fisher (ed.), p. 812.
47. Saenger, K. L. (1981). *J. Chem. Phys.* **75**, 2467; Saenger, K. L., and Fenn, J. B. (1983). *J. Chem. Phys.* **79**, 6043.
48. Kuswa, G., Stallings, C., and Stam, A. (1970). *Rev. Sci. Instrum.* **41**, 1362.
49. Inutake, M., and Kuriki, K. (1972). *Rev. Sci. Instrum.* **43**, 1670.
50. Riley, J. A., and Giese, C. F. (1970). *J. Chem. Phys.* **53**, 146; Watters, R. L., Jr., and Walters, J. P. (1977). *Rev. Sci. Instrum.* **48**, 643.
51. Inoue, N., and Uchida, T. (1968). *Rev. Sci. Instrum.* **39**, 1461.
52. Liverman, M. G., Beck, S. M., Monts, D. L., and Smalley, R. E. (1979). *J. Chem. Phys.* **70**, 192.
53. R. M. Jordan Company, 990 Golden Gate Terrace, Grass Valley, CA 95945.
54. Gentry, W. R., and Giese, C. F. (1978). *Rev. Sci. Instrum.* **49**, 595.
55. Auerbach, A., and McDiarmid, R. (1980). *Rev. Sci. Instrum.* **51**, 1273; Cross, J. B., and Valentini, J. J. (1982). *Rev. Sci. Instrum.* **53**, 38.
56. Lasertechnics, 5500 Wilshire Avenue NE, Albuquerque, NM 87113.
57. Köhler, K.-A. (1973). *Rev. Sci. Instrum.* **44**, 73.
58. Behlen, F. M., and Rice, S. A. (1981). *J. Chem. Phys.* **75**, 5672; Tusa, J., Sulkes, M., Rice, S. A., and Jouvet, C. (1982). *J. Chem. Phys.* **76**, 3513.
59. Otis, C. E., and Johnson, P. M. (1980). *Rev. Sci. Instrum.* **51**, 1128.
60. General Valve Corporation, 19 Gloria Lane, Fairfield, NJ 07004.
61. Adams, T. E., Rockney, B. H., Morrison, R. J. S., and Grant, E. R. (1981). *Rev. Sci. Instrum.* **52**, 1469.
62. Hopkins, J. B., Langridge-Smith, P. R. R., Morse, M. D., and Smalley, R. E. (1983). *J. Chem. Phys.* **78**, 1627.
63. Lemire, G. W. (1989). Ph.D. thesis, University of Utah, Salt Lake City, Utah.
64. Adriaens, M. R., Allison, W., and Feuerbacher, B. (1981). *J. Phys. E: Sci. Instrum.* **14**, 1375.
65. Amirav, A., Even, U., and Jortner, J. (1981). *Chem. Phys. Lett.* **83**, 1.
66. Lovejoy, C. M., and Nesbitt, D. J. (1987). *Rev. Sci. Instrum.* **58**, 807.
67. Smalley, R. E., Ramakrishna, B. L., Levy, D. H., and Wharton, L. (1974). *J. Chem. Phys.* **61**, 4363; Smalley, R. E., Wharton, L., and Levy, D. H. (1975). *J. Chem. Phys.* **63**, 4977.
68. Engelking, P. C. (1991). *Chem. Rev.* **91**, 399.
69. Foster, S. C., and Miller, T. A. (1989). *J. Phys. Chem.* **93**, 5986.
70. Ahmad-Bitar, R., Lapatovich, W. P., Pritchard, D. E., and Renhorn, I. (1977). *Phys. Rev. Lett.* **39**, 1657; Smalley, R. E., Auerbach, D. A., Fitch, P. S., Levy, D. H., and Wharton, L. (1977). *J. Chem. Phys.* **66**, 3778.
71. Monts, D. L., Dietz, T. G., Duncan, M. A., and Smalley, R. E. (1980). *Chem. Phys.* **45**, 133; Xie, W., Harkin, C., and Dai, H.-L. (1990). *J. Chem. Phys.* **93**, 4615.
72. Farthing, J. W., Fletcher, I. W., and Whitehead, J. C. (1983). *J. Phys. Chem.* **87**, 1663.
73. Heaven, M., Miller, T. A., and Bondybey, V. E. (1984). *J. Chem. Phys.* **80**, 51.
74. Song, X.-M., and Cool, T. A. (1992). *J. Chem. Phys.* **96**, 8664; Cool, T. A., and Song, X.-M. (1992). *J. Chem. Phys.* **96**, 8675.

75. Getty, J. D., Burmeister, M. J., Westre, S. G., and Kelly, P. B. (1991). *J. Am. Chem. Soc.* **113**, 801; Getty, J. D., Liu, X., and Kelly, P. B. (1992). *J. Phys. Chem.* **96**, 10155.

76. Searcy, J. Q. (1974). *Rev. Sci. Instrum.* **45**, 589; Leasure, E. L., Mueller, C. R., and Ridley, T. Y. (1975). *Rev. Sci. Instrum.* **46**, 635; Brutschy, B., and Haberland, H. (1977). *J. Phys. E* **10**, 90; Ganteför, G., Siekmann, H. R., Lutz, H. O., and Meiwes-Broer, K. H. (1990). *Chem. Phys. Lett.* **165**, 293.

77. Droege, A. T., and Engelking, P. C. (1983). *Chem. Phys. Lett.* **96**, 316; Carrick, P. G., and Engelking, P. C. (1984). *Chem. Phys. Lett.* **108**, 505.

78. Engelking, P. C. (1986). *Rev. Sci. Instrum.* **57**, 2274.

79. Brazier, C. R., Carrick, P. G., and Bernath, P. F. (1992). *J. Chem. Phys.* **96**, 919.

80. Sharpe, S., and Johnson, P. (1984). *Chem. Phys. Lett.* **107**, 35; Sharpe, S., and Johnson, P. (1986). *J. Mol. Spectrosc.* **116**, 247; Sharpe, S., and Johnson, P. (1986). *J. Chem. Phys.* **85**, 4943.

81. Dunlop, J. R., Karolczak, J., and Clouthier, D. J. (1988). *Chem. Phys. Lett.* **151**, 362.

82. Kohn, D. W., Clauberg, H., and Chen, P. (1992). *Rev. Sci. Instrum.* **63**, 4003.

83. Chen, P., Colson, S. D., Chupka, W. A., and Berson, J. A. (1986). *J. Phys. Chem.* **90**, 2319.

84. Minsek, D. W., and Chen, P. (1990). *J. Phys. Chem.* **94**, 8399.

85. Minsek, D. W., Blush, J. A., and Chen, P. (1992). *J. Phys. Chem.* **96**, 2025; Blush, J. A., Minsek, D. W., and Chen, P. (1992). *J. Phys. Chem.* **96**, 10150.

86. Blush, J. A., and Chen, P. (1992). *J. Phys. Chem.* **96**, 4138.

87. Clouthier, D. J., and Karolczak, J. (1991). *J. Chem. Phys.* **94**, 1.

88. Karolczak, J., and Clouthier, D. J. (1993). *Chem. Phys. Lett.* **201**, 409.

89. Karolczak, J., Zhuo, Q., and Clouthier, D. J. (1993). *J. Chem. Phys.* **98**, 60.

90. Karolczak, J., Clouthier, D. J., and Judge, R. H. (1991). *J. Mol. Spectrosc.* **147**, 61; Clouthier, D. J., Karolczak, J., Rae, J., Chan, W.-T., Goddard, J. D., and Judge, R. H. (1992). *J. Chem. Phys.* **97**, 1638.

91. Kohn, D. W., and Chen, P. (1993). *J. Am. Chem. Soc.* **115**, 2844.

92. Dietz, T. G., Duncan, M. A., Powers, D. E., and Smalley, R. E. (1981). *J. Chem. Phys.* **74**, 6511.

93. O'Brien, S. C., Liu, Y., Zhang, Q., Heath, J. R., Tittel, F. K., Curl, R. F., and Smalley, R. E. (1986). *J. Chem. Phys.* **84**, 4074.

94. Weidele, H., Frenzel, U., Leisner, T., and Kreisle, D. (1991). *Z. Phys. D* **20**, 411.

95. Morse, M. D. (1986). *Chem. Rev.* **86**, 1049; Morse, M. D. (1993). *Adv. Metal Semiconductor Clusters* **1**, 83.

96. Bourne, O. L., Humphries, M. R., Mitchell, S. A., and Hackett, P. A. (1986). *Opt. Commun.* **56**, 403; Simard, B., Mitchell, S. A., Humphries, M. R., and Hackett, P. A. (1988). *J. Mol. Spectrosc.* **129**, 186; Simard, B., Mitchell, S. A., Hendel, L. M., and Hackett, P. A. (1988). *Faraday Disc. Chem. Soc.* **86**, 163.

97. Bechtold, P. S., Parks, E. K., Weiller, B. H., Pobo, L. G., and Riley, S. J. (1990). *Z. f. Phys. Chem.* **169**, 101; Kaldor, A., Cox, D. M., and Zakin, M. R. (1988). *Adv. Chem. Phys.* **70**, 211.

98. Armentrout, P. B., Hales, D. A., and Lian, L. (1994). *Adv. Metal Semiconductor Clusters* **2**, 1.

99. Moazzen-Ahmadi, N., McKellar, A. R. W., and Amano, T. (1989). *J. Chem. Phys.* **91**, 2140; Heath, J. R., and Saykally, R. J. (1990). *J. Chem. Phys.* **93**, 8392; Heath,

J. R., Saykally, R. J. (1991). *J. Chem. Phys.* **94,** 3271; Arnold, D. W., Bradforth, S. E., Kitsopoulos, T. N., and Neumark, D. M. (1991). *J. Chem. Phys.* **95,** 8753.

100. Kroto, H. W., Heath, J. R., O'Brien, S. C., Curl, R. F., and Smalley, R. E. (1985). *Nature* **318,** 162.

101. Guo, B. C., Kerns, K. P., and Castleman, A. W., Jr. (1992). *Sciences* **255,** 1411; Chen, Z. Y., Walder, G. J., and Castleman, A. W., Jr. (1992). *J. Phys. Chem.* **96,** 9581; Castleman, A. W., Jr., Guo, B., and Wei, S. (1992). *Int. J. Mod. Phys. B* **6,** 3587.

102. Bloomfield, L., Geusic, M. E., Freeman, R., and Brown, W. L. (1985). *Chem. Phys. Lett.* **121,** 33; Liu, Y., Zhang, Q.-L., Tittel, F. K., Curl, R. F., and Smalley, R. E. (1986). *J. Chem. Phys.* **85,** 7434.

103. Zheng, L.-S., Brucat, P. J., Pettiette, C. L., Yang, S., and Smalley, R. E. (1985). *J. Chem. Phys.* **83,** 4273.

104. Johnson, M. A., Alexander, M. L., and Lineberger, W. C. (1984). *Chem. Phys. Lett.* **112,** 285; Posey, L. A., and Johnson, M. A. (1988). *J. Chem. Phys.* **89,** 4807.

105. Armbruster, M., Haberland, H., and Schindler, H.-G. (1981). *Phys. Rev. Lett.* **47,** 323; Haberland, H., Schindler, H.-G., and Worsnop, D. R. (1984). *Ber. Bunsenges. Phys. Chem.* **88,** 270; Haberland, H., Langosch, H., Schindler, H.-G., and Worsnop, D. R. (1984). *J. Phys. Chem.* **88,** 3903; Haberland, H., Ledewigt, C., Schindler, H.-G., and Worsnop, D. R. (1984). *J. Chem. Phys.* **81,** 3742.

106. Steimle, T. C., Fletcher, D. A., Jung, K. Y., and Scurlock, C. T. (1991). *Chem. Phys. Lett.* **184,** 379.

107. Schulz, C. P., Haugstätter, R., Tittes, H. U., and Hertel, I. V. (1986). *Phys. Rev. Lett.* **57,** 1703.

# 3. FAST BEAM SOURCES

## A. Chutjian and O. J. Orient

Jet Propulsion Laboratory, California Institute of Technology
Pasadena, California

## 3.1 Introduction

Energetic neutral beams of well-defined quantum state are needed in a variety of basic and applied physics applications. The fast neutral projectile can undergo an inelastic, energy-loss collision with an atomic or molecular target, or participate in a gas-phase chemical reaction. Collision of the fast beam with a surface can lead to single or multiple charge transfers, and erosion or chemical reaction with the surface or surface-adsorbed species. Reaction-excitation channels which are closed at thermal energies will be exoergic at higher beam energies.

A beam arrangement is usually the most convenient laboratory method for carrying out single-collision excitation, ionization, or reaction studies. Production of neutral beams having energies $E$ above thermal (say, $E > 1.0$ eV) is difficult because the charge "handle" is missing: The species must be accelerated as a positive or negative ion, followed by electron detachment (either by electron or photon), charge exchange or collisional detachment, or surface neutralization. Alternatively, the neutral species can be accelerated directly through a momentum-transfer step such as in a seeded supersonic beam, or in a plasma-ball explosion.

The various techniques used to date for producing hyperthermal-energy beams have trade-offs. Factors such as quantum state, beam purity, kinetic energy range, kinetic energy spread, Coulomb broadening, space-charge, flux, angular divergence, beam diameter, and beam detection method have to be considered within each application. Several diagnostic tests must be incorporated within each source type to characterize the output. One must have definite knowledge, for example, that ions or metastable states are not admixed if a ground-state neutral beam is desired; or one must know ultraviolet photons from a source are not causing spurious (even synergistic) effects at the target. Given these considerations, it is not surprising that the 1–10 eV range has sometimes been referred to as the no-man's land in beam work [1]. Yet it is in this energy range that surface and gas-phase reaction thresholds lie.

Driven by an important, practical need to simulate low-earth orbit (LEO) phenomena, several of the systems discussed in this chapter have been applied to,

EXPERIMENTAL METHODS IN THE PHYSICAL SCIENCES
Vol. 29B

or developed for, the production of fast atomic oxygen beams. Depending on the source type, other fast atoms can be made by a suitable choice of feed gas.

## 3.2 Types of Sources

### 3.2.1 Photodetachment in the Negative Ion

The efficient production of a negative ion, followed by acceleration to the desired final energy and electron photodetachment to a known final quantum state of the neutral atom, is the most straightforward technique for producing well-characterized, fast beams. This technique has been used to produce fast $H(^2S)$ [2–6] and $O(^3P)$ [7–11] atoms via the $H^-\,(^1S)$ and $O^-\,(^2P)$ ion states, respectively. In the case of $H(^2S)$, the final kinetic energy range was 10–3000 eV, while the energy range utilized in $O(^3P)$ was 2.0–25 eV.

The experimental techniques used can rely on either electrostatic focusing of the negative ion, or magnetic confinement. Shown in Figure 1 is the detachment region of a source for fast hydrogen atoms [2]. $H^-$ ions are formed in a biased duoplasmotron source, extracted electrostatically, and momentum-analyzed in

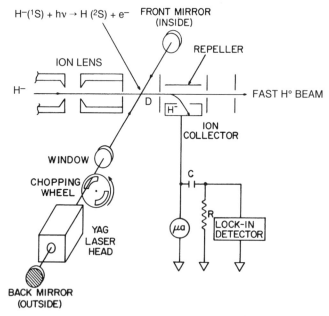

FIG. 1. Electrostatically focused source used to produce $H(^2S)$ by photodetachment of $H^-\,(^1S)$ [2–6]. Ions are focused to a detachment point D where they are crossed with a modulated YAG laser (30 W, cw) at 1060 nm. Undetached $H^-$ ions are deflected away after D. The gain in fast $H^0$ is given by the loss in $H^-$.

the 45° magnetic sector. The ions are decelerated/accelerated by the potential between the ion source and the final (grounded) potential of the ion lens. They are focused to the point D and crossed with a photon beam from a YAG laser operating at 1060 nm in an intracavity mode. A slight 9° "jog" between the ion lens and D prevents fast neutrals stripped upstream from reaching the target region. The detachment efficiency was found to be 5% at 1000 eV, with the expected increase as (ion velocity)$^{-1}$ towards lower energies. At energies less than 200 eV, space-charge spreading of the H$^-$ beam decreased the ion density, which canceled the expected velocity gain. Detachment efficiencies of greater than 50% should easily be realizable with higher-power lasers. Care must be exercised to incorporate space charge in the focusing. One must define a collimated ion beam at D which will continue, as a neutral beam, towards the experimental region with small divergence losses.

Magnetic confinement of the negative-ion beam allows one to attain much higher currents, and to perform several operations (extraction, confinement, transport, mass selection, and detachment) with the assistance of the confining **B** field. Shown in Figure 2 is a recent approach to the generation of fast $O(^3P)$ neutral beams utilizing dissociative electron attachment (DA) and a solenoidal magnetic field to confine the electrons and negative ions [7, 9]. Electrons are generated from a spiral-wound tungsten filament (F), extracted, then decelerated to 8 eV. NO molecules effusing from a gas nozzle (G) dissociatively attach to form $O^-(^2P)$ ions via the step

$$e + NO(X^2\Pi) \rightarrow N(^4S^\circ, {}^2D^\circ, {}^2P^o) + O^-(^2P). \quad (3.1)$$

NO was chosen as the feed gas because of its large DA cross section at a relatively high (8 eV) electron energy, allowing larger space-charge limited currents at G. Also, only the O$^-$ ion is formed via this channel, with low kinetic energy at onset to minimize divergence in the final O-beam [7]. The confined ions and electrons are accelerated to the desired final atomic energy (5 eV, say). Using a trochoidal deflector ($T_1$), the faster electrons are separated from the slower ions, and the electrons are trapped in a Faraday cup. $T_1$ also serves as a low-resolution mass analyzer and deflects the O$^-$ beam off-axis so that photons and other species from F and the lens region cannot reach the target. Currents as high as 50 µA of $O^-(^2P)$ have been measured after $T_1$. Using all visible lines from a 20 W argon-ion laser L, the electron in O$^-$ is detached at M in a multiple-pass mirror geometry to produce exclusively $O(^3P)$ atoms traveling at 5 eV energy. Undetached ions can be separated from the neutral beam either by a second trochoidal deflector after M, or by biasing the target region (or collection mirror C) negative with respect to the ion energy. (Their current may also be measured in a suitably placed Faraday cup.) With a second trochoidal deflector one may monitor the detachment efficiency with ion energy. This detachment efficiency g can be written in terms of the ion mass $m$, energy $E$,

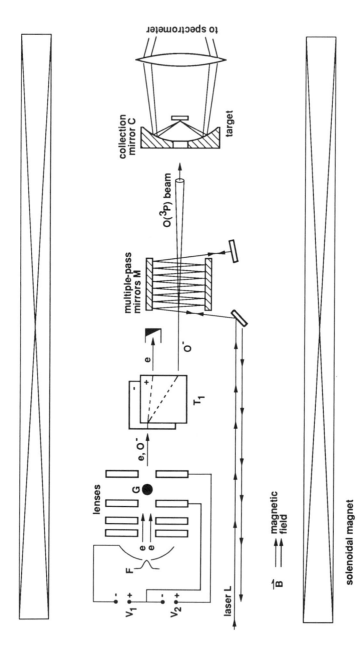

FIG. 2. Magnetically confined source used to produce fast O($^3P$) atoms (2–40 eV) by electron photodetachment [7–11]. The O$^-$($^2P$) ions are produced by dissociative electron attachment in NO, accelerated to the desired final energy, then separated from the electrons by the trochoidal monochromator T$_1$. Photodetachment takes place in the multiple-pass mirrors M using all visible lines from a 20 W cw argon-ion laser.

mirror reflectivity $\varepsilon$, detachment region length $l$, and photon-ion beams filling factor $\mathcal{F}$ as

$$g = 1 - \exp\left[ -\frac{\varepsilon\mathcal{F}l}{hc} \sqrt{\frac{m}{2E}} \sum_{i=0}^{n} \lambda_i \sigma_i W_i \right], \qquad (3.2)$$

where $W_i$ and $\sigma_i$ are the laser power and detachment cross section at each laser wavelength $\lambda_i$ (there are, for example, typically about 10 visible lines in an argon-ion laser output); $h$ is Planck's constant; and $c$ is the speed of light. The efficiency $g$ has been measured between 2.5 and 5.0 eV, and ranges between 16% and 8.5%, respectively, with the expected $\sqrt{E}$ dependence. Scattering results obtained using this apparatus have been obtained for both surface-adsorbed [7, 9, 10] and gaseous [8, 11] targets.

Significant advantageous to the photodetachment approach are (a) there is usually only one bound state of the negative ion, (b) there is a broad range of photodetachment wavelengths which produce only the ground electronic state of the projectile, and (c) measurement of the final beam flux is simple: Because there is only one detachment channel, every negative ion lost results in production of one ground-state atom. *Current* depletion measurements are easily made via a Faraday cup; *flux* measurements require additional knowledge of the neutral beam diameter at the target. This can be measured at low energies by surface erosion (using silver or osmium films, for example), or at high (kilo–electron volt) energies using a Faraday cup through secondary-electron ejection or a particle multiplier with scanning aperture. Energy measurements can be made by retarding potential or time-of-flight techniques, with the former requiring knowledge of contact-potential shifts.

Disadvantages to the photodetachment technique are that [see Eq. (3.2)] the detachment efficiency $g$ depends upon the magnitude of $\sigma$ (relatively small, $\sim 6 \times 10^{-18}$ cm$^2$ for O$^-$ [12]). It also varies inversely with ion velocity, or the time the ion spends in the detachment region. Hence, efficiencies are less at the higher kinetic energies $E$. This can limit the highest energy at which the detachment approach is practical, although some intensity is gained back by the higher space-charge limited currents available. Also, there may be applications where high neutral fluxes ($10^{18}$–$10^{20}$ atoms/cm$^2$-sec) are desired, for example, in so-called "accelerated testing" of spacecraft materials. Within present technology, no source can meet this requirement and still offer control over beam purity, energy, and quantum state (see later discussion). In LEO, typical O($^3P$) fluxes are in the range $10^{14}$–$10^{16}$ atoms/cm$^2$-sec at 5 eV energy. Such LEO-type fluxes for laboratory simulations can be attained with magnetic confinement of the negative ion [7]. Due respect must be given to ion space-charge effects, leading to final beam divergence, and to Coulomb–Coulomb exchange, leading to a broadening of the longitudinal energy distribution of the O$^-$ ions, and hence of the neutral beam [7, 13]. The beam divergence can be estimated from a balance between the repulsive force $\mathbf{F}_r = e\mathbf{E}_r$ of the radial electric field $\mathbf{E}_r$, and the centripetal force $\mathbf{F}_c = e\mathbf{v}_\theta \times \mathbf{B}$ of the spiraling ions, where $\mathbf{v}_\theta$ is the ion tangential velocity. The quantity $\frac{1}{2}mv_\theta^2$ is the rotational kinetic energy $T_\theta$ and can be estimated by $2T_\theta/r = |\mathbf{F}_c - \mathbf{F}_r|$, from which an angular divergence can be calculated. Under a typical set of beam conditions, the divergence half-angle was estimated to be 9°, in reasonable agreement with a 5° divergence measured from Ag film erosion profiles [14].

By proper choice of the target gas, channel, and/or type of ion source, photodetachment is also capable of producing, in addition to O and H, the fast neutral species Li, C, F, Na, Al, Si, Cu, Rb, Cs, Al, Si, Cl, K, Ge, Br, Rb, I, Ta, W, $C_2$, OH, $SF_6$, and others [15]. Negative ions of many of these species can be generated in sputter-type sources by charge-exchange of positive ions at a cesium-coated metal surface, followed by extraction of the metal negative ion [16].

### 3.2.2 Positive- and Negative-Ion Charge Exchange

Charge-capture neutralization of singly charged positive ions by a target gas has been extensively used to generate fast neutral beams [1]. As with other methods, a successful application requires diagnostics of the neutral beam. DC discharge, radio-frequency (RF), microwave, or bucket-type sources of $O^+$, for example, can produce the ground $^4S^\circ$ state and the metastable $^2D^\circ$ and $^2P^\circ$ levels. In principle, all three states may charge exchange with the target gas to produce a distribution of final neutral states as given by the branching ratio and $O^+$ kinetic energy [17]. A low-lying metastable level in $C^+$ would be $^4P$, and in $N^+$ the $^1D$, $^1S$, and $^5S^\circ$ levels.

The quest to attain moderate to high beam fluxes in the 1–10 eV range invariably leads to the problem of space charge and beam divergence. In the conventional approach to ion extraction [17], one can extract from a DC-type discharge a milliampere current of beam (to be charge-exchanged downstream) at 3000 eV with a "narrow" beam divergence, say 2° half angle. In subsequent deceleration to 10 eV, the space-charge limited current is reduced by a factor $(3000/10)^{3/2} = 5200$; by the Helmholtz–Lagrange law, the beam divergence must increase by a factor of $(3000/10)^{1/2} = 17$ (assuming unity magnification). Hence, both beam space charge and beam phase space diminish the current available at the lower energy prior to charge exchange. Some relief is gained by use of a volume-type positive ion source [18–20] because the positive ions are created in a larger volume, and their space charge is partially neutralized by a magnetically confined electron cloud within the source plasma. Both this larger volume and any spreading of the beam after extraction or charge exchange will lower the available flux at the target.

Advantages to charge exchange are that one can conveniently tune the neutral energy over a broad range (30 eV to 10 keV, say). Moreover, cross sections, especially in the case of resonant charge exchange, can be large, and hence neutral beam production can be efficient. However, one must examine the momentum-transfer cross section to see whether the final neutral beam will remain reasonably collimated (forward-scattered) after the exchanging collision [21]. A divergent beam after exchange would result in loss of neutral flux during subsequent recollimation, for example. Differential charge-exchange scattering intensities measured for the $H^+ + H_2(v = 0)$ reaction have shown fairly isotropic scattering in the angular range 0–14° at 30 eV laboratory energy [22]. One could expect an increasingly isotropic behavior at lower charge-exchange energies, and hence even greater beam loss.

Examples of the use of resonant charge exchange at 3 keV ion energy are given in a recent series of measurements of atomic electron-impact ionization cross sections [23–25]. A representative positive-ion charge-exchange apparatus is shown in Figure 3 [23]. Positive ions are produced in a DC-discharge source, accelerated (L1), velocity-selected at about 1 keV energy, and charge-exchanged. With suitable choice of charge-transfer gas, one may generate nanoampere currents of many atomic species in the 0.1–1.0 keV range, often with elimination or minimization of metastable levels in the final beam. After the cell, nonexchanged ions are deflected from the beam path, and high-Rydberg states of the neutral are removed via field-ionization. The subsequent beam line in Figure 3 is used to ionize the target, measure the electron–neutral overlap, and measure neutral and ion currents. Ions of different mass-to-charge ratio (from multiple ionization) are energy-analyzed with the hemispherical electrostatic analyzer.

Extension of the charge-exchange approach to the 10–100 eV energy range almost certainly requires an additional lens system between the velocity filter and charge-transfer cell for focusing. Potential-and-trajectory codes exist for testing a design to fit the instrument geometry [26, 27]. One may use shaped electrodes and/or a five-element decelerating lens system recently proposed [28]. In any case, the greatest divergence, ion loss, and angst will be experienced in this step.

Electron-impact ionization cross section measurements [25] provide one diagnostic for the presence of metastables via the lowered threshold for ionization from the higher-energy metastable state(s). Alternatively, measurement of the charge-transfer efficiency vs. thickness $\mu$ of the target gas can lead to attenuation efficiencies that can be resolved into effects of two or more independent charge-exchanging states. This technique has been used by several groups to measure metastable contents in ion beams [29–32]. It is illustrated in a higher-energy application through Figures 4 and 5 [29]. Here, a 60 keV $He^+$ beam is partially neutralized in a gas or alkali metal vapor. The beam then traverses a distance between the neutralizer and target cell during which short-lived excited states decay, and ionic components and high-Rydberg levels are removed by the purifier. The neutral $He^0$ prior to the target cell consists of a

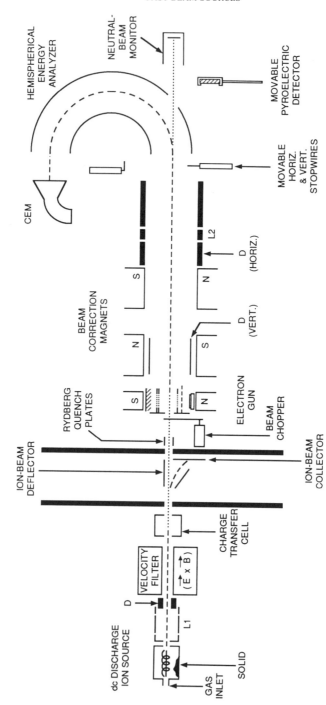

FIG. 3. A charge-exchange source used to generate neutral beams at kilo–electron volt energies [23] and to subsequently measure electron-impact ionization cross sections. Attention is given to the removal of nonexchanged ions and ions which have exchanged into high Rydberg levels. Operation at low energies (1–100 eV) will depend upon how carefully the ion extraction and focusing lens systems prior to the charge-exchange cell are designed.

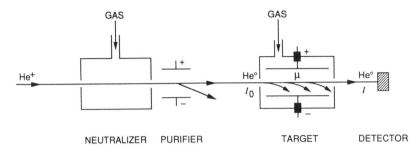

FIG. 4. Technique to detect the presence of metastable states in a charge-exchanged beam [29]. The transmitted current $I$ of fast (10–500 keV) He atoms is measured as a function of target thickness $\mu$. Ions generated within the target cell are swept away prior to reaching the detector and cannot contribute to the transmitted neutral current.

fraction $f$ in several metastable ($*$) states, and a fraction $(1 - f)$ in the ground state. Within the target cell the incident beam (intensity $I_0$) is attenuated, and the attenuation is measured as a function of target-gas thickness $\mu$. The charge-exchanged ions are swept out continuously, and only the transmitted $He^0$

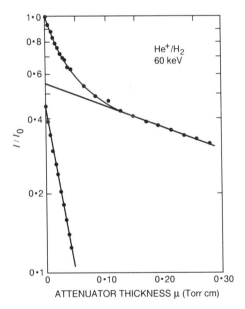

FIG. 5. Variation of the attenuation fraction $I/I_0$ with target thickness $\mu$ in a 60 keV He$^+$ beam [29]. A single break in slope indicates the presence of two independently charge-exchanging states. Extrapolation of the lower cross section (usually ground-state) slope to zero thickness gives a fractional concentration here of 45%.

(intensity $I$) strikes the secondary emission detector. The measured attenuation $I/I_0$ is given by

$$\frac{I}{I_0} = (1-f)\exp\left[-\mu\sum_n \sigma_{0n}\right] + f\exp\left[-\mu\sum_n \sigma_0*_n\right], \qquad (3.3)$$

where $\sigma_{0n}$ and $\sigma_{0*n}$ are cross sections for charge exchange from the ground and metastable $He^0$ states, respectively, leading to final states $n$. A plot of $I/I_0$ vs. $\mu$ (Figure 5) can often be resolved into two or more separate linear parts whose slopes give the total exchange cross sections, and the intercepts the population fractions in the ground and metastable neutral states.

Finally, limits to metastable population can also be established via laser-induced fluorescence (LIF) by searching for optical emissions uniquely coupled with the metastables, and via resonant multiphoton ionization (REMPI) by detecting the resulting ion or electron at photon excitation energies lower than the ground-state onset [19, 33].

One may also start with the negative ion of the atomic species to be produced, and charge-exchange ($\sigma_{10}$) or collisionally detach ($\sigma_D$) the electron with a suitable gas. Total cross sections ($\sigma_{10} + \sigma_D$, with $\sigma_{10} \gg \sigma_D$) for $H^-$, $D^-$, and $O^-$, have been measured [34–37]. This total can be quite large, in the range 50–100 $Å^2$ for energies less than 100 eV. Work on the $O^-$, $F^-$, and $S^-$ systems indicates that the final neutral state may be the ground state (F), or a mixture of the ground and metastable $^1D$ states (O, S) [38, 39]. Positive ions are produced via autoionization of the intermediate, doubly excited state of the neutral [40]. These should be removed along with undetached negative ions prior to the target.

### 3.2.3 Plasma Sources with Surface Neutralization

Surface neutralization of positively or negatively charged ions produced in an RF plasma has been used to generate species of energy in the range 4–32 eV (see Figure 6) [41–43]. Here, one generates the appropriate positive ion in an RF discharge ($N^+$, $O^+$, $Ne^+$, $Ar^+$, and $Kr^+$ have been produced). The ions are magnetically confined, accelerated to the desired final energy (including the effect of the plasma sheath), then neutralized and reflected by a metal plate. The resulting neutral beam retains about 40–50% of the incident ion energy, with an energy spread of 8–14 eV (FWHM), depending on the final energy [42].

The neutral beam diverges after the plate. A fraction is intercepted by a differential-pumping aperture/baffle and directed towards the target. The method is capable of producing fluxes of the order $10^{14}$–$10^{15}$ atoms/$cm^2$-sec in a target chamber located ~30 cm from the plate. The energy width is large and depends on instrumental parameters. Also, much remains to be understood about surface neutralization: One must have for each species a diagnostic of the final quantum state(s), and even of the species produced in the neutral beam. There is a

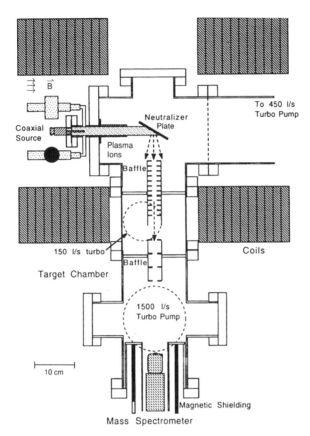

FIG. 6. Plasma/surface-neutralization source used to generate fast atoms [42]. The microwave-generated plasma particles are magnetically confined and directed towards a neutralization plate. A fraction of the reflected, charge-exchanged neutrals are intercepted by a pumping–collimating baffle and directed towards the target. The kinetic energy of neutrals can be adjusted by the incident ion energy, and to a lesser extent by the angle of surface incidence/reflection, and surface material.

dependence of the final energy and ratio beam energy/beam width on incident ion energy, on angle of incidence and reflection with respect to the surface normal, and on surface material. These should be quantified in each apparatus to determine the output energy and width.

### 3.2.4 Laser-Generated Plasma Sources

In order to fill a gap in the 1–10 eV beam energy range, and to utilize the high peak power levels available in pulsed laser sources, a laser-surface "blow-off" scheme was demonstrated for the production of pulsed, fast beams [44–46]. The target can be either a metal or a thin film on a glass substrate, or a dense "puff" of gas in vacuum [47–49]. While peak fluxes of the order $10^{17}$ atoms/cm²-sec have been attained [44], one does not have a priori knowledge of the species, quantum state, or energy spread without further diagnostics. Peak fluxes can be extended to the range $10^{19}$–$10^{21}$ atoms/cm²-sec using a 10 J laser pulse at 10.6 μm from a $CO_2$ laser [48]. It was pointed out for the laser blow-off source that, because of the high temperatures involved, atoms or molecules could be formed in excited states [44]. In the recent studies [48, 49], the presence of highly excited states in the neutral beam has been monitored through the decay with distance of the OI 777.3 nm transition, corresponding to the transition $3p\,^5P \rightarrow 3s\,^5S^\circ$ [50], as a probe of the neutral beam velocity. The energies of these states are 10.7 eV and 9.15 eV, respectively, above the $O(^3P)$ ground state. This line is thought to arise from the dissociative recombination $O_2^+ + e$. If so, one is assuming that the $O_2^+$ velocity is a measure of the neutral O-atom velocity within the beam. However, the $O_2^+$ ion may be subject to plasma potentials and stray accelerating fields which do not affect the neutral component(s).

The simultaneous presence of $O^+$ ions, electrons, metastables, and UV photons in this plasma [49, 51] makes the extraction of unambiguous information difficult. These particles can strike the target, causing concurrent excitation, dissociation, recombination, heating or, for dense targets (surfaces), synergistic effects.

### 3.2.5 Laser and Arc-Jet Sources

A continuous-wave laser has been used to sustain a plasma arc in a rare-gas mixture with the appropriate feed gas [52, 53]. A schematic diagram of the laser-sustained source is shown in Figure 8. The cw $CO_2$ laser beam is focused by lens L into a region of pressure $P_0 > 10^3$ torr of the $O_2$/rare gas mixture. Breakdown is initiated by means of a lower-power, pulsed $CO_2$ laser. The position of the plasma ball B relative to the nozzle is adjusted by moving lens L. Hydrodynamic expansion through the nozzle produces an energetic (1–10 eV) beam of the atomic species, with an energy width (FWHM) of 50–60% of the peak energy. Two chopper wheels are subsequently used to measure beam velocity through time of flight [53]. Because of the high laser powers deposited in the gas, the plasma ball is also a source of infrared-to-UV radiation. This broad radiation distribution, any collisionally unquenched metastable states, and possibly nozzle material can strike the target.

Use has been made of a 1.2–2 kW DC arc in He or Ar [54, 55], without a laser. The feed gas is injected into the hot gas downstream of the discharge. It dissociates to the desired atomic (and other) species, and the nonequilibrium flow is expanded out of a nozzle and the composition frozen. The central core of the

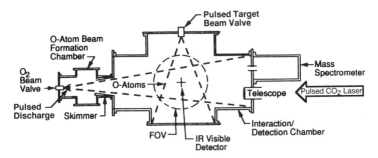

FIG. 7. Pulsed laser-plasma source used to generate fast atomic oxygen atoms and ions [48]. A pulsed $CO_2$ laser initiates breakdown in a pulsed $O_2$ beam. The plasma containing atoms and approximately 1% ions enters the interaction chamber. The target is either a surface or a second pulsed beam. Optical emissions are detected normal to the beam direction. A later version of this source utilizes a magnetic field to attenuate the charged-particle components of the plasma [49].

beam is skimmed off to give a centerline flux estimated to be of the order $10^{15}$ atoms/$cm^2$-sec. Beam velocities can be generated in the range $(1.5-4.0) \times 10^5$ cm/sec (0.2–1.3 eV for 0 atoms) by choice of the carrier gas. Two slotted, rotating disks serve to block the intense photon emission from the discharge, and to transmit the selected beam velocity. Fast beams of N, $N_2$, O, $O_2$, and $N_2O$ of various purities have been generated [54]. The flux can presumably be estimated through an intensity-calibrated quadrupole mass

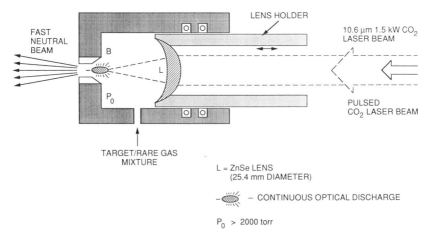

FIG. 8. Continuous-wave laser-sustained fast O-atom source [52, 53]. A cw $CO_2$ laser is focused by lens L into a high-pressure region of $O_2$ and Xe. Breakdown is initiated by the pulsed $CO_2$ laser. A plasma ball at B is generated, and fast atoms expand through the nozzle toward the target region.

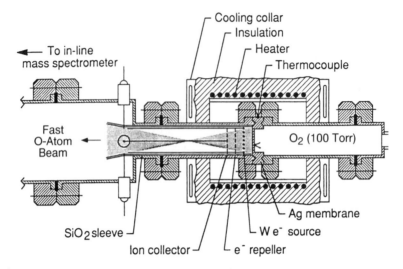

FIG. 9. Appratus for generating fast O-atoms through electron-simulated desorption [57]. Molecular $O_2$ on the high-pressure side adsorbs, dissociates, and diffuses through a thin Ag–ZrO$_2$ membrane. Desorption of a fast beam of the adsorbed species takes place through bombardment by 1 keV electrons produced by acceleration from a simple tungsten (W) emitter [57, 59].

analyzer, or through a calorimetric measurement. A related approach using an electrodeless RF discharge has been used to generate fast N and O atoms in a supersonic beam of $N_2$ or $O_2$ seeded in He [56].

### 3.2.6 Electron-Stimulated Surface Desorption

Surface desorption by an electron beam has been found to produce a hyperthermal beam of atomic oxygen [57–61], with a flux of the order of $10^{10}$–$10^{13}$ atoms/cm$^2$-sec [57, 61]. The technique (Figure 9) involves a supply of $O_2$ from a high-pressure reservoir to a heated (up to 800°C) Ag membrane with a 1000 Å ZrO$_2$ coating. Molecular oxygen adsorbs, dissociates, and diffuses through the membrane. The ultrahigh-vacuum side of the membrane is irradiated by a 1 keV electron beam. Desorbed species include O, O$^+$, and H$_3$O$^+$ or F. The energy distribution of O has not been reported, although the O$^+$ distribution is peaked at ~$5 \pm 4$ eV (FWHM) [57] or ~$6 \pm 3.6$ eV (FWHM) [61]. The population in excited states of O or O$^+$ has not been measured: One would expect a 1 keV electron beam to excite the long-lived, low-lying metastable levels in the neutral and ion as they desorb.

In a somewhat similar vein, electron-stimulated desorption of fast O and CO has been measured for a CO/Ru(001) surface using a pulsed electron gun and time-of-flight [62]. Kinetic energies of CO (with no or little internal electronic excitation) in the range 0.1–2 eV were measured. Neutral O and CO with energies in the range 0.5–5 eV, and with internal excitation, were also detected.

## 3.2.7 Photodissociation Sources

Molecular photodissociation has been used to generate fast atoms and to study in situ chemical reactions. As examples, fast H atoms have been produced through photodissociation of HBr and have subsequently been used to study the H + $N_2O$ reaction [63]. The H-atom energy distribution is bimodal, peaking at 2.21 eV and 2.67 eV for photodissociation at 193 nm. These energies correspond to the Br atoms in the $^2P_{1/2}$ and $^2P_{3/2}$ states, respectively. Fast $O(^3P)$ have been generated through photodissociation of $O_3$ [64, 65], and the energy distribution of the O-atoms measured through time-of-flight (TOF). The TOF results show a broad distribution peaking at $\sim$2 eV for photodissociation at 226.06 nm. The energy width is set by the rotational and vibrational distribution in the $O_2(X)$ fragment.

In general, the types of beams experiments possible via photodissociation are rather specific. One has limited control over the energy distribution of the atoms (set by the kinematics of the dissociation), and over their mean energy. For the latter, one must have photodissociation proceeding over a wide laser-wavelength range in order to study an energy dependence or threshold behavior of a reaction over a several electron–volt range. The photodissociation energy divides between the fragments, with only some fraction appearing as translational energy of the desired atom. The fragments—arising with rotational, vibrational and/or electronic excitation—can also enter into chemical reaction.

## Acknowledgments

Thanks are due to E. Murad for many discussions of the relative merits of fast-atom sources. This work was carried out at the Jet Propulsion Laboratory, California Institute of Technology, and was supported by the AFOSR and Phillips Laboratory through agreement with the National Aeronautics and Space Administration.

## References

1. Anderson, J. B., Andres, R. P., and Fenn, J. B. (1965). In *Advances in Atomic and Molecular Physics*, D. R. Bates and I. Estermann (eds.), Vol. 1. Academic Press, New York.

2. Van Zyl, B., Utterback, N. G., and Amme, R. C. (1976). *Rev. Sci. Instr.* **47**, 814.
3. Van Zyl, B., Le, T. Q., Neumann, H., and Amme, R. C. (1977). *Phys. Rev. A* **15**, 1871.
4. Van Zyl, B., Neumann, H., Rothwell, H. L., Jr., and Amme, R. C. (1980). *Phys. Rev. A* **221**, 716.
5. Gealy, M. W., and Van Zyl, B. (1987). *Phys. Rev. A* **36**, 3091.
6. Gealy, M. W., and Van Zyl, B. (1987). *Phys. Rev. A* **36**, 3100.
7. Orient, O. J., Chutjian, A., and Murad, E. (1990). *Phys. Rev. A* **41**, 4106. For details of the states and kinematics, see Orient, O. J., and Chutjian, A. (1995). *Phys. Rev. Letters* **74**, 5017.
8. Orient, O. J., Chutjian, A., and Murad, E. (1990). *Phys. Rev. Letters* **65**, 2359.
9. Orient, O. J., Martus, K. E., Chutjian, A., and Murad, E. (1992). *Phys. Rev. A* **45**, 2998.
10. Orient, O. J., Martus, K. E., Chutjian, A., and Murad, E. (1992). *J. Chem. Phys.* **97**, 4111.
11. Orient, O. J., Chutjian, A., Martus, K. E., and Murad, E. (1993). *Phys. Rev. A* **48**, 427; Orient, O. J., Chutjian, A., and Murad, E. (1994). *J. Chem. Phys.* **101**, 8297; Orient, O. J., Chutjian, A., and Murad, E. (1995). *Phys. Rev. A* **51**, 2094.
12. Massey, H. (1976). *Negative Ions*, 3rd edition. Cambridge University Press, New York.
13. Jansen, G. H. (1990). *Coulomb Interactions in Particle Beams*, Academic Press, New York; Zimmermann, B. (1970). In *Advances in Electronics and Electron Physics*, L. Marton (ed.), Vol. 29. Academic Press, New York.
14. Orient, O. J., and Chutjian, A., unpublished results.
15. Smirnov, B. M. (1982). *Negative Ions*, McGraw-Hill, Inc., New York.
16. Mori, Y. (1992). *Rev. Sci. Instr.* **63**, 2357; Ishikawa, J. (1992). *Rev. Sci. Instr.* **63**, 2368.
17. Sjolander, G. W. (1990). In *Materials Degradation in Low Earth Orbit*, V. Srinivasan and B. A. Banks (eds.). The Minerals, Metals and Materials Society, Warrendale, Pennsylvania.
18. Kaufman, H. R., Cuomo, J. J., and Harper, J. M. E. (1982). *J. Vac. Sci. Technol.* **21**, 725; Matsubara, Y., Tahara, H., Takahashi, M., Nogawa, S., and Ishikawa, J. (1992). *Rev. Sci. Instr.* **63**, 2595.
19. Becker, C., Copeland, R. A., and Slanger, T. G. (1993). Private communication.
20. Walther, S. R., Leung, K. N., and Kunkel, W. B. (1990). *Rev. Sci. Instr.* **61**, 315.
21. Gianturco, F. A., Palma, A., Semprini, E., Stefani, F., and Baer, M. (1990). *Phys. Rev. A* **42**, 3926.
22. Niedner, G., Noll, M., Toennies, J. P., and Schlier, Ch. (1987). *J. Chem. Phys.* **87**, 2685.
23. Freund, R. S., Wetzel, R. C., Shul, R. J., and Hayes, T. R. (1990). *Phys. Rev. A* **41**, 3575.
24. Shul, R. J., Freund, R. S., and Wetzel, R. C. (1990). *Phys. Rev. A* **41**, 5856.
25. Freund, R. S., Wetzel, R. C., and Shul, R. J. (1990). *Phys. Rev. A* **41**, 5861.
26. Herrmannsfeldt, W. B. (1988). Stanford Linear Accelerator Report No. SLAC-Report-331, Stanford Univ., Stanford, California; see also Becker, R. (1989). *Nucl. Instr. Methods* **B42**, 162.
27. Bernius, M. T., and Chutjian, A. (1989). *J. Appl. Phys.* **66**, 2783.
28. Foo, K. K., Lawson, R. P. W., Feng, X., and Lau, W. M. (1991). *J. Vac. Sci. Technol.* **A9**, 312.
29. Gilbody, H. B. (1978). *Inst. Phys. Conf. Ser. No. 38*, Ch. 4.

30. Aumayr, F., and Winter, H. (1989). *Phys. Scripta* **T28**, 96.
31. Wolfrum, E., Schweinzer, J., and Winter, H. (1992). *Phys. Rev. A* **45**, R4218.
32. Zuo, M., Smith, S. J., Chutjian, A., Williams, I. D., Tayal, S. S., and McLaughlin, B. M. (1995). *Astrophys. J.* **440**, 421.
33. Bamford, D. J., Jusinski, L. E., and Bischel, W. K. (1986). *Phys. Rev. A* **34**, 185.
34. Iluels, M. A., Champion, R. L., Doverspike, L. D., and Wang, Y. (1990). *Phys. Rev. A* **41**, 4809.
35. Gauyacq, J. P., Wang, Y., Champion, R. L., and Doverspike, L. D. (1988). *Phys. Rev. A* **38**, 2284.
36. Huq, M. S., Doverspike, L. D., Champion, R. L., and Esaulov, V. A. (1982). *J. Phys. B* **15**, 951.
37. Huq, M. S., Scott, D., Champion, R. L., and Doverspike, L. D. (1985). *J. Chem. Phys.* **82**, 3118.
38. Boumsellek, S., and Esaulov, V. A. (1990). *J. Phys. B* **23**, 279.
39. Boumsellek, S., and Esaulov, V. A. (1990). *J. Phys. B* **23**, 1303.
40. Boumsellek, S., Tuan, V. N., and Esaulov, E. (1990). *Phys. Rev. A* **41**, 2515.
41. Cuthbertson, J. W., Langer, W. D., and Motley, R. W. (1990). In *Materials Degradation in Low Earth Orbit,* V. Srinivasan and B. A. Banks (eds.). The Minerals, Metals and Materials Society, Warrendale, Pennsylvania.
42. Cuthbertson, J. W., Motley, R. W., and Langer, W. D. (1992). *Rev. Sci. Instr.* **63**, 5279.
43. Cuthbertson, J. W., Langer, W. D., and Motley, R. W. (1992). *J. Nucl. Mater.* **196–198**, 113.
44. Friichtenicht, J. F. (1974). *Rev. Sci. Instr.* **45**, 51.
45. Utterback, N. G., Tang, S. P., and Friichtenicht, J. F. (1976). *Phys. Fluids* **19**, 900.
46. Bakos, J. S., Bürger, G., Ignácz, P. N., Szigeti, J., and Kovács, J. (1988). *J. Phys. E: Sci. Instr.* **21**, 1095.
47. Caledonia, G. E., Krech, R. H., and Green, B. D. (1987). *AIAA J.* **25**, 59.
48. Upschulte, B. L., and Caledonia, G. E. (1992). *J. Chem. Phys.* **96**, 2025.
49. Sonnenfroh, D. M., Caledonia, G. E., and Lurie, J. (1993). *J. Chem. Phys.* **98**, 2872.
50. Baskin, S., and Stoner, J. A., Jr. (eds.) (1975). *Atomic Energy Levels and Grotrian Diagrams*, American Elsevier Publ. Co., New York.
51. Upschulte, B. L., Oakes, D. B., Caledonia, G. E., and Blumberg, W. A. M. (1992). *Geophys. Res. Letters* **19**, 993.
52. Cross, J. B., Cremers, D. A.., Sprangler, L. H., Hoffbauer, M. A., and Archuleta, F. A. (1986). *Proc. 15th Int. Symp. Rarefied Gas Dynamics*, V. Boffi and C. Cercignani (eds.). B. G. Teubner Publ., Stüttgart, Germany.
53. Cross, J. B., Blais, N. C. (1989). In *Rarefied Gas Dynamics: Space-Related Studies*, E. P. Muntz, D. P. Weaver, and D. H. Campbell (eds.). AIAA, Washington, D.C.
54. Greer, W. A. D., Stark, J. P. W., and Pratt, N. H. (1990). In *Rarefied Gas Dynamics, Proc. 17th Int. Symp.,* A. E. Beylich (ed.). VCH Publ., Weinheim, Germany.
55. Greer, W. A. D., Pratt, N. H., and Stark, J. P. W. (1993). *Geophys. Res. Letters* **20**, 731.
56. Pollard, J. E. (1992). *Rev. Sci. Instr.* **63**, 1771.
57. Outlaw, R. A., Hoflund, G. B., and Davidson, M. R. (1989). *J. Vac. Sci. Technol.* **A7**, 2087; NASA Tech Briefs (May 1988), p. 42; Outlaw, R. A., and Davidson, M. R. (1994). *J. Vac. Sci. Technol.* **A12**, 854.
58. Hoflund, G. B., Davidson, M. R., and Outlaw, R. A. (1992). *Interface Anal.* **19**, 325.

59. Davidson, M. R., Hoflund, G. B., and Outlaw, R. A. (1993). *J. Vac. Sci. Technol.* **A11,** 264.
60. Davidson, M. R., Hoflund, G. B., and Outlaw, R. A. (1993). *Surf. Sci.* **281,** 111.
61. Hoflund, G. B., and Weaver, J. F. (1994). *Meas. Sci. Technol.* **5,** 201.
62. Feulner, P. (1975). In *Electronic Transition, DIET II, Proc. 2nd Int. Workshop,* W. Brenig (ed.). Springer, Heidelberg.
63. Hoffman, G., Oh, D., Iams, H., and Wittig, C. (1989). *Chem. Phys. Letters* **155,** 356.
64. Kinugawa, T., Sato, T., Arikawa, T., Matsumi, Y., and Kawasaki, M. (1990). *J. Chem. Phys.* **93,** 3289.
65. Slanger, T. G., and Copeland, R. A. (1993). *Advances in Physical Chemistry: Current Problems and Progress in Atmospheric Chemistry,* **281,** 111.

# 4. VAPOR CELLS AND HEAT PIPES

## C. R. Vidal

Max Planck Institut für Extraterrestrische Physik
Garching bei München, Germany

## 4.1 Introduction

If some vapor is kept at a temperature $T$, its density $N$ is given by the ideal gas relation $N = P/kT$, where $k$ is the Boltzmann constant and where the static pressure $P$ is given by the vapor pressure curve of the enclosed element. $P$ is the pressure of the saturated vapor which is in equilibrium with its liquid phase and is only a function of the temperature $T$. In order to obtain a well-defined density in optical measurements, one has to provide an isothermal heater over the entire optical path length.

One special, extremely useful device that satisfies this requirement is the so-called heat pipe. In general, it can be designed either as a closed device or as an open device terminated by windows which are protected from the vapor of interest by inert gas boundaries. In the latter system, the vapor is accessible to optical investigations. These devices have been used in a number of applications in which the total and partial pressures have to be accurately known:

1. Measurements of vapor pressure curves,
2. measurements of oscillator strengths,
3. measurements of pressure broadening parameters and of various collisional transfer cross sections,
4. investigations of nonlinear processes in metal vapor/inert gas mixtures for sum frequency mixing, Raman processes, and other nonlinear effects, and
5. generation of metal vapor plasmas.

## 4.2 Thermal Vapor Cells

If a saturated vapor is enclosed in some volume $V$, the pressure is governed by the vapor pressure corresponding to the coldest point in the system. This determines the integrated column density $\int_0^L N\,dx$. Hence, it is important to supply a constant temperature of the heater to have a well-defined density over the entire optical path length. Because of the steepness of the vapor pressure curve, a small change in temperature gives rise to a significantly larger change in pressure, and the temperature requirements are generally more stringent than the

EXPERIMENTAL METHODS IN THE PHYSICAL SCIENCES
Vol. 29B

density requirements. For this reason it is better to stabilize the pressure than the temperature. The homogeneity requirements have been greatly improved by the invention of the heat pipe.

## 4.3 The Closed Heat Pipe

Since the original paper by Grover, Cotter, and Erickson [1], heat pipes have served the purpose of providing elements of very large heat conductivity [2, 3]. The arrangement in Figure 1 shows the original closed heat pipe. In such a device the vapor is driven from a heat source to a heat sink and returns as a liquid inside the wick back to the source because of capillary forces. In this manner continuous evaporation and condensation occurs which physically carries the heat of vaporization down the tube. The direction of the vapor flow, as indicated by the arrows in Figure 1, is only determined by the temperature difference between the heat source and the heat sink, which can, of course, be interchanged, whereby the direction of the heat flow in Figure 1 is reversed. With lithium as a working material, the temperature difference between the source and the sink has been shown to be less than 1°C at about 1000°C, depending on the heat transfer of the pipe. Such a device can also be shaped as a hollow tube and can thus be used as a liner around a conventional vapor cell to provide the necessary isothermal heating for establishing a well-defined vapor pressure in the cell. Other applications of the closed heat pipe range from very large devices such as heat exchangers in air conditioners down to very small devices for cooling semiconductors and solid-state lasers.

## 4.4 The Open-Ended Heat Pipe Oven

This system is used for spectroscopic observations if one wants to investigate a sample of homogeneous vapor and well-defined optical path length. In this case

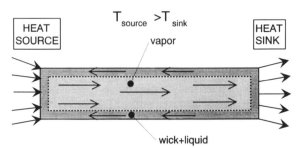

FIG. 1. Sectional view of a closed heat pipe, indicating the heat source and heat sink and the wick structure connecting them. Source and sink may be interchanged, reversing the direction of the vapor flow.

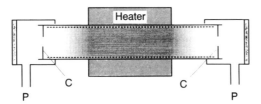

FIG. 2. Sectional view of an open-ended heat pipe as used in the heat pipe oven. Windows at its ends allow optical investigations, and apertures C eliminate convection of the confining inert gas. From Figure 1 of reference [4].

the heat pipe has to be modified as a *heat pipe oven* [4]. This is a heat pipe with windows at its ends which allow spectroscopic investigations of the vapor contained inside.

Similar to the closed heat pipe, the normal, open-ended heat pipe oven shown in Figure 2 consists of a tube with a mesh structure on its inner surface which acts as a wick. Each end of the tube is terminated by windows which are separated from the vapor by inert gas boundaries. It again has a heat source and a heat sink and contains the working material to be investigated. Because of the power supplied by the external heater, the working material of interest is evaporated inside the central portion of the tube. It is particularly important to note that, as will be discussed, its vapor pressure is equal to the externally adjusted inert gas pressure. The external heater may be an ohmic heater or an induction heater or some other heating element. Induction heaters are generally preferred at elevated temperatures, and the heating zone is typically surrounded by vacuum to avoid corrosion of the pipe at these temperatures. The pipe itself may be manufactured from some refractory metal such as tungsten or molybdenum.

The vapor flows out of the center toward the ends of the tube and condenses in the outer, slightly colder parts of the heat pipe oven. The condensed liquid then returns through the mesh back to the heater section because of capillary forces. In this regard, the wetting properties of the wick are very important because they determine whether the liquid can return to all areas of the tube. If this does not occur because of changes in the surface properties of the wick, areas of the pipe may be superheated, forming so-called "hot spots," because no new vapor can be formed and these areas are not properly cooled.

The vapor zone finally extends over a length of the heat pipe oven for which the power supplied by the heater is balanced by the losses due to radiation and due to heat conduction through the walls of the tube. Because of the continuous evaporation and condensation of the working material, a large amount of heat is transferred through the heat pipe oven. This gives rise to the very large heat conductivity of a heat pipe oven, which can exceed the thermal conductiviy of solids by orders of magnitude. Hence, it provides an extremely homogeneous temperature and density distribution over the length of the vapor column. Finally, a stable equilibrium is reached in which the central vapor column is confined by inert gas boundaries whose pressure determines the vapor pressure inside the heat pipe oven. It is important to note that a heat pipe oven provides a "dynamic equilibrium" in which the continuous transport of the vapor pushes the confining inert gas away from the central portion of the heat pipe oven. It does not provide a "static equilibrium" in which the pressure is given by the sum of the partial pressures which would be the same over the entire system. The inert gas has the additional virtue of protecting the windows at the end of the tube from any vapor deposition, and hence from any corrosion associated with it. Any residual vapor which is transported by the convection of the confining inert gas due to the temperature gradient of the enclosing tube can be minimized by inserting small apertures C which prevent the circulation of the inert gas carrying some small droplets of vapor (fog formation). They are indicated at the end of the tube in Figure 2. The fog formation in the transition zone between vapor and inert gas can be observed by irradiating the transition zone with a laser and observing the Mie scattering.

In contrast to conventional vapor cells, in which one tries to maintain a constant temperature distribution over the entire cell, the heat pipe oven provides a vapor column of constant pressure which is equal to the confining inert gas pressure. To model a heat pipe oven, a one-dimensional flow model has been described in reference [5] where the vapor parameters are determined using the Bernoulli equation. The onset of sonic flow limits the ultimate flow velocity of the vapor inside the heat pipe, and hence the total heat conductivity. Heat pipes should typically be operated at pressures of a few millibars or more, to minimize the vapor flow velocity inside the Bernoulli equation. Because the ultimate flow velocity is equal to the sound velocity, as demonstrated in laser-induced fluorescence of $Li_2$ molecules [5] in an open-ended heat pipe oven, the confining inert gas pressure at room temperature should be large enough to avoid start-up problems. They occur if the vapor pressure at room temperature is very small, so that the vapor has to approach the sound velocity to maintain the proper heat conductivity and hence the overall power balance. In closed heat pipes, this may be circumvented by inserting some inert gas whose pressure determines the minimum vapor pressure for operating the closed heat pipe.

Heat pipe ovens can also be operated with vapors such as barium confined by reactive gases such as oxygen [6]. In this case chemical reactions take place, and the generated molecules in the transition zone can only be investigated qualitatively, because the partial pressure of the reacting vapor in the transition zone is only poorly defined.

The heat pipe oven has also been built in the shape of a disk [7] with a cylindrical glass ring as a window which provides the spacing between the upper and the lower disk. In this design the vapor can be viewed from all angles, a property which might be valuable in nonlinear optics. Closed disk-shaped devices have also been used to distribute the heat from a small spot over the entire diameter of the disk. This may be necessary for operating power meters at the very large input powers provided by the focused beams of high-power lasers.

The need for a wick inside the heat pipe oven may be eliminated altogether in a rotating heat pipe oven which can be sealed with magnetic fluids [8]. Instead of the wick, it uses centrifugal forces for returning the liquid. However, for optical studies, vapor cells without any moving parts are to be preferred. The rotating heat pipe oven is only of interest for materials with wetting problems which provide a small surface tension with the material of the wick.

Open-ended heat pipe ovens have been used in a large number of studies, and these are briefly summarized next.

### 4.4.1 Vapor Pressure Measurements

One of the most fundamental applications of the heat pipe oven is the accurate measurement of vapor pressure curves. In the system used by Bohdansky and Schins [9, 10], one end of the oven contains a thin tube that is connected to a reservoir containing the inert gas. Measurement of the inert gas pressure, which is identical to the vapor pressure, gives directly the vapor pessure of interest.

### 4.4.2 Crossed Heat Pipes

In order to investigate the vapor of interest spectroscopically, the vapor is frequently irradiated by a laser beam which enters through one of the windows. The beam has to be well collimated to avoid difficulties with scattered light. In another method the entire heat pipe oven is built in the shape of a cross, as shown in Figure 3, where the laser enters the vapor through one arm of the cross, and the laser-induced fluorescence is observed at right angles through the other arm of the heat pipe oven by means of a detector $D$. In this case, the wicks of the two arms should be well connected to obtain a homogeneous and identical vapor pressure in all arms.

In both cases, reabsorption by the thermal vapor can be a problem for quantitative measurements. To avoid reabsorption, one should choose fluorescence transitions that do not end on thermally populated levels of the vapor. In those cases where windows are available that do not react with the vapor, it is possible to insert a tube with a window at the end which reduces the reabsorbing vapor column in the side arm.

### 4.4.3 Superheated Heat Pipes

There is another problem in generating well-defined samples of vapor. In providing, for example, alkali vapors, one generates besides the atomic species sizeable amounts of dimers and large clusters [11]. This can be overcome to

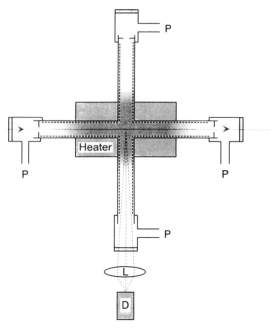

Fɪɢ. 3. Crossed heat-pipe oven for the observation of laser-induced fluorescence with a detector D.

some extent by superheating the vapor by inserting a separate heating element inside the vapor [12]. The vapor parameters are then no longer determined by the vapor pressure curve of the saturated element in equilibrium with its liquid phase, but are rather determined by the temperature of some volume element whose temperature lies above the vapor pressure curve and has been raised at a fixed density to an unsaturated value. In this manner, the partial pressure, and hence the concentration, of molecules or clusters can be lowered. This is important if molecular absorption is undesirable and must be avoided.

### 4.4.4 Measurement of Oscillator Strengths

Another very important application is the measurement of oscillator strengths. This requires the determination of the optical depth $\tau = \kappa N L$, where $\kappa$ is the desired absorption coefficient. For this purpose, a constant density $N$ and a well-defined length $L$ must be achieved. The homogeneity over the optical depth is provided by the large heat conductivity of the heat pipe oven. However, the optical path length is not well defined because of the transition zones between the vapor and the inert gas due to the mutual diffusion of the two components.

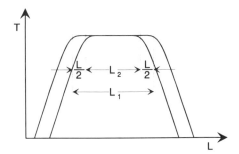

FIG. 4. Temperature distribution along an open-ended heat pipe oven measured at two different input powers, $H_1$ and $H_2$, which provide identical transition zones between the metal vapor and the confining inert gas. Taken from Figure 2 of reference [4].

This problem can be overcome in a very elegant way by using measurements at different heater powers, which provides vapor columns of different lengths, but with identical transition zones if the pressure is maintained constant, as indicated in Figure 4. By measuring at different powers, the effect of the transition zones can be cancelled. A more elegant approach is indicated in Figure 5, where two heat pipes are interconnected and hence are operated at the same vapor pressure $P$, but with different heater powers $H_1$ and $H_2$, respectively. In this manner, vapor columns of different lengths, $L_1$ and $L_2$, but identical transition zones, are provided. Hence, by measuring the optical depth $\tau$ in a double-beam experiment, the transition zones are automatically compensated and the effective column density $L$ is given by the difference in length, $(L_2 - L_1)$, at a given density in the two pipes. This can be accurately measured using an

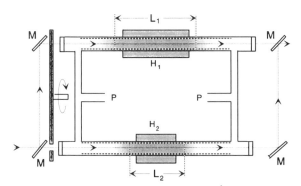

FIG. 5. Double-beam arrangement with two heat pipe ovens in which the transition zones from the vapor to the confining inert gas cancel and a simple and accurate measurement of oscillator strengths can be performed. The chopper wheel passes the beam alternately through one or the other of the two heat pipe ovens. From Figure 4 of reference [17].

optical pyrometer to determine the temperature profiles on the surface of the pipe. In this manner an optical depth $\tau$ given by $\tau_1 - \tau_2 = \kappa N(L_1 - L_2) = \kappa NL$ is defined.

Figure 5 shows a simple setup for a double-beam experiment. The chopper wheel alternately passes the beam through one or the other of the two heat pipes.

### 4.4.5 Metal Vapor Plasmas

Metal vapor plasmas can be generated by exciting the whole vapor column inside the heat pipe oven using either a low-pressure arc or a positive column or a negative glow discharge [13]. (Other methods such as electron beam excitation can, of course, also be applied.) To establish a low-pressure arc, one can insert a heated tungsten wire down the axis of the heat pipe oven. If the heat pipe is filled with alkali or alkaline earth metals which lower the work function of the cathode, large emission currents can be obtained. Eventually one does not even have to provide the external heating as the discharge itself acts as the heat source. Currents of the order of several hundred amps at a few volts can easily be drawn across the vapor column. In the limit of vanishing currents, the system approaches the behavior of a thermionic diode [14]. Another method of achieving an arc inside the vapor was suggested by Boyd and Harter [15] using a disk-shaped heat pipe oven [7].

For a glow discharge, the mesh structures should be split into at least two sections which are mutually insulated. This can be done by using a ceramic tube for the heat pipe with separate, insulated sections of the mesh inside. Depending on the polarity and the design of the different mesh sections, one can either run a negative glow discharge or a positive column in the metal vapor section. In this manner, emission spectra of the neutral or the ionized element can be generated. In another method, a flute-type hollow cathode discharge generating a negative glow discharge was run inside a heat pipe oven [16].

## 4.5 The Concentric Heat Pipe Oven

To provide extremely homogeneous metal vapor/inert gas mixtures, a new device, the concentric heat pipe oven, was developed [17]. It is based on the principle of the normal, open-ended heat pipe oven. It is required for many situations in spectroscopy. The schematic arrangement of the concentric heat pipe oven is shown in Figure 6. An inner heat pipe oven A, containing the metal vapor/inert gas mixture, is surrounded by an outer heat pipe oven B, acting as an isothermal heater which imposes an extremely homogeneous temperature distribution on the inner heat pipe oven A. The temperature of the outer heat pipe oven B, and hence also of the inner heat pipe A, is determined by the confining

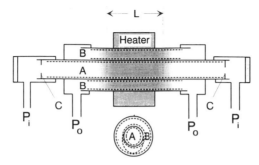

FIG. 6. Schematic arrangement of the concentric heat pipe oven which contains an outer heat pipe oven B as a well-defined heater. It provides a homogeneous metal vapor/inert gas mixture of adjustable length in the inner heat pipe A. The overall optical path length depends on the heater power supplied. The loop-shaped structures in the outer heat pipe B connect the outer and inner mesh structures avoid the depletion of the liquid in one of the wicks and avoid the formation of any hot spots. From Figures 1 and 3 of reference [17].

inert gas pressure $P_o$ according to the vapor pressure curve of the working material. Frequently, the inner and outer heat pipe ovens are operated with the same working material. Then the inert gas pressure $P_i$ of the inner heat pipe A should be larger than $P_o$, and the total pressure of the outer heat pipe oven B is equal to the partial pressure of the working material in the inner heat pipe oven A. Hence, the inner heat pipe oven is filled with a metal vapor/inert gas mixture where the partial pressure of the metal vapor is given by $P_o$ and the partial pressure of the inert gas by $(P_i - P_o)$. This requires $P_i > P_o$. Similar to the heat pipe oven discussed earlier, the concentric system should have apertures C to eliminate the convection of the confining inert gas. They are shown in Figure 6 at the end of the inner heat pipe A.

Compared with static cells containing metal vapor/inert gas mixtures, the advantage of a concentric heat pipe oven is that the partial pressures can be easily and independently adjusted at any time by adjusting the pressures $P_i$ and $P_o$. This aspect is particularly valuable for the very stringent phase-matching requirements in nonlinear optics exploiting the wavelength-dependent refractive indices of the constituents according to the Sellmeier equations.

For technical reasons, it can sometimes be advantageous to operate the outer heat pipe oven B with a different working material which has a higher vapor pressure for a particular desired temperature than the working material of the inner heat pipe oven A. This can eliminate start-up problems in the outer heat pipe oven and is generally recommended if the metal vapor in the inner heat pipe oven A has to be operated at rather low pressures to maintain a sufficiently small optical depth. As an example, Sr–Xe or Mg–Kr mixtures in four-wave frequency-mixing experiments [18] have frequently been operated in a concentric heat pipe oven where the outer heat pipe system B is operated with sodium and an inert gas such as argon. This can, of course, be different from the inert gas used in the inner heat pipe oven A. The larger vapor pressure in the outer heat pipe oven leads to reduced flow velocities in the vapor zone for maintaining a particular power balance [5]. It leads to greater stability and homogeneity of the whole heat pipe oven, as indicated by smaller short-term pressure fluctuations. The partial pressure of the metal vapor inside the iner heat pipe oven is no longer given by the total pressure $P_0$ of the outer heat pipe oven. However, it can be easily calibrated against the pressure in the outer heat pipe.

## 4.6 Crossed Concentric Heat Pipe Oven

The length of the metal vapor/inert gas mixture for the concentric heat pipe oven in Figure 6 depends on the heater power supplied. Small changes in the power balance will lead to corresponding changes of the total column density inside the heat pipe oven. For maintaining the phase-matching condition in situations with a pronounced oscillatory structure of the nonlinear process [18], it may be necessary to stabilize the heater power. Furthermore, for establishing a stable equilibrium, a long warm-up time is generally required.

In order to circumvent these problems, a modified concentric heat pipe oven was developed [19]. It combines the virtue of the accurately defined partial pressures with a well-defined column density. This has been achieved with a system shown in Figure 7, where the length of the vapor column in the heat pipe oven A is defined by the diameter $L$ of the vertical heat pipe oven B, which acts again as a well-defined isothermal heater. Any changes in the power balance will immediately move the transition region between the metal vapor and the confining inert gas inside the vertical heat pipe oven B up or down. However, it will not affect the mechanically defined, fixed length of the heating zone L which is imposed on the inner heat pipe oven A. In addition, stable operation of the modified concentric heat pipe system is achieved already long before the vapor column in the vertical heat pipe oven B has reached its final equilibrium position.

In practice it is advantageous to insert a helical piece of sheet metal in the upper vertical tube B above the horizontal tube A. This has no influence on the pressure balance, but prevents droplets of the condensed liquid falling directly onto the horizontal tube A, which would locally cool the heat pipe A until the liquid has completely evaporated again.

For Gausian laser beams, it is primarily the pressure ratio and not so much the length that, according to the confocal parameter, must be maintained, as long as the beam waist stays inside the metal vapor/inert gas mixture. In practical applications, the modified concentric heat pipe oven has so far provided the most

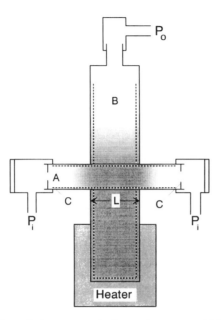

FIG. 7. A modification of the concentric heat pipe oven which provides a stable, well-defined column density that is independent of the heater power supplied and only depends on the diameter of the outer heat pipe L. From Figure 1 of reference [19].

stable and well-defined metal vapor/inert gas mixtures. In such applications, it can also be operated as a sealed-off system.

The concentric heat pipe oven has been extremely useful in a number of applications which require accurate partial pressures of the metal vapor/inert gas mixture and which will be mentioned in the following sections.

### 4.6.1 Measurement of Collisional Cross Sections

Collisional cross-sections in mixtures of vapors and inert gases can be measured directly. They give rise either to pressure broadening of spectral lines whose line width determines the broadening parameters, or to an inelastic energy transfer to other levels of the system whose pressure dependence yields the corresponding transfer cross sections.

### 4.6.2 Four-Wave Frequency Mixing

One of the most important applications of the concentric heat pipe oven has been the generation of coherent VUV or IR radiation by means of four-wave frequency mixing where energy and momentum have to be conserved [18, 20]. This requires the following conditions:

1. The nonlinear susceptibility $\chi_T^{(n)}$ has to be sufficiently large. This may be achieved by exploiting one or more suitable multiphoton resonances of the nonlinear medium.

2. For large conversion efficiencies, it is necessary to maintain proper phase matching such that the fundamental and product waves travel with the same phase velocity through the nonlinear medium.

3. According to the fundamental equations of nonlinear optics [18], the optical depths at the incident and the generated wavelengths $j$ should be sufficiently small that $\tau_j < 1$.

4. In practical applications, one should operate the nonlinear medium at laser powers which just approach the onset of saturation. Any higher laser power merely saturates the system and does not raise the efficiency, and indeed can lead to complicated behavior due to higher-order nonlinear susceptibilities which affect the refractive indices and hence the phase matching.

Because of the third requirement, gaseous media inside a concentric heat pipe oven are generally used for four-wave frequency mixing in the VUV. The first two requirements are satisfied using a two-component system where the first component (metal vapor) is selected to provide a large nonlinear susceptibility due to a suitable multiphoton resonance, and the second component (inert gas) is responsible for achieving the phase matching. Because the linear and nonlinear susceptibilities are determined by the transition moments and the positions of the atomic or molecular energy levels, the performance of a particular gaseous nonlinear medium depends strongly on the frequency of the incident radiation. Over a wide range of frequencies, the linear and nonlinear properties are dominated by the properties of the first few resonance transitions. For this reason, one uses metal vapor/inert gas mixtures in the near VUV ($\lambda < 200$ nm), whereas in the far vacuum ultraviolet ($\lambda < 100$ nm), also labeled XUV, it may frequently be more appropriate to employ pure inert gas mixtures such as those provided in pulsed nozzle beams [21, 22]. In order to meet the very stringent homogeneity requirements imposed by the phase-matching condition, the metal vapor inert gas mixtures are generally produced using a concentric heat pipe oven [17, 19].

## 4.7 Modified Concentric Heat Pipe Oven for Mixtures of Different Vapors

In order to achieve phase matching in metal vapor/inert gas mixtures, one may sometimes need rather high inert gas pressures at a particular wavelength for achieving the desired refractive indices. This gives rise to a large homogeneous line width of the nonlinear medium and may lead to a reduction of the conversion efficiency, in particular for situations with a small laser line width

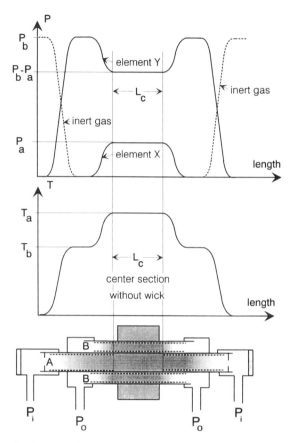

FIG. 8. Schematic diagram of the temperature distribution and of the partial pressures along the inner heat pipe of the modified concentric heat pipe oven shown in the bottom of the figure operated with two different metal vapors $X$ and $Y$. From Figure 3 of reference [26].

[23]. It may therefore be advantageous to use mixtures of different metal vapors for phase matching, as shown by Wynne *et al.* [24] in the infrared using a Na–K system and by Bloom *et al.* [25] in the ultraviolet region using a Na–Mg system. For a nonlinear medium of this kind, a different type of concentric heat pipe oven can be used that generates mixtures of a *saturated* and an *unsaturated* metal vapor [26], where the individual pressures can be independently adjusted at any time.

The important difference in a two-metal heat pipe is that the wick in the inner heat pipe is in two sections with a gap of length $L_c$ in its center, as indicated at the bottom of Figure 8. Hence, the liquid of the condensate cannot return to this

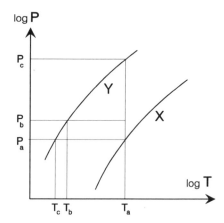

FIG. 9. Vapor pressure curves of two elements X and Y, showing the different pressures and temperatures of a heat pipe oven containing the two elements. For purposes of clarity, the temperature scale has been greatly expanded.

center section after a stable temperature distribution has been established. It purposely generates a large "hot spot" in the central section.

While the outer heat pipe is filled with an element X, the inner heat pipe contains both materials, X and Y. The pressure of the confining inert gas is $P_o$ in the outer heat pipe and $P_i$ in the inner heat pipe, where $P_i > P_o$. If the outer heat pipe is heated at a power level where it again acts like a conventional heat pipe oven, the outer heat pipe serves as a homogeneous heater. For a particular vapor pressure curve, its temperature $T_o$ is completely specified by the confining inert gas pressure $P_o$. As indicated in Figure 6, the outer wick and the inner wick of the outer heat pipe are both connected by loop-shaped wick structures to avoid depletion of the liquid material in one of the wicks, which gives rise to unwanted hot spots.

To understand the behavior of the inner heat pipe, we must realize that it contains two materials with different vapor pressure curves, as shown in Figure 9. In an ordinary heat pipe oven, the temperature–pressure characteristic is determined by the vapor pressure curve of the element Y with the larger vapor pressure if that element can be saturated anywhere inside the tube. If $P_o < P_c$, then the vapor Y is pushed out of the heat pipe into the colder areas which have a lower vapor pressure. This can, by the way, also be exploited for a fractionated distillation of two different vapors, X and Y.

When the center section of the outer heat pipe containing element X is heated, the confining inert gas pressure $P_o$ is equal to some pressure $P_a$, which gives rise to a temperature $T_a$ which is identical in the outer and inner heat pipes. Because initially both elements X and Y are located in the center section of the inner heat pipe, the partial pressure of X is $P_a$ and of Y is $P_c$ for a temperature $T_a$, as shown in Figure 9, as long as both vapors are saturated.

If, however, the confining inert gas pressure of the inner heat pipe $P_i$ is equal to $P_b$, where $P_a < P_b < P_c$, the vapor of element Y will be driven out to the cooler areas of the pipe, while element X remains in equilibrium with its liquid phase. Because there is no wick in the center section, the liquid of element Y vaporizes and becomes depleted because it cannot return. Element Y is finally unsaturated in the center section and has a partial pressure $(P_i - P_a)$. In this manner we have achieved in the center section of the inner heat pipe an unsaturated pressure $(P_b - P_a)$ of element Y, together with a saturated vapor X at a pressure $P_a$. The partial pressures are completely specified by the inert gas pressures $P_o = P_a$ of the outer heat pipe and $P_i = P_b$ of the inner heat pipe.

Without the wick inside the inner heat pipe, both elements X and Y would eventually be deposited in the colder parts of the pipe. To avoid this, both materials are returned by the capillary forces of the wick-covered sections adjacent to the center section. Hence, we have an outer heat pipe operated with element X at $P_a$ and an inner heat pipe with a mixture of a saturated element X and an unsaturated element Y in the center section. As a consequence, the inner heat pipe cools the outer heat pipe very effectively, such that for the outer heat pipe the length of the zone at temperature $T_a$ is just as long as the length $L_c$ of the inner heat pipe without a mesh.

This system is more difficult to operate than a system containing a metal vapor/inert gas mixture; it has therefore only been used in a few cases. The system must be operated at an optimum heater power that has to be adjusted in such a manner that the heater power supplied establishes a vapor zone in the outer sections of the pipe of the desired length.

The behavior of the inner heat pipe is illustrated in Figure 9, which shows schematically the distribution of the temperature as well as the various partial pressures along the inside of the inner heat pipe for the optimum input power mentioned earlier. The center section of length $L_c$ has a temperature $T_a$ which is determined by the confining inert gas pressure $P_o = P_a$ of the outer heat pipe. The partial pressure of the saturated element X is therefore $P_a$, while the partial pressure of the unsaturated element Y is $(P_b - P_a)$ for $P_i = P_b$. Outside the center region, the temperature drops rapidly, and its value $T_b$ is determined by the vapor pressure curve of element Y at the confining inert gas pressure $P_b$.

As an alternative method Engelke et al., have suggested an injection heat pipe oven [27], where the second metal is injected through a thin capillary tube into the main vapor column of the heat pipe oven. Devices of this kind have a lower homogeneity and a finite time of operation depending on the size of the reservoir containing the second metal.

### 4.7.1 Applications of the Two-Metal Concentric Heat Pipe Oven

Modified two-metal concentric heat pipe ovens can be used to undertake spectroscopy of heteronuclear molecules such as NaLi [26, 28] and MgCa [29]. This is particularly interesting if the partial pressures of the vapors are rather far away from the vapor pressure curves of one of the individual, saturated elements.

As noted previously, two-metal ovens are used in situations where vapor/inert gas systems require an extremely large inert gas pressure and a metal vapor/inert gas mixture becomes therefore impractical.

## 4.8 General Technical Considerations for the Heat Pipe Oven

The principle of the heat pipe oven imposes several technical constraints on its design. For stable operation, the metal vapor should have a vapor pressure in excess of a few millibars at temperatures below 1200°C, where high temperature–resistant stainless steel can be used for constructing the oven. Otherwise, one can use an induction heater in conjunction with a tube made of some refractory metal such as tungsten or molybdenum, to achieve higher temperatures. One should also avoid metals forming a eutectic with the tube material [30], which changes properties of the heat pipe oven such as its temperature resistivity, and hence its mechanical stability. As a further consideration, one should select mesh materials that have no wetting problems to avoid the formation of hot spots. The welding seams in crossed heat-pipe ovens must be designed with great care because they also have to resist the vapor and avoid any fatigue. If possible, one should try to use the same material as the material of the tube.

## References

1. Grover, G. M., Cotter, T. P., and Erickson, G. F. (1964). *J. Appl. Phys.* **35**, 1990.
2. Dunn, P., and Reay, D. A. (1976). *Heat Pipes*, Pergamon Press, New York.
3. Ivanovskii, M. N., Sorokin, V. P., and Yagodkin, I. V., translated by R. Berman (1982). *The Physical Principles of Heat Pipes*, Clarendon Press, Oxford.
4. Vidal, C. R., and Cooper, J. (1969). *J. Appl. Phys.* **40**, 3370.
5. Vidal, C. R. (1973). *J. Appl. Phys.* **44**, 2225.
6. Sakurai, K., and Broida, H. P. (1976). *J. Chem. Phys.* **65**, 1138.
7. Boyd, R. W., Dodd, J. G., Krasinski, J., and Stroud, C. R. (1980). *Opt. Lett.* **5**, 117.
8. Hessel, M. M., and Lucatorto, T. B. (1973). *Rev. Scient. Instr.* **44**, 561.
9. Bohdansky, J., and Schins, H. E. J. (1965). *J. Appl. Phys.* **36**, 3683.
10. Bohdansky, J., and Schins, H. E. J. (1967). *J. Phys. C* **71**, 215.
11. Nesmeyanov, A. N. (1963). *Vapor Pressure of the Elements*, Academic Press, New York.
12. Milošević, S., Beuc, R., and Pichler, G. (1986). *Appl. Phys. B* **41**, 135.

13. Vidal, C. R. (1972). *Physik in unserer Zeit* **3**, 174.
14. Niemax, K. (1985). *Appl. Phys. B* **38**, 147.
15. Boyd, R. W., and Harter, J. (1980). *Appl. Opt.* **19**, 2660.
16. Veža, D., and Pichler, G. (1983). *Opt. Commun.* **45**, 39.
17. Vidal, C. R., and Haller, F. B. (1971). *Rev. Scient. Instr.* **42**, 1779.
18. Vidal, C. R. (1987). "Four-Wave Frequency Mixing in Gases," in *Tuneable Lasers*, L. F. Mollenauer and J. C. White (eds.), *Topics in Appl. Phys.* **59**, 57–113.
19. Scheingraber, H., and Vidal, C. R. (1981). *Rev. Scient. Instr.* **52**, 1010.
20. Hanna, D. C., Yuratich, M. A., and Cotter, D. (1979). *Nonlinear Optics of Free Atoms and Molecules*, Springer Series in Optical Sciences, Vol. 17. Springer, Berlin.
21. Marinero, E. E., Rettner, C. T., and Zare, R. N. (1983). *Chem. Phys. Lett.* **95**, 486.
22. Bethune, D. S., Rettner, C. T. (1987). *IEEE J. Quant. Electron.* **QE-23**, 1348.
23. Leubner, C., Scheingraber, H., and Vidal, C. R. (1981). *Opt. Commun.* **36**, 205.
24. Wynne, J. J., Sorokin, P. P., and Lankard, J. R. (1974). In *Laser Spectroscopy*, R. G. Brewer and A. Mooradian (eds.), pp. 103–111. Plenum, New York.
25. Bloom, D. M., Young, J. F., and Harris, S. E. (1975). *Appl. Phys. Lett.* **27**, 390.
26. Vidal, C. R., and Hessel, M. M. (1972). *J. Appl. Phys.* **43**, 2776.
27. Engelke, F., Ennen, G., and Meiwes-Broer, K. H. (1984). *Chem. Phys.* **83**, 187.
28. Hessel, M. M., and Jankowski, P. (1972). *J. Appl. Phys.* **43**, 209.
29. Atmanspacher, H., Schreingraber, H., and Vidal, C. R. (1985). *J. Chem. Phys.* **82**, 3491.
30. Hansen, M., and Anderko, K. (1958). *Constitution of Binary Alloys*, McGraw-Hill, New York; Elliot, R. P. (1965). First Supplement. McGraw-Hill, New York; Shunk, F. A. (1969). Second Supplement, McGraw-Hill, New York.

# 5. FREE RADICAL SOURCES

## K. M. Evenson

National Institute of Standards and Technology
Boulder, Colorado


## J. M. Brown

Department of Physical Chemistry
Oxford University

In the early days of understanding molecular composition, it was known that stable molecules were built up from characteristic parts, which were called radicals. When the bond in a molecule is broken, the components can have a fleeting existence of their own; these fragments are called *free radicals*. For example, water can be broken into two free radicals: a hydroxyl radical, OH, and atomic hydrogen. Free radicals are atoms or molecules which have one or more unpaired electrons; consequently, they have a great propensity to react with other species and have a very short lifetime in most environments. They react with other free radicals or with stable species to form other stable molecules. Much of the interest in free radicals derives from the roles they play as intermediates in chemical reactions.

Free radicals are often created by dissociating stable molecules. In the simplest case, atomic radicals are created by dissociating diatomic gases such as $O_2$, $N_2$, $H_2$, $F_2$, HCl, or $Cl_2$; or, if a polyatomic molecule is used, then both atomic and molecular radicals are created, as in dissociating $H_2O$, $NH_3$, or $CH_4$ to produce OH and H, N and H and NH and $NH_2$, or H and $H_2$ and C and CH and $CH_2$ and $CH_3$, respectively. Dissociation can occur in electrical discharges, by photolysis, in ovens, by laser ablation, and in chemical reactions.

## 5.1 Electrical Discharges

Much of the progress in the spectroscopy of transient molecules over the last 20 years has depended on the use of electrical discharges (either AC or DC) to generate free radicals, ions, or metastables. Electrical discharges are particularly useful in forming ionic species.

EXPERIMENTAL METHODS IN THE PHYSICAL SCIENCES
Vol. 29B

### 5.1.1 The Microwave Discharge

A microwave discharge operating at 2450 MHz is one of the most commonly used sources for producing free radicals and ions. This frequency is set aside by international agreement for use in microwave ovens, medical diathermy units, and scientific apparatus. Water and many oils have absorptions at this frequency; hence, it is useful for heating many biological materials. The radiation, with a wavelength of 12.25 cm, permits the construction of conveniently sized microwave cavities for running electrodeless, and hence "clean," discharges.

A review and test of many discharge cavities was made in 1964 [1]. Cavity 5 in that review is still the one most commonly used and is shown in Figure 1. This cavity is a quarter-wavelength coaxial cavity tuned by varying the length of the center electrode (the resonant frequency is not dependent on the diameter of the cavity); it also has a variable coupling adjustment on the coaxial input line. The addition of one-half-wavelength-long metal tubes covering the discharge tube on

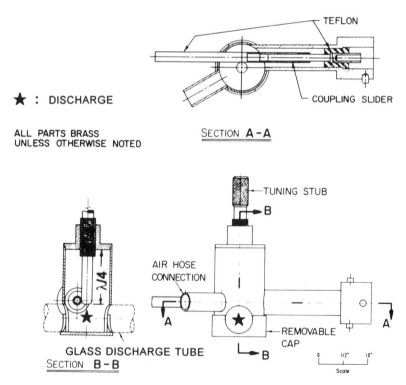

FIG. 1. Quarter-wavelength coaxial cavity operating at 2450 MHz. (Cavity 5 from reference [1].) The discharge location is shown with a star in the lower views.

each side of the cavity prevents radiation from leaking from the cavity [2]. This is very important, for this radiation can produce cataracts [3]. This shielding also dramatically reduces electromagnetic pickup in microwave detectors due to microwave leakage. A high-$Q$, $TMO_{010}$ cylindrical cavity which operates at high pressures has also been developed [4], but its use is inconvenient because it cannot be removed from the vacuum system without dismantling the system; consequently, it is seldom used.

Cavity 5 was used to produce atomic oxygen from molecular oxygen for the direct measurement of the fine structure frequencies of the three isotopes of atomic oxygen using far infrared laser magnetic resonance [5]. Recently, we have operated a modified microwave discharge cavity with its discharge parallel to a magnetic field for the production of ions [6]. The combination of the discharge and magnetic field produced a considerable increase in the brightness of the discharge, along with a significant pinching and localization of the plasma. Work on this type of plasma is now in progress at the Boulder Laboratories of NIST.

## 5.1.2 The DC Discharge

A DC electric discharge provides a method for extending the discharge length and generating transient molecules continuously over several meters of path length. Consequently, it is easy to achieve long absorption paths and hence high sensitivity; most other methods of generation, including the RF discharge, produce the species in very confined volumes.

Gaseous electric discharges have been used for a variety of purposes for a very long time, and the principles of their operation are available in many treatises such as reference [7]. The last decade has seen some attempt at characterizing them from the point of view of the formation of both charged and neutral molecules. This has led to the invention of two new discharge configurations: the hollow cathode discharge [8] and the magnetically extended positive column discharge [9].

### 5.1.2.1 The Conventional DC Discharge. A conventional DC discharge is shown in Figure 2. The concentration of positive ions is greatest in the negative glow region, which is relatively short under normal conditions. Its length is equal to the distance traveled by the high-energy electrons which have been accelerated cross the cathode drop region and is proportional to the electron energy. It is typically a few centimeters long at a pressure of 135 Pa (1 torr). The ions are formed in this region by electron ($e$) bombardment of neutrals (M):

$$e + M \rightarrow M^+ + 2e.$$

The characteristics of a glow discharge are determined principally by cathode phenomena. In a normal discharge, the voltage across the cathode drop remains

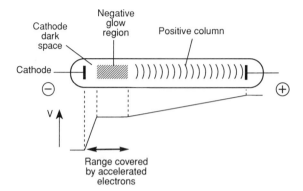

FIG. 2. A diagram showing the different regions in an electric discharge through a low pressure gas. The potential varies nonuniformly through the discharge as shown, the largest fall occurring in the cathode dark space. Positive ions are formed predominantly in the negative glow region.

essentially constant as the current increases; hence, there is a need for a series ballast resistor. At the same time, the cathode area covered by the discharge plasma increases. In the "anomalous discharge," the plasma covers the entire cathode area, and the voltage across the cathode drop rises roughly in proportion to the current. As a result both the energy of the electrons and the distance they travel increase with current. For more details, see Chapter VIII in reference [7].

A DC discharge 4 m long was used to create atomic oxygen [10] for the accurate measurement of its fine structure spectrum in a tunable far infrared spectrometer.

5.1.2.2 Hollow Cathode Discharge Cell. The principle of the hollow cathode discharge is to wrap the cathode around as much of the discharge volume as possible, thereby increasing the extent of the negative glow region and hence the number of ions formed in the discharge. Van den Heuvel and Dymanus published an effective design [8] which has since been copied by many other groups. It is shown in Figure 3. The discharge is struck between the hollow cathode and an anode positioned at the end of a perpendicular glass side arm in the center of the main tube, as is shown in Figure 3. Stable operation is possible over a large range of currents and pressures. The positive column is confined to the glass insert between anode and cathode only, and the negative glow is concentrated along the axis of the cathode tube. A few millimeters from the wall, the glow changes from the cathode dark space to the negative glow which extends along the axis of the tube. The length of the glow is directly related to the voltage applied to the anode. The cathode is cooled efficiently by a slow flow of liquid nitrogen through a helical copper tube soldered around the cathode tube.

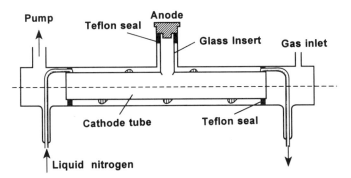

FIG. 3. A hollow cathode discharge cell for the generation of positive ions, after the design of van den Heuvel and Dymanus [8]. The copper cathode first inside a glass cell and is cooled by flowing nitrogen through a spiral tube soldered to its outside surface.

This cell was used in Dymanus's group [11] to produce $NH^+$ to observe the first rotational transition of that ion.

### 5.1.2.3 Magnetic Field Enhancement of the Negative Glow Region.
In a conventional DC discharge tube (Figure 2), the electrons which are accelerated through the cathode gap spread out; consequently, many are lost through collisions with the wall of the tube. The distance which they travel down the discharge tube is therefore much less than it might otherwise be. The length of the negative glow can be extended by the application of a longitudinal magnetic field [12]. De Lucia and his group at Duke University exploited this observation to achieve dramatic increases in the signals recorded from ions formed in a discharge [9]. Using a solenoid magnet around their discharge tube, they achieved magnetic field strengths of several tens of milliteslas. Fields of this magnitude are sufficient to confine electrons with several hundred volts of transverse energy to a cyclotron radius of the order of 1 cm, roughly the cross-section of the discharge probed by the laser beam. The discharge can be forced to run in the anomalous mode described in Section 5.1 by restricting the size of the electrodes. The increase in positive ion concentration achieved in this way is about 100-fold.

The group of Destombes et al. [13] used this type of discharge to produce the ion complex $ArH_3^+$ and measure its millimeter-wave spectrum.

## 5.2 Photolysis

The photolytic generation of free radicals with UV flash lamps has been used for a long time. It provided the basis for a large number of papers for electronic spectroscopy in the 1950s, the technique being referred to as flash photolysis [14]. However, the concentrations of transient species generated in this way were not high enough to permit the detection of rotational or vibrational spectra. Now the situation has been changed with the availability of high-power excimer lasers. It is usual to photolyze the stable parent molecule in the bulk phase. For example, the methyl radical can be generated efficiently by the photolysis of $CH_3I$ [15]:

$$CH_3I + h\nu \rightarrow CH_3 + I.$$

Radicals and even ions have also been formed by photolysis or photoionization of a stable precursor seeded into a rare gas free-jet expansion [16] (to be described later). The molecules are thereby generated cold, considerably simplifying the resultant spectrum.

## 5.3 Chemical Formation

Free radicals are produced in electrical discharges from reactions between the reactive species produced in the dissociation. They can also be produced downstream from the discharge when one of the neutral reactive species reacts with another reactant. Most methods of generating transient molecules involve flow systems in which streams of gas are pumped continously through tubes. In the discharge/flow system, two such streams of gases are mixed in a reaction zone to generate the molecule of interest. The principle behind these reactions in the gas phase is to generate a reactive species A that is relatively easy to handle, such as an atom, and to react it with a parent molecule BR to form the free radical of interest, ideally in a fast single-step process:

$$A + BR \rightarrow R + AB.$$

A diagram of the production scheme is shown in Figure 4. Although the

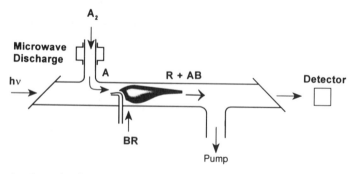

FIG. 4. A schematic diagram of the discharge/flow apparatus for the generation of short-lived molecules. The products of a discharge through flowing gas, producing A, are mixed with a continous flow of a secondary gas BR and react downstream to generate the desired product.

discharge can be operated DC (or at low AC frequencies), it is preferable to use either an RF or a microwave discharge because these do not require electrodes within the gaseous volume (electrodes provide very efficient sites for the removal of transient molecules). An example of the use of this method is the production of the HCO radical [17]:

$$F + H_2CO \rightarrow HCO + HF,$$

where the fluorine atoms are formed by discharging $CF_4$ or 10% $F_2$ in helium (the second source produces significantly more fluorine atoms) and reacting them with formaldehyde.

## 5.4 Laser Ablation

This method is used to vaporize less volatile materials, such as metals. Radiation from a powerful, pulsed laser (either a Nd:YAG or an excimer laser) is focused onto a solid sample. The material is thereby vaporized, often to form a gas plasma above the surface of the solid. The atoms or molecules in the gas phase then react with a secondary species to form the desired product, for example [18]:

$$2Al + N_2O \rightarrow Al_2O + N_2.$$

This method for the formation of transient species is routinely used in conjunction with supersonic nozzle expansion to form the species for observation at low rotational and vibrational temperatures [16]. Often the solid surface needs to be fresh if the material is to be vaporized efficiently. In these circumstances, the material is constructed in the form of a rod which is then slowly rotated to present a new surface for each pulse of laser light. The choice of wavelength of laser light depends on the material to be ablated; this technique is still somewhat of a "black art."

## 5.5 Thermal Generation

At first glance, the use of high temperatures to generate molecules for the study of their infrared spectra, does not look like a promising approach, because a furnace is a strong source of blackbody radiation. However, a laser beam is easy to collimate, so spatial separation of signal beam and furnace background can easily be achieved. Several groups have recorded infrared spectra of transient molecules by generating them in a furnace [19, 20], and Jones and his group at Ulm University have exploited this technique in studying a whole series of diatomic hydride molecules [21]. Their apparatus is shown schematically in Figure 5.

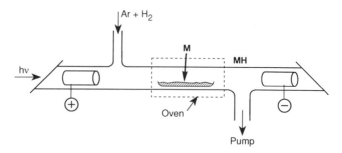

FIG. 5. Apparatus for the generation of metal hydrides by flowing discharged $H_2$ over metal heated in a furnace indicated by the dashed outline.

The cell is made of an aluminum oxide ceramic body and has two cylindrical water-cooled stainless steel electrodes fitted coaxially to the ends of the tube. The cell is placed inside a furnace capable of producing temperatures up to 1300°C. A suitable metal, M, is placed in the center of the cell in a ceramic boat, and the cell is filled with a 20% mixture of hydrogen in helium at a total pressure of 506 Pa (5 mbar). The gas is discharged to generate hydrogen atoms which combine with the hot metal to produce MH. The technique works particularly well for the more volatile metals for which significant vapor pressure can be generated.

## 5.6 Supersonic Expansion

Supersonic expansion is not necessarily a way of creating free radicals, but rather a method of cooling the molecules and isolating them for spectroscopic investigation. However, new, weakly bound complexes including free radicals can be formed in this way. When a gas is forced through a small hole from a high- to a low-pressure region, a diverging jet of molecules for which the equipartition of energy is no longer applicable is formed. The translational velocity of the molecules is increased at the expense of the internal (vibrational and rotational) degrees of freedom [16]. With low vibrational "temperatures," it is possible to form and maintain weakly bound complexes such as van der Waals molecules [22]. Much work has been done in the past 5 years on the infrared spectroscopy of such species [23, 24], using an apparatus like that shown in Figure 6.

The compound from which the complex is to be formed is seeded into an inert gas carried at moderately high pressures, 100 kPa (1 atm), and the whole is expanded through a pulsed nozzle into a high-vacuum region. The region just in front of the nozzle (where the complexes are formed) is sampled with a beam from an infrared laser. The laser absorption is enhanced by the use of a multipass

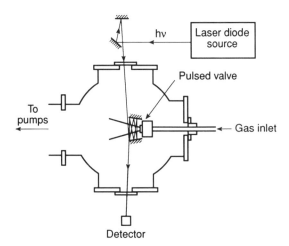

FIG. 6. Apparatus for the study of infrared spectra of van der Waals molecules, formed by supersonic expansion of the component gases through a pulsed nozzle. The system of tilted plane mirrors causes the laser beam to make several passes through the sample, thereby increasing the absorption.

system; the plane mirrors are not parallel, so successive reflections of the beam "walk" toward the nozzle and then out again. Greater path lengths and hence sensitivity can be achieved by using a slit nozzle [25]. The slit nozzle was used very effectively in the study of $(HCl)_2$ [26].

## References

1. Fehsenfeld, F. C., Evenson, K. M., and Broida, H. P. (1995). *Rev. Sci. Inst.* **36**, 294.
2. McCarroll, Bruce (1970). *Rev. Sci. Instrum.* **41**, 279.
3. Parmeggiani, Luigi (1983). In *Encyclopaedia of Occupational Health and Safety*, 3rd Edition, p. 426.
4. Beenakker, C. I. M. (1976). *Spectrochimica Acta* **31B**, 483.
5. Brown, John M., Evenson, K. M., and Zink, Lyndon R. (1993). *Phys. Rev. A* **48**, 3761.
6. Varberg, T. D., Evenson, K. M., and Brown, John M. (1994). *J. Chem. Phys.* **100**, 2487.
7. Cobine, J. D. (1958). *Gaseous Conductors*, Dover.
8. van den Heuvel, F. C., and Dymanus, A. (1982). *Chem. Phys. Letts.* **92**, 219.
9. De Lucia, F. C., Herbst, E., Plummer, G. M., and Blake, G. A. (1983). *J. Chem. Phys.* **78**, 2312.
10. Zink, L. R., Evenson, K. M., Matsushima, F., Nelis, T., and Robinson, R. (1991). *Astro. Phys. J.* **371**, L85.
11. Verhoeve, P., ter Meulen, J. J., Meerts, W. L., and Dymanus, A. (1986). *Chem. Phys. Lett.* **132**, 213.

12. Thomson, J. J., and Thomson, G. P. (1928). *Conductivity of Electricity through Gases*, Cambridge University Press.
13. Destombes, J. L., Demuynck, C., and Bogey, M. (1988). *Phil. Trans. Roy. Soc. Ser. A* **324,** 147.
14. Porter, G. (1950). *Proc. Roy. Soc.* **200A,** 284.
15. Hall, G. E., Sears, T. J., and Frye, J. M. (1989). *J. Chem. Phys.* **90,** 6234.
16. Foster, S. C., Kennedy, R. A., and Miller, T. A. (1988). In *Frontiers of Laser Spectroscopy of Gases*, A. C. P. Alves, J. M. Brown, and J. M. Hollas (eds.), p. 421. Kluwer Academic.
17. Bowater, I. C., Brown, J. M., and Carrington, A. (1971). *J. Chem. Phys.* **54,** 4957.
18. Cai, M., Carter, C. C., Miller, T. A., and Bondybey, V. E. (1991). *J. Chem. Phys.* **95,** 73.
19. Haese, N. N., Liu, D. J., and Altman, R. S. (1984). *J. Chem. Phys.* **81,** 3700.
20. Lemoine, B., Demuynck, C., Destombes, J. L., and Davies, P. B. (1988). *J. Chem. Phys.* **89,** 673.
21. Magg, U., and Jones, H. (1988). *Chem. Phys. Lett.* **146,** 415, and subsequent papers.
22. Levy, D. H. (1980). *Ann. Rev. Phys. Chem.* **31,** 197.
23. Hayman, G. D., Hodge, J., Howard, B. J., Muenter, J. S., and Dyke, T. R. (1987). *J. Chem. Phys.* **86,** 1670.
24. Lovejoy, C. M., Schuder, M. D., and Nesbitt, D. J. (1986). *Chem. Phys. Lett.* **127,** 374.
25. Lovejoy, C. M., and Nesbitt, D. J. (1987). *Rev. Sci. Instrum.* **58,** 807.
26. Schuder, M. D., Lovejoy, C. M., Lascola, R., and Nesbitt, D. J. (1993). *J. Chem Phys.* **99,** 4346.

# 6. SOURCES OF METASTABLE ATOMS AND MOLECULES

## Timothy J. Gay

Behlen Laboratory of Physics, University of Nebraska
Lincoln, Nebraska

## 6.1 Introduction

This chapter reviews the various techniques used to produce beams of neutral metastable atoms and molecules. Metastable excited states cannot decay quickly to a state of lower energy because the necessary transition is forbidden by dipole and/or spin–flip selection rules. Because of their large internal energy ($\sim 10$ eV) and long natural lifetimes ($\gtrsim 10^{-5}$ s), such species play an important role in a variety of situations involving planetary atmospheres, discharges, plasmas, and vapor deposition techniques [1–4] and are essential in many basic studies of atomic collisions [5–19], laser cooling [20, 21], atomic spectroscopy and interferometry [21–23], nuclear physics [24, 25], and quantum electrodynamics [26, 27]. They are also valuable surface-specific probes of condensed matter phenomena [28–31]. We also consider methods of metastable beam modification and characterization; the detection of metastables is the subject of Chapter 11. Metastable sources have been discussed by a number of other authors [32–37]. Moreover, we have found several papers reporting the details of specific sources [8, 9, 11, 18, 26, 28, 38–48] to be particularly comprehensive and useful.

Metastable sources belong generally to one of five categories:

1. Electron-beam bombardment
2. Discharge
3. Charge transfer
4. Optical pumping
5. Thermal

The primary metastable production mechanism in discharge and electron bombardment sources is the impact excitation of ground states by electrons to metastable states. In discharge sources, this occurs in the active plasma of an electrical discharge or arc. Electron bombardment sources, on the other hand, use a well-defined electron beam to excite an effusive or supersonic beam of ground state atoms or molecules. Some sources involve two discharge regions or extremely high electron beam fluxes and are best viewed as hybrids between the two categories.

EXPERIMENTAL METHODS IN THE PHYSICAL SCIENCES
Vol. 29B

In charge-transfer sources, a fast beam of ions is partially neutralized in a target cell, with both ground and excited states being produced. Most excited states which are not metastable decay in a drift region or can be field-ionized, leaving primarily ground and metastable states remaining in addition to the residual ion beam. Optical pumping sources are similar conceptually to those involving electron-beam bombardment, except they use photons to excite the ground state to a level above the desired metastable state. Subsequent decay produces the metastable species. This two-step process is unavoidable with photon excitation because dipole selection rules must be obeyed, whereas electron bombardment excitation is not subject to these rules, and metastable excitation can proceed directly. Thermal sources are based on the fact that a number of elements, including Tl and Ga, have metastable states very close in energy to their ground states. When they are heated in an oven to temperatures sufficient to produce high vapor pressures, the Boltzmann factor for the upper level is large enough to yield an effusive beam rich in metastables.

Sources of a wide variety of metastable atoms and molecules have been developed to date. A representative number of these, listed by the respective metastable species they produce, are listed in Table I. (Table I does not include reference to metastable experiments in which little or no source information is given. Virtually every element in the periodic table and most molecules have metastable states; see references [32–37] and references therein for reports on work with other metastable species.) The large majority of sources has been developed for the production of the metastable states of hydrogen ($2^2S_{1/2}$), helium ($2^1S_0$ or $2^3S_1$), and the heavy noble gases neon, argon, krypton, and xenon ($n^3P_0$ and $n^3P_2$, where $n = 3 \cdots 6$ for Ne through Xe, respectively). Metastable fluxes, as reported in the various references, are given in Table I to provide a general comparison of relative source performance. The reader is cautioned, however, that such fluxes are determined in a variety of ways and, on occasion, have proven to be unreliable. This issue is taken up in Section 6.7 and also in Chapter 11.

Each of the source types listed above has advantages and disadvantages, both generically and with regard to the production of specific metastable species. Discharge sources are much simpler to construct than the others, but much of the beam they produce consists of ground states. Ground-state-to-metastable ratios are typically between $10^5$ and $10^4$ [39, 41, 91, 92]. The thermal or near-thermal velocity beams produced by discharge and electron bombardment (and optical pumping) sources are generally easier to analyze and work with than the fast beams from charge-transfer sources. However, charge-transfer sources produce beams with much lower ground-state backgrounds. Discharge and electron bombardment devices typically require large vacuum pumps because of their high gas loads.

TABLE I. Sources of Metastable Atoms and Molecules

| Atom/ Molecule (Atomic Number) | Primary Metastable States | Types of Sources[a] | Typical Quoted/ Implied Flux/Intensity | References |
|---|---|---|---|---|
| H (1) | $2^2S_{1/2}$ | O, E, C | $10^6$–$10^8$ s$^{-1}$ (O, E); $\sim 10^{11}$ s$^{-1}$ (C) | 22, 24, 26, 38, 49 |
| He (2) | $2^1S_0$, $2^3S_1$ | D, E, C | $10^{13}$–$10^{15}$ sr$^{-1}$ (D, E); $10^{11}$–$10^{13}$ s$^{-1}$ (C) | 7, 9, 11, 12, 18, 29, 30, 39, 43–47, 50–59 |
| N (7) | $2^2D_{3/2, 5/2}$ | D | — | 60 |
| Ne (10) | $3^3P_{0, 2}$ | D, E, C | $10^{13}$–$10^{15}$ s$^{-1}$ sr$^{-1}$ (D, E); $10^{11}$–$10^{13}$ s$^{-1}$ (C) | 8, 14, 16, 29, 39, 41, 42, 48, 51, 54, 61–67 |
| Mg (12) | $3^3P_{0, 1, 2}$ | D, E | $\sim 10^{15}$ s$^{-1}$ cm$^{-2}$ (D) | 68, 69, 70 |
| Al (13) | $3^2P_{3/2}$ | D | $10^{10}$–$10^{12}$ s$^{-1}$ sr$^{-1}$ | 71[b] |
| Ar (18) | $4^3P_{0, 2}$ | D, E, C | $10^{13}$–$10^{15}$ s$^{-1}$ sr$^{-1}$ (D, E); $10^{11}$–$10^{13}$ s$^{-1}$ (C) | 13, 14, 39, 41, 45, 47, 54, 59, 64, 66, 72–74 |
| Ca (20) | $4^3P_{0, 1, 2}$ | D, E | $\sim 10^{15}$ s$^{-1}$ cm$^{-2}$ (D) | 40, 68, 75, 76 |
| Sc (21) | $4^4F_{5/2, 7/2}$ | O | — | 23 |
| Zn (30) | $4^3P_{0, 2}$ | E | — | 69 |
| Ga (31) | $4^2P_{3/2}$ | T | — | 6, 77 |
| Kr (36) | $5^3P_{0, 2}$ | D, E | $\sim 10^{13}$ s$^{-1}$ sr$^{-1}$ (D) | 41, 63 |
| Sr (38) | $5^3P_{0, 2}$ | D | $\sim 10^{15}$ s$^{-1}$ cm$^{-2}$ | 70 |
| Zr (40) | $4^3F_{3, 4}$ | D | $10^{10}$–$10^{12}$ s$^{-1}$ sr$^{-1}$ | 71[b] |
| Nb (41) | $4^5S_2$ | D | $10^{10}$–$10^{12}$ s$^{-1}$ sr$^{-1}$ | 71[b] |
| Mo (42) | $4^5S_2$ | D | $10^{10}$–$10^{12}$ s$^{-1}$ sr$^{-1}$ | 71[b] |
| In (49) | $5^2P_{3/2}$ | T | —[c] | 78 |
| Xe (54) | $6^3P_{0, 2}$ | D | [c] | 73, 79 |
| Ba (56) | $6^3P_{0, 2}$; $5^1D_2$; $5^3D_{1, 2, 3}$ | D | $\sim 10^{15}$ s$^{-1}$ cm$^{-2}$ | 80 |
| Gd (64) | $4^{11}F, 4^{11}G$ | O | $\sim 10^{15}$ s$^{-1}$ cm$^{-2}$ | 81 |
| W (74) | $5^5D_{0-4}$ | D | $10^{10}$–$10^{12}$ s$^{-1}$ sr$^{-1}$ | 71[b] |
| Hg (80) | $6^3P_{0, 2}$ | D, E | $\sim 10^{12}$ s$^{-1}$ sr$^{-1}$ (D) | 5, 82–84 |
| Tl (81) | $6^2P_{3/2}$ | O, T | — | 85, 86 |
| Pb (82) | $6^3P_{1, 2}$ | D, T | — | 85, 87 |
| Bi (83) | $6^2D_{3/2, 5/2}$ | D | — | 88 |
| H$_2$ | $C^3\Pi_u$ | E | — | 63 |
| N$_2$ | $A^3\Sigma_u^+$ | D, E | $\sim 10^{14}$ s$^{-1}$ sr$^{-1}$ (D) | 47, 54, 63, 89 |
| O$_2$ | $A^1\Delta_g$ | D | — | 63 |
| CO | $a^3\Pi, b^3\Sigma^+$ | E | — | 90 |

[a] O—optical pumping; E—electron bombardment; C—charge transfer; D—discharge; T—thermal.

[b] Flux estimates: D. W. Duquette, private communication.

[c] Although references [71] and [78] report no Xe* flux, it can be expected that standard discharge sources would produce Xe* fluxes comparable to those of other noble gases.

Metastable atoms of the metallic or alkaline earth elements are produced most easily in a discharge source, while noble gas metastables are readily made in any source other than one utilizing optical pumping. Hydrogen metastables can be destroyed relatively easily by external fields and are thus usually made in electron bombardment sources. Thermal sources are limited to those elements with low-lying metastable states.

Finally, the velocity profiles of the various source types are very different, and this is often a major consideration when picking a source for a given experiment. Reactions at typical "chemical physics" energies require the lower velocities of discharge or electron beam sources. Between these two, the electron bombardment sources offer better velocity control and resolution than do discharge sources. Charge-transfer sources have by far the best energy resolution, but produce fast metastables that are unsuitable for the study of chemical reactions or the probing of condensed-matter surface phenomena.

Next we discuss in detail the various source types, and then consider techniques for metastable beam modification and characterization. Practical experimental concerns will be addressed throughout, particularly with regard to metastable-associated backgrounds.

## 6.2 Electron-Beam Bombardment Sources

### 6.2.1 Overview

Electron bombardment sources make use of a conceptually simple scheme: the excitation of a beam of atoms to metastable states by a well-defined beam of electrons. Insofar as they effect the metastable intensity and velocity distribution, the important design parameters for such a source are the electron beam energy and current, and its overlap with the atomic beam. Given the rather small cross-section values for metastable excitation ($\lesssim 10^{-17}$ cm$^2$ [93]), intensity considerations demand large electron-beam currents and efficient beam overlap. Unfortunately, the excitation cross-sections are maximal just above threshold, where space charge limitations on the electron beam current density are the most severe. The relative populations of various metastable components within a given atomic beam are also energy-dependent, because triplet excitation falls much more rapidly with electron energy than does singlet excitation. Thus, by 100 eV, virtually all of the metastables produced in, for example, a helium target are in $2^1S$ states.

The electron and atomic beams can be either coaxial or transverse; neither geometry is clearly superior to the other, and other experimental factors not associated with the source often determine which is used. Coaxial sources offer generally larger overlap volumes, but are more complex, typically requiring more electrodes for electron-beam focusing. Magnetic fields are often used in either configuration to confine the electron trajectories to a well-defined overlap region. Momentum transfer to the atomic beam, in combination with the initial atomic-source velocity profile, is crucial in determining the velocity distribution of the metastables. This issue has been discussed extensively in the literature [44–46, 73, 89]. Early transverse excitation designs suffered from poor velocity resolution; the already broad atomic ground-state distribution characteristic of their effusive sources was further broadened by the momentum transfer perpendicular to the beam resulting from electron bombardment. The first coaxial sources improved velocity resolutions somewhat. Recent sources of both transverse and coaxial design, using supersonic atom beams, geometric or mechanical velocity selection, pulsing of the atomic and/or electron beams, or a combination of these, have succeeded in producing $\Delta v/v$ values approaching 3%. Velocity spreading due to electron bombardment is less when heavy species such as Ar are used instead of H or He. Mean velocities can be modified by using resistive heating or conductive cooling [11, 46, 74] of the atomic beam source. Indeed, one of the major advantages of electron bombardment sources is that velocity distributions can be controlled with relative ease.

### 6.2.2 Coaxial Geometry

A state-of-the-art coaxial electron-beam source, used for producing metastable noble gas beams, is shown schematically in Figure 1 [45]. It uses a supersonic nozzle atomic beam source, downstream ($\sim$15 mm) from which is a 0.6 mm diameter skimmer. The atomic stagnation temperature and pressure are nominally 300 K and several atmospheres, respectively. To avoid overloading the vacuum pumps, the atomic source is pulsed using an automobile fuel injector, with typical pulse-time widths of tens of milliseconds.

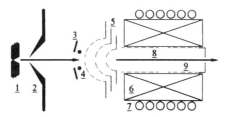

Fig. 1. Schematic diagram of the coaxial electron-beam bombardment metastable source of Kohlhase and Kita [45]. Shown are (1) pulsed supersonic nozzle ground-state atom source; (2) skimmer; (3) electron repeller; (4) electron filament; (5) extractor, control, and acceleration grids; (6) electromagnet; (7) cooling water coil; (8) first anode; (9) second anode.

The atomic beam enters an excitation volume whose first element prevents electrons from migrating upstream. Downstream, a thermionic cathode is looped to just circumscribe the atomic beam, and three hemispherical mesh electrodes accelerate and pulse the electron beam. The main collision volume, surrounded by a water-cooled solenoid, has a two-part mesh anode and an exit aperture. The electron beam pulsing is used in place of a mechanical chopper to allow metastable time-of-flight (TOF) information to be obtained.

The natural velocity percentage width of supersonic nozzle sources is generally less than 10%, and the coaxial excitation does not affect this distribution significantly. This source is reported to have a velocity width of between 3 and 4% for He, comparable to the best results reported to date. It is also reported to produce an extremely high instantaneous metastable flux: with Ar the authors quote a value of $5 \times 10^{15}$ atoms s$^{-1}$ sr$^{-1}$. It should be noted that coaxial sources can suffer velocity broadening due to production of fast metastable negative ions which are neutralized in collisions with ground state atoms before exiting the source. This process yields a second velocity component of the output beam [72].

### 6.2.3 Transverse Geometry

Tommasi *et al.* [44] have described a metastable source with transverse excitation geometry. In their apparatus, shown schematically in Figure 2, the momentum transfer due to electron-impact excitation is turned to an advantage by using it to substantially separate the supersonic ground-state beam from the metastable component produced in the electron gun.

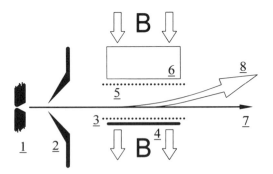

Fig. 2. Schematic diagram of the transverse electron-beam bombardment metastable source of Tommasi *et al.* [44]. Shown are (1) pulsed supersonic nozzle ground-state atom source; (2) skimmer; (3) electron control grid; (4) indirectly heated planar cathode; (5) electron control grid; (6) parallel-plate slotted anode; (7) undeflected ground-state beam; (8) collisionally deflected metastable atoms. The electron bombardment volume is immersed in a uniform magnetic field.

A piezoelectric valve produces pulses with a repetition rate of $\sim$2 Hz and a width of <1 ms, corresponding to $\sim$10$^{18}$ atoms/pulse. The duty cycle is again limited by pumping speed. After collimation by a skimmer, the atom beam enters a region where it is crossed by a 5 mm $\times$ 25 mm electron beam, defined by the emitting area of the cathode. The cathode is specially treated with barium–strontium carbonate, using a recipe developed to provide a very robust emitting surface [46, 47]. Electron energies are controlled by two grids and are kept near the metastable excitation threshold to minimize collisionally induced spatial spread of the beam. The electron beam is dumped into an anode comprising an array of parallel plates, designed to minimize the effects of space charge. This design permits electron currents greater than 0.1 A to be produced, even though the intergrid gap of 11 mm, dictated by the necessity of preventing metastable–grid collisions, is rather large.

The He or Ar metastable beam thus produced is reported to have extremely high instantaneous flux ($>$10$^{15}$ s$^{-1}$ sr$^{-1}$) and has a FWHM divergence of less than 5°. The percentage velocity width of the beam is 7%, due almost entirely to the ground-state source velocity spread. Moreover, the deflected beam population is actually inverted, i.e., over a restricted angular range, the flux of metastables exceeds that of ground state atoms. In addition, Tommasi *et al.* have observed spatial separation within the He* beam of the $2^1S$ and $2^3S$ states. This is due to the different excitation thresholds of the two species, which cause kinematic shifts in their trajectories. The $^3P_2$ and $^3P_0$ states of the heavy noble gases would exhibit a similar separation, although it would be smaller than in He because of the correspondingly smaller energy splittings.

## 6.3 Discharge Sources

### 6.3.1 Overview

A large variety of discharge sources have been developed, based on a broad range of geometries and discharge types. While low-current ($\leq$1 A) cold-cathode sources are the most commonly used, other types, involving pulsed, RF, and even audio frequency discharges have been reported. Often they are outgrowths of molecular dissociators or arc-heated fast atom sources, in which metastable fractions are considered a nuisance. While discharge sources are generally simpler to build and operate than electron bombardment devices, they suffer from broader velocity distributions which are less easily controlled and which are dependent on discharge polarity, current, and voltage. Another potential problem is the large ground-state fraction present in beams from discharge sources; the metastable-to-ground-state ratio is typically less than 10$^{-4}$. Discharge-induced sputter erosion is the limiting factor in source lifetime, as opposed to gun filament life in electron beam sources. In atomic scattering experiments, photon backgrounds from the discharge can be particularly severe. Sources of this type can be classified very roughly as belonging to one of the following categories.

### 6.3.2 Hot-Filament-Cathode Effusive Sources

These are used for gaseous samples. The metastable atoms effuse from an aperture in the plate anode opposite the cathode (Figure 3a). Fluxes are low ($\lesssim 10^{10}$ metastables s$^{-1}$ sr$^{-1}$), with velocities determined by the discharge temperature, which is determined in turn by power input to the cathode. Fast atom components can result from collisional neutralization of metastable negative ions which have been accelerated toward the anode.

### 6.3.3 Effusive Hollow-Cathode Discharge/Sputter Sources

In these sources (Figures 3b–3d), a DC discharge is struck between a hollow cathode and an anode which may or may not define the output beam. Two disparate examples of this type are that of Theuws *et al.* [41], in which the cathode acts as the tube from which the parent gas effuses, and that of Duquette and Lawler [71], which has been used to produce metastable metal atoms. In the latter source, argon is used to maintain a discharge between anode and cathode, and argon ions sputter a metal sample which coats the cathode. The free metal atoms are subsequently excited in the discharge, and some effuse from an exit aperture in the anode. A variant on this type of source, used to produce alkali-earth metastables, has been reported by Ureña *et al.* [40]. Ca is heated in a cylindrical crucible to ~1400 K and effuses into a discharge region between the crucible and the concentric, electrically isolated cylindrical oven. There it is excited and emerges from the oven. The discharge can be pulsed if desired. Large Ca* fluxes have been reported for this source. A similar source with an axial beam geometry has been reported by Brinkmann *et al.* [75].

### 6.3.4 Gas-Dynamic Sources

A unique source, developed by Brinkmann and co-workers [76], was also used for Ca (Figure 3e). The Ca is vaporized in a cylindrical oven and diffuses into a hollow stainless-steel cathode through holes drilled in the cathode wall. The discharge runs between the cathode and an anode made of wire wrapped in a spiral whose axis is coincident with that of the cathode. Along this axis is directed a "carrier" beam of argon, which is passed through a filament preheater prior to entering the hollow cathode. The metastable atoms are thus carried out of the discharge and attain a velocity distribution characteristic of the argon,

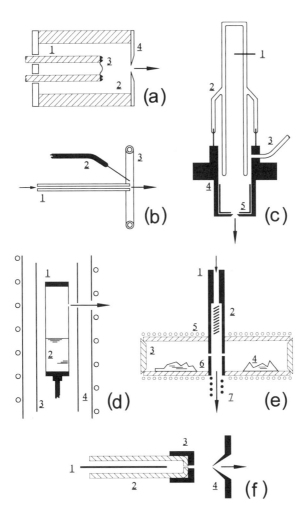

FIG. 3. Various configurations of discharge metastable sources. (a) Hot-cathode effusive source [7, 8] showing (1) insulating cathode supports; (2) insulating discharge container; (3) hot cathode; (4) anode. (b) Hollow-cathode arc source [41] showing (1) hollow cathode; (2) water-cooled ring anode; (3) ignition electrode. (c) Hollow-cathode sputter source [60] showing (1) anode; (2) argon column; (3) argon inlet; (4) hollow cathode; (5) sputter target. (d) Effusive oven source [40] showing (1) crucible (anode); (2) metal to be vaporized; (3) hollow cathode; (4) water-cooled heat shield. (e) Gas-dynamic jet source [76] showing (1) argon inlet; (2) preheater; (3) oven; (4) sample to be vaporized; (5) oven heater coils; (6) hollow cathode with effusive entrance apertures; (7) coil anode. (f) Supersonic cold cathode discharge source of Brand *et al.*, incorporating improvements on the design of reference [64] showing (1) W cathode rod; (2) alumina tube with laser-drilled nozzle; (3) aluminium cap anode; (4) skimmer and secondary anode.

which can be supersonic when an appropriate nozzle is used. Electrons carried away from the discharge can be used to seed a second discharge downstream. The additional discharge dramatically increases the metastable production [43].

### 6.3.5 Supersonic DC Discharge Sources

These sources, which are among the simplest to build and operate, are based on the original designs of Searcy [52] and Leasure et al. [53]. They have been developed extensively in a number of laboratories [9, 25, 28, 29, 39, 42, 50, 51] and are the most widely used source type today. An advantage of supersonic discharge sources over effusive ones is the relatively narrow velocity spread of the metastables. One version of this source type, representing modifications and improvements on the designs of Fahey et al. [54] and of Brand et al. [61], is shown in Figure 3f. Noble gas with a stagnation pressure of 1–2 kPa emerges in supersonic flow from a 0.15 mm diameter nozzle that has been laser-drilled in an alumina tube. Emerging atoms are excited by a discharge struck between a tungsten cathode rod coaxial with the alumina tube and an aluminum-cap anode. This source is thus similar in concept and excitation geometry to the electron-beam bombardment source of Kohlhase and Kita [45]. Earlier designs relied on a discharge between the skimmer and the cathode. The present configuration allows for easier initiation of the discharge (no high-voltage pulser or Tesla coil is needed), and a second discharge can, if desired, be maintained between the anode and the skimmer, which is electrically isolated. A second in-line discharge has been used both in the hot arc source of Ferkel et al. [43] and, as mentioned earlier, the gas dynamic source of Brinkmann et al. [76] to significantly increase metastable production. Liquid nitrogen cooling of the discharge region has also been used effectively to reduce metastable velocities and velocity spreads while increasing flux [20, 42]. When the source's polarity is reversed, positive ions are accelerated towards the skimmer, and fast ground-state or metastable neutrals can populate the beam [52, 72, 91].

Various authors have reported the metastable intensity and fine-structure distribution (e.g., the ratio of $2^3S_1$ to $2^1S_0$ populations in He) to depend on stagnation pressure and discharge current and voltage [28, 39, 41–43, 51, 61, 72, 95]. Some designs use coaxial or radial magnetic fields to enhance the discharge and hence metastable intensity. For this type of source, however, the two most crucial factors appear to be the vacuum in the region between the nozzle and skimmer, and the nozzle/skimmer alignment [39, 61]. The vacuum is important because elastic collisions with background gas remove metastables from the beam. Thus, skimmer geometry and nozzle/skimmer distance can also be important, because any supersonic beam reflected from the upstream skimmer walls will act as residual background gas. The source in Figure 3f has a nozzle operated in a confined chamber (<1 liter volume) pumped by a 300 liter/s turbomolecular pump. The aluminum cylinder is mounted on a robust precision $x$–$y$ translational stage for alignment. These design features have been instrumental in obtaining high metastable fluxes.

## 6.4 Charge-Transfer Sources

A serious problem that can occur with electron-beam and discharge sources is background due to the large ground-state component accompanying the metastable fraction. For example, in electron-metastable scattering experiments, such a ground-state component means that incident electron energies must be kept below (depending on the specific experiment) the first excitation threshold for optical decay or the ionization threshold of the ground-state atoms. In the case of He targets, this limits incident electron energies to the range below 25 eV. Charge-transfer sources produce ground-state-to-metastable ratios closer to $10^0$ than to $10^5$, and thus eliminate a large part of this problem [49, 59, 65, 66].

Virtually all of the charge-transfer sources reported have been used for the production of hydrogen or noble gas metastable atoms, with the alkali metals, parent noble gases, or $H_2$ used as the charge-transfer target. The charge-transfer and metastable production process proceeds most efficiently when two conditions are met: the energy defect, $\Delta E$, in the production process is as small as possible, and the incident ion velocity is such that electron velocity matching between the initial and final states occurs [36, 59]. The defect $\Delta E$ is defined as the difference in ionization energy between the target ground state and the projectile metastable state. Velocity matching simply means that the electron to be transferred need not undergo large acceleration during the collision.

The velocities required for the best metastable production thus imply the use of high-voltage accelerators, which, in conjunction with the use of ion sources, means that efficient charge-transfer sources are relatively complicated. Moreover, the fast metastables and the accompanying ground-state component can excite background gas in the experimental interaction region, leading to another source of background. Also, the fast beam is more difficult to manipulate with lasers (Section 6.7).

A recently reported state-of-the-art source, which has produced very high Ne* fluxes, is shown schematically in Figure 4 [36, 67]. $Ne^+$ is produced in a standard water-cooled Penning ion source, which is floated at $+800$ V with respect to ground. The extracted, space-charge-limited beam of about 35 µA is accelerated and periodically focused by a series of Einzel lenses which transport the beam to the charge-transfer cell and serve to decouple the rest of the experiment from the effusive gas load of the ion source. The ion beam traverses a Na vapor target 1 cm thick at a nominal pressure and temperature of 0.03 Pa and 500 K, respectively. The neutral flux leaving the charge exchange cell is $\sim 10^6 \text{ s}^{-1} \text{ sr}^{-1}$, of which 50% are in a metastable state. The metastable density

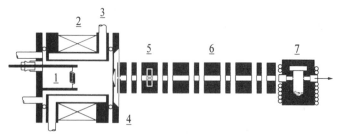

FIG. 4. Schematic diagram of high-flux charge-transfer metastable source [36, 95] showing Penning ion source floated at +800 V with respect to ground: (1) hot anode; (2) electromagnet; (3) cooling water. Also shown are (4) extraction cathode; (5) electrostatic deflection electrodes; (6) transport Einzel lenses; (7) grounded charge–transfer cell.

at the interaction region 25 cm downstream from the charge exchange cell is $\sim 10^6$ cm$^{-3}$. If Ne is used in the charge exchange cell instead of sodium vapor, the source can produce a fast ground-state beam. Source lifetime is limited by Na coating of Einzel lens element insulators, which causes electrical breakdown after several hundred hours of operation.

An interesting variation on the standard charge-transfer vapor target is the use of a graphite multichannel "converter," developed by Gostev et al. [57, 58]. They describe a device in which He$^+$ from a Penning source is neutralized with 25% efficiency into the $2^1S$ and $2^3S$ metastable states. The graphite array also serves to collimate the atomic beam and control its velocity by acting as the extraction electrode for the ion source. This apparatus was reported to produce extremely high steady-state fluxes of He* ($\sim 10^{20}$ s$^{-1}$ sr$^{-1}$), although more recent analyses have cast doubt on this number [36]. Moreover, attempts to use it to produce metastable species of the heavier noble gases led to very short source lifetimes due to sputtering of the converter material. Discussions of other charge-transfer sources are contained in references [13–19, 24, 48, 49, 59, 65, 66].

## 6.5 Optical-Pumping Sources

Metastable atoms can be produced in two-step "optical pumping" processes involving excitation of the ground state to an excited state by photon impact, followed either by collisional "quenching" or photon decay to a metastable level. The latter process has the advantage that neither the excitation nor the decay processes involve significant momentum transfer, and the resultant metastable velocity profiles can thus be very narrow. Optical production of metastables has the general advantage over bombardment and discharge sources that by tuning or filtering the photon source, specific metastable species can be produced. (This is particularly simple if dye lasers are used.) Sources using optical pumping have been reported for Hg [5, 82], Sc [23], Tl [86], H [38], and Gd [81]. With the exception of the Sc and Gd sources, which used dye lasers, discharges containing the parent atom were used to provide the pumping photons.

In the Hg source of Haberman *et al.* [82], Hg vapor produced in an oven is combined with nitrogen and the mixture is passed coaxially at a pressure of $\sim 10$ kPa over a cylindrical Hg resonance lamp. The 254 nm radiation from the lamp excites Hg $6^3P_1$ states. Some of these are subsequently converted to metastable $6^3P_0$ states through collisions with $N_2$ molecules. At the end of the lamp, the Hg-seeded nitrogen is formed into a beam by passing through a nozzle and skimmer. In addition to its role in collisional production of the $6^3P_0$ metastables, the $N_2$ also serves to accelerate and narrow the velocity distribution of the Hg as it passes through the nozzle. The $N_2$ must be kept very pure to prevent $O_2$ and hydrocarbon quenching of the Hg metastables. With this source, Hg* fluxes of $4 \times 10^{12}$ s$^{-1}$ sr$^{-1}$ were observed.

Beams of metastable hydrogen atoms have been produced in a similar fashion, but with somewhat more difficulty, given that atomic hydrogen must first be produced from $H_2$ (e.g., by RF discharge) and formed into a beam [38]. Lyman $\beta$ radiation (103 nm) from a second RF discharge is used to pump the beam to the $3P$ state, with subsequent decay to the metastable $2^2S_{1/2}$ level. Particular care must be used in the case of hydrogen to avoid external stray fields that can quench the metastable levels by motional or static Stark mixing with the $2P$ state. The use of windows for Ly$\beta$ radiation and of Ly$\alpha$ detectors to monitor the metastable production complicate matters further. Harvey [38] reports production of a metastable H beam with intensity $10^6$ s$^{-1}$ using this technique. This value is comparable to that typical for electron-bombardment sources.

## 6.6 Thermal Sources

Many atoms have ground states which are the lowest components of fine-structure multiplets. If other components of this multiplet are metastable and are not too far above the ground-state in energy, the effusive beam from an oven source can have a significant metastable fraction. Hishinuma and Sueoka [6], for example, describe a graphite oven source for Ga $(4^2P_{3/2})$ metastable effusive beams which operates at 1600 K. The splitting between the ground $4^2P_{1/2}$ and $4^2P_{3/2}$ metastable states in Ga is $\Delta E = 0.10$ eV. Thus, a Boltzmann distribution with appropriate statistical weights yields a ratio for metastable-to-ground-state atoms as

$$\frac{N^*}{N_g} = \frac{g^*}{g_g} \exp\left( + \frac{\Delta E}{kT} \right) = 0.47, \qquad (6.1)$$

where $g^*$ and $g_g$, the statistical weighting factors, are 4 and 2 for the metastable and ground states, respectively. Geesmann *et al.* [85] and Bartsch *et al.* [78] have reported observing metastables in Th and Pb from a thermal source operated between 1000 K and 1400 K. Other elements which have the appropriate multiplet structure for such sources include, B, Al, In, Cl, Br, I, and Lu.

## 6.7 Beam Modification and Analysis

Beams of particles emerging from metastable sources generally contain ions, fast or slow neutrals in a variety of excited states, ground-state neutrals, and electrons, as well as the metastable species. In addition, more than one metastable state may be produced, as is the case with noble gas beams. It is thus important to characterize and possibly alter the emergent beam. This can be done in a variety of ways, which depend primarily on the metastable atom or molecule of interest. A diagram illustrating a generic gauntlet of diagnostic devices is shown in Figure 5. We note that fast metastable beams, such as those produced by charge transfer, are more difficult to alter by means of photons or inhomogeneous (Stern–Gerlach or six-pole) magnetic fields than are thermal beams [48, 96].

### 6.7.1 Collimation and Charged Particle Removal

Discharge and thermal sources require some collimation immediately following the beam production region. This is typically accomplished with a skimmer, whose geometric profile can be quite important [39], and which often serves as one of the discharge electrodes. In addition to reducing the divergence of the beam, collimators also substantially reduce background photons. Electron-beam, optical, and charge-transfer sources more often rely on collimation of the parent beam prior to excitation. Electrons and ions can then be removed from the beam by electric field plates or a charged grid [18]. This technique has the ancillary benefit that atoms in highly excited Rydberg levels, which might not decay to a ground or metastable level over the normal flight path of the apparatus, are field-ionized [97]. A simple drift region after the source can also be effective in removing many of the shorter-lived excited states.

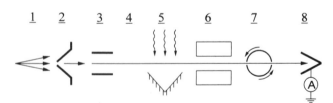

FIG. 5. Schematic gauntlet of beam modification and diagnostic devices (see text): (1) divergent metastable beam from source; (2) skimmer; (3) field quenching/ion deflection plates; (4) drift region; (5) optical pumping/quenching region; (6) inhomogeneous magnetic field for state selection or analysis; (7) velocity selector/analyzer; (8) metastable detector.

## 6.7.2 Metastable Quenching and State Selection

If more than one metastable species is produced by the source, it is often desirable to "quench" all but one of these states. This is accomplished most readily by photon excitation of the levels to be eliminated. The excited states can decay subsequently either directly or indirectly to the ground state. Helium $2^1S$ states are usually quenched using 2.06 μm light from a He discharge tube wrapped in a spiral about the beam [9, 30, 43, 46, 98]. This light induces transitions to the $2^1P$ state, which can decay to the ground state. A high-power Ti-sapphire laser could also be used for this purpose. The $2^3S$ states cannot be destroyed except by collision processes. Tommasi et al. [44] use the different momentum transferred to the He atom upon electron impact creation of $2^1S$ and $2^3S$ states to separate the two metastable components spatially. Lasers have been used to eliminate either the $^3P_2$ or $^3P_0$ metastables in heavy noble gas beams [48, 51, 61, 99].

Once a single metastable variety has been isolated, further state selection can be accomplished either by additional optical pumping or by manipulation with inhomogeneous magnetic fields. A simple Stern–Gerlach magnet can be used to select and/or analyze metastables with a given magnetic quantum number, albeit with significantly reduced intensity [7, 8, 28, 30]. Six-pole magnets have been used very effectively to produce highly polarized He ($2^3S$) beams [25]. Baum et al. [28] used a supersonic DC discharge source of He* succeeded immediately by a skimmer and six-pole magnet with a central "stop," or plug placed on the beam axis. Ground-state atoms (which make up the vast majority of the emergent beam) and $2^1S$ states, both of which have $M_J = 0$, are undeflected in the magnet and are largely stopped by the axial plug. He ($2^3S$) states are not only polarized by the six-pole [96], but are also focused past the stop, so that their intensity on-axis is enhanced. Baum et al. report metastable polarizations of 90%, with ground-state-to-metastable ratios as low as $10^2$. Six-poles are generally ineffective with fast beams from charge transfer sources.

Optical pumping can also be used to polarize the metastables [100]. (See also Chapter 9.) Consider as an example the $^3P_2$ states of the heavy noble gases. If circularly polarized radiation is used to drive multiple transitions between the $np^5(n + 1)s\ ^3P_2$ and $np^5(n + 1)p\ ^3D_3$ levels, the $^3P_2$ state will become oriented in the direction of the incident photon angular momentum. Once a beam has been spin-polarized, the extent of its polarization can be measured using, e.g., a Stern–Gerlach magnet.

## 6.7.3 Beam Compression

In addition to their utility for state selection, photon beams and six-pole magnets can also be used for beam compression. As mentioned in Section 6.7.2, the use of a six-pole magnet for this purpose has been demonstrated [28].

Recently, laser cooling and compression of noble gas metastable beams has been demonstrated [20, 21]. Detailed discussions of such techniques are presented in these references and in Chapter 8 of this volume.

### 6.7.4 Velocity Selection and Analysis

An emergent beam's velocity profile can be analyzed in a variety of ways. For a DC source, a mechanical slotted chopper in conjunction with time-of-flight (TOF) analysis is often used [9, 43, 51, 54, 72]. This method is useful for establishing the beam fraction due to photons and fast neutrals. For pulsed beams, TOF techniques are also applicable [40, 45, 64]. A number of authors have reported using laser absorption measurements to determine velocity distributions [42]. By monitoring the light absorption as a function of photon incidence angle, and hence Doppler shift, the metastable velocity distribution can be mapped.

Velocity selection can be done completely mechanically or with a combination of pulsing and mechanical chopping with good resolution, but with some intensity diminution [6–8, 101]. The pulsed rotor source of Simons et al. [73] provides the simplest way to accomplish crude mechanical velocity selection. Generally, velocity selection is best accomplished by adjustment of the velocity profile of the initial (parent) ground state atoms, prior to excitation [11, 20, 46]. Such velocity control is sufficient for most applications.

### 6.7.5 Intensity Measurements

The topic of intensity and flux measurement is taken up on detail in Chapter 11. We simply note here that a variety of techniques can be used, including chemical reactions [3, 66, 98], laser-induced fluorescence and quenching [36, 42, 102], static electromagnetic field quenching [22, 38], and Auger electron emission from solid surfaces [31, 36, 82, 103–106]. These disparate techniques must be carefully calibrated if two of them are used to compare source intensities. Specifically, absolute fluxes are difficult to measure using the Auger method because of the uncertainties in secondary electron emission coefficients and the wide range of assumptions made in using them to estimate source strength.

## Acknowledgments

The author acknowledges numerous useful discussions with Dr. M. E. Johnston, and support from the Physics Division of the National Science Foundation.

## References

1. See, e.g., Allen, L. H. (1984). *Astrophysics and Space Science Library* **112**, 64; 116.

2. See, e.g., Makabe, T., Nakano, N., and Yamaguchi, Y. (1992). *Phys. Rev. A* **45**, 2520, and references therein.

3. See, e.g., Massey, H. S. W., McDaniel, E. E., and Bederson, B. (1982). *Applied Atomic Collision Physics; Atmospheric Physics and Chemistry*, Vol. 1, eds. Academic Press, New York.

4. See, e.g., Boyd, I. W., and Krimmel, E. F. (1989). *Photon, Beam and Plasma Assisted Processing*, eds. North Holland, Amsterdam.

5. Hayashi, S., Mayer, T. M., and Bernstein, R. B. (1978). *Chem. Phys. Lett.* **53**, 419.

6. Hishinuma, N., and Sueoka, O. (1983). *Chem. Phys. Lett.* **98**, 414.

7. Rothe, E. W., Neynaber, R. H., and Trujillo, S. M. (1965). *J. Chem. Phys.* **42**, 3310.

8. Tang, S. Y., Marcus, A. B., and Muschlitz, E. E., Jr. (1972). *J. Chem. Phys.* **56**, 566.

9. Ohno, K., Takami, T., Mitsuke, K., and Ishida, T. (1991). *J. Chem. Phys.* **94**, 2675.

10. Ben Arfa, M., Le Coz, G., Sinou, G., Le Nadan, A., Tuffin, F., and Tannous, C. (1994). *J. Phys. B* **27**, 2541.

11. Martin, D. W., Gregor, R. W., Jordan, R. M., and Siska, P. E. (1978). *J. Chem. Phys.* **69**, 2833.

12. Chen, C. H., Naberland, H., and Lee, Y. T. (1974). *J. Chem. Phys.* **61**, 3095.

13. Schall, H., Beckert, T., Alvariño, J. M., Vecchiocattivi, F., and Kempter, V. (1981). *Il Nuovo Cimento* **63**, 378.

14. Alvariño, J. M., Hepp, C., Kreinsen, M., Staudenmayer, B., Vechiocattivi, F., and Kempter, V. (1984). *J. Chem. Phys.* **80**, 765.

15. Rebick, C., and Dubrin, J. (1971). *J. Chem. Phys.* **55**, 5825.

16. Coleman, M. L., and Hammond, R. (1973). *Chem. Phys. Lett.* **19**, 271.

17. Reynaud, C., Pommier, J., Tuan, V. N., and Barat, M. (1979). *Phys. Rev. Lett.* **43**, 579.

18. Dixon, A. L., Harrison, M. F. A., and Smith, A. C. H. (1976). *J. Phys. B* **9**, 2617.

19. Morgenstern, R., Lorents, D. C., Peterson, J. R., and Olson, R. E. (1973). *Phys. Rev. A* **8**, 2372.

20. Hoogerland, M. (1993). Ph.D. thesis, Eindhoven University.

21. Metcalf, H., and van der Straten, P. (1994). *Phys. Rep.* **244**, 203.

22. Biraben, F., Garreau, J. C., Julien, L., and Allegrini, M. (1990). *Rev. Sci. Instrum.* **61**, 1468.

23. Zeiske, W., Meisel, G., Gebauer, H., Hofer, B., and Ertmer, W. (1976). *Phys. Lett.* **55A**, 405.

24. Madansky, L., and Owen, G. E. (1959). *Phys. Rev. Lett.* **2**, 209.

25. Slobodrian, R. J., Rioux, C., Giroux, J., and Roy, R. (1986). *Nucl. Instrum. Methods* **A244**, 127.

26. Lamb, W. E., Jr., and Retherford, R. C. (1950). *Phys. Rev.* **79**, 549.

27. Izrailov, E. K. (1994). *Op. Spectrosc.* **77**, 182.

28. Baum, G., Raith, W., and Steidl, H. (1988). *Z. Phys. D* **10**, 171.

29. Bozso, F., Yates, J. T., Jr., Arias, J., Metiu, H., and Martin, R. M. (1983). *J. Chem. Phys.* **78**, 4256.

30. Riddle, T. W., Onellion, M., Dunning, F. B., and Walters, G. K. (1981). *Rev. Sci. Instrum.* **52**, 797.

31. Johnson, P. D., and Delchar, T. A. (1977). *J. Phys. E* **10**, 428.

32. Lew, H. (1967). In *Methods of Experimental Physics*, Vol. 4, Pt. A: *Atomic Sources and Detectors*, V. W. Hughes and H. L. Schultz (eds.), pp. 183–198. Academic Press, New York.
33. Pauly, H. (1988). In *Atomic and Molecular Beam Methods*, Vol. 1, G. Scoles (ed.), pp. 113–117. Oxford University Press, Oxford.
34. Lin, C. C., and Anderson, L. W. (1992). *Adv. At. Mol. Opt. Phys.* **29**, 1.
35. Rundel, R. D., and Stebbings, R. F. (1972). In *Case Studies in Atomic Collisions Physics II*, E. W. McDaniel and M. R. C. McDowell (eds.), North Holland, Amsterdam.
36. Trajmar, S., and Nickel, J. C. (1993). *Adv. At. Mol. Opt. Phys.* **30**, 45.
37. Childs, W. J. (1972). *Case Stud. Atom. Phys.* **3**, 215.
38. Harvey, K. C. (1982). *J. Appl. Phys.* **53**, 3383.
39. Verheijen, M. J., Beijerinck, H. C. W., v. Moll, L. H. A. M., Driessen, J., and Verster, N. F. (1984). *J. Phys. E* **17**, 904.
40. Ureña, A. G., Costales, E. V., and Rábanos, V. S. (1990). *Meas. Sci. Technol.* **1**, 250.
41. Theuws, P. G. A., Beijerinck, H. C. W., Verster, N. F., and Schram, D. C. (1982). *J. Phys. E* **15**, 573.
42. Kawanaka, J., Hagiuda, M., Shimizu, K., Shimizu, F., and Takuma, H. (1993). *Appl. Phys. B* **56**, 21; Ishibashi, Y., Irikura, K., and Matsuzawa, H. (1989). *J. Vac. Sci. Technol. A* **7**, 2818.
43. Ferkel, H., Feltgen, R., and Pikorz, D. (1991). *Rev. Sci. Instrum.* **62**, 2626.
44. Tommasi, O., Bertuccelli, G., Francesconi, M., Giammanco, F., Romanini, D., and Strumia, F. (1992). *J. Phys. D* **25**, 1408.
45. Kohlhase, A., and Kita, S. (1986). *Rev. Sci. Instrum.* **57**, 2925.
46. Brutschy, B., and Haberland, H. (1977). *J. Phys. E* **10**, 90; MacNair, D., Lynch, R. T., and Hannay, N. D. (1953). *J. Appl. Phys.* **24**, 1335.
47. Freund, R. S. (1970). *Rev. Sci. Instrum.* **41**, 1213.
48. Gaily, T. D., Coggiola, M. J., Peterson, R., and Gillen, K. T. (1980). *Rev. Sci. Instrum.* **51**, 1168.
49. Donnally, B. L., Clapp, T., Sawyer, W., and Schultz, M. (1964). *Phys. Rev. Lett.* **12**, 502.
50. Hotop, H., Kolb, E., and Lorenzen, J. (1979). *J. Electron Spectrosc. Rel. Phenom.* **16**, 213.
51. Kraft, T., Bregel, T., Ganz, J., Harth, K., Ruf, M.-W., and Hotop, H. (1988). *Z. Phys. D* **10**, 473.
52. Searcy, J. Q. (1974). *Rev. Sci. Instrum.* **45**, 589.
53. Leasure, E. L., Mueller, C. R., and Ridley, T. Y. (1975). *Rev. Sci. Instrum.* **46**, 635.
54. Fahey, D. W., Parks, W. F., and Schearer, L. D. (1980). *J. Phys. E* **13**, 381.
55. Onellion, M., Riddle, T. W., Dunning, F. B., and Walters, G. K. (1982). *Rev. Sci. Instrum.* **53**, 257.
56. Fry, E. S., and Williams, W. L. (1969). *Rev. Sci. Instrum.* **40**, 1141.
57. Gostev, V. A., Elakhovskii, D. V., Zaitsev, Yu. V., Luizova, L. A., and Khakhaev, A. D. (1980). *Opt. Spectrosc.* **48**, 251.
58. Gostev, V. A., Elakhovskii, D. V., and Khakhaev, A. D. (1980). *Sov. Phys. Tech. Phys.* **25**, 1258.
59. Peterson, J. R., and Lorents, D. C. (1969). *Phys. Rev.* **182**, 152.
60. Jackson, L. C., and Broadway, L. F. (1930). *Proc. Roy. Soc.* **A127**, 678.
61. Brand, J. A., Furst, J. E., Gay, T. J., and Schearer, L. D. (1992). *Rev. Sci. Instrum.* **63**, 163.

62. Verheijen, M. J., Beijerinck, H. C. W., and Verster, N. F. (1982). *J. Phys. E* **15,** 1198.
63. Olmsted, J., III, Newton, A. S., and Street, K., Jr. (1965). *J. Chem. Phys.* **42,** 2321.
64. Rundel, R. D., Dunning, F. B., and Stebbings, R. F. (1974). *Rev. Sci. Instrum.* **45,** 116.
65. Coggiola, M. J., Gaily, T. D., Gillen, K. T., and Peterson, J. R. (1979). *J. Chem. Phys.* **70,** 2576.
66. Neynaber, R. H., and Magnuson, G. D. (1976). *J. Chem. Phys.* **65,** 5239.
67. Johnston, M. E. (1993). Ph.D. Thesis, University of California-Riverside; also submitted to *J. Phys. B* for publication.
68. Giusfredi, G., Miguzzi, P., Strumia, F., and Tonelli, M. (1975). *Z. Phys. A* **274,** 279.
69. Lurio, A. (1959). *Bull. Am. Phys. Soc.* **4,** 419, 429.
70. Kowalski, A., and Heldt, J. (1978). *Chem. Phys. Lett.* **54,** 240.
71. Duquette, D. W., and Lawler, J. E. (1982). *Phys. Rev. A* **26,** 330.
72. Hardy, K. A., Gillman, E., and Sheldon, J. W. (1990). *J. Appl. Phys.* **67,** 7240.
73. Simons, J. P., Suzuki, K., and Washington, C. (1984). *J. Phys. E* **17,** 581.
74. Foreman, P. B., Parr, T. P., and Martin, R. M. (1977). *J. Chem. Phys.* **67,** 5591.
75. Brinkmann, U., Kluge, J., and Pippert, K. (1980). *J. Appl. Phys.* **51,** 4612.
76. Brinkmann, U., Busse, H., Pippert, K., and Telle, H. (1979). *Appl. Phys.* **18,** 249.
77. Berkling, K., Schlier, C., and Toschek, P. (1962). *Z. Physik* **168,** 81.
78. Bartsch, M., Geesman, H., Hanne, G. F., and Kessler, J. (1992). *J. Phys. B* **25,** 1511.
79. Friedburg, H., and Kuiper, H. (1957). *Naturwiss.* **44,** 487.
80. Ishii, S., Ohlendorf, W. (1972). *Rev. Sci. Instrum.* **43,** 1632.
81. Ogura, K., and Shibata, T. (1994). *J. Phys. Soc. Jpn.* **63,** 834.
82. Haberman, J. A., Wilcomb, B. E., Van Itallie, F. J., and Bernstein, R. B. (1975). *J. Chem. Phys.* **62,** 4466.
83. McDermott, M. N., and Lichten, W. L. (1960). *Phys. Rev.* **119,** 134.
84. Burrow, P. D. (1967). *Phys. Rev.* **158,** 65.
85. Geesmann, H., Bartsch, M., Hanne, G. F., and Kessler, J. (1991). *J. Phys. B* **24,** 2817.
86. Gould, G. (1956). *Phys. Rev.* **101,** 1828.
87. Garpman, S., Lido, G., Rydbert, S., and Svanberg, S. (1971). *Z. Physik* **241,** 217.
88. Svanberg, S. (1972). *Physica Scripta* **5,** 73.
89. Zavilopulo, A. N., Shpenik, O. B., Zhukov, A. I., and Snergurskii, A. V. (1933). *Tech. Phys.* **38,** 42.
90. Belić, D. S., and Hall, R. I. (1981). *J. Phys. B* **14,** 365.
91. Fahey, D. W., Schearer, L. D., and Parks, W. F. (1978). *Rev. Sci. Instrum.* **49,** 503.
92. Haberland, H., and Oschwald, M. (1988). *J. Phys. B* **21,** 1183.
93. Fabrikant, I. I., Shpenik, O. B., Snegursky, A. V., and Zavilolpulo, A. N. (1988). *Phys. Repts.* **159,** 1.
94. Pearl, J. C., Donnelly, D. P., and Zorn, J. C. (1969). *Phys. Lett.* **30A,** 145.
95. Ignat'ev, A. V., Kozyrev, A. V., and Novoselov, Yu. N. (1993). *Tech. Phys.* **38,** 529.
96. Hughes, V. W., Long, R. L., Jr., Lubell, M. S., Posner, M., and Raith, W. (1972). *Phys. Rev. A* **5,** 195.
97. Koch, P. M., and Mariani, D. R. (1981). *Phys. Rev. Lett.* **46,** 1275.

98. Ruf, M.-W., Yencha, A. J., and Hotop, H. (1987). *Z. Phys. D* **5,** 9.
99. Dunning, F. B., Cook, T. B., West, W. P., and Stebbings, R. F. (1975). *Rev. Sci. Instrum.* **46,** 1072.
100. Bußert, W. (1986). *Z. Phys. D* **1,** 321; see also Butler, W. H., Hammond, M. S., Lynn, J. G., Dunning, F. B., and Walters, G. K. (1988). *Rev. Sci. Instrum.* **59,** 2083, and references therein.
101. Kinsey, J. L. (1966). *Rev. Sci. Instrum.* **37,** 61.
102. Kinsey, J. L. (1977). *Ann. Rev. Phys. Chem.* **28,** 349; Schohl, S., Klar, D., Kraft, T., Meijer, H. A. J., Ruf, M.-W., Schmitz, U., Smith, S. J., and Hotop, H. (1991). *Z. Phys. D* **21,** 25.
103. Dunning, F. B., Smith, A. C. H., and Stebbings, R. F. (1971). *J. Phys. B* **4,** 1683.
104. Dunning, F. B., and Smith, A. C. H. (1971). *J. Phys. B* **4,** 1696.
105. Dunning, F. B., Rundel, R. D., and Stebbings, R. F. (1975). *Rev. Sci. Instrum.* **46,** 697.
106. Rundel, R. D., Dunning, F. B., Howard, J. S., Riola, J. P., and Stebbings, R. F. (1973). *Rev. Sci. Instrum.* **44,** 60.

# 7. PRODUCTION OF RYDBERG ATOMS

## T. F. Gallagher

Department of Physics, University of Virginia
Charlottesville, Virginia

## 7.1 Introduction

Rydberg atoms, those in states of high principal quantum number $n$, have an excited valence electron which is weakly bound in a large orbit. The former property is evident from the energy $W$, given in atomic units by

$$W = -\frac{1}{2n^2}.$$ (7.1)

The atomic unit of energy is twice the Rydberg constant. For an electron moving in a Coulomb potential, the energy of Eq. (7.1) implies that the expectation value of the electron's distance from the ionic core, $\langle r \rangle$, is given by

$$\langle r \rangle \sim 3n^2/2,$$ (7.2)

where $\langle r \rangle$ is given in $a_0$, the Bohr radius. Because the excited electron is in a large, weakly bound orbit, Rydberg atoms have exaggerated properties, making them interesting systems for a variety of experiments. For example, the fact that the Rydberg electron is nearly free has been used to do low-energy electron–molecule scattering [1].

This chapter is devoted to commonly used methods of producing usable samples of neutral Rydberg atoms for further experiments.

## 7.2 Overview of the Methods

There are two general ways of producing Rydberg atoms: adding a loosely bound electron to an ion by charge exchange, and exciting one of the electrons from the ground state of an atom. Correspondingly, Rydberg atoms have been produced by charge exchange, electron impact, and photoexcitation, the processes

$$A^+ + B \rightarrow A_{n\ell} + B^+,$$ (7.3a)

$$e^- + A \rightarrow A_{n\ell} + e^-,$$ (7.3b)

and

$$h\nu + A \rightarrow A_{n\ell}.$$ (7.3c)

EXPERIMENTAL METHODS IN THE PHYSICAL SCIENCES
Vol. 29B

Since collisional excitation, shown by Eqs. (7.3a) and (7.3b), is not state-selective, pure collisional excitation is becoming less common, and optical excitation is often used in conjunction with it. We shall for the moment consider the three processes of Eq. (7.3) separately. Charge exchange, represented by Eq. (7.3a), has been used extensively to make fast beams of Rydberg atoms. The starting point is a fast ion beam, as described by Alton [2] and Chutjian [3]. Electron impact or optical excitation requires thermal atoms in a beam, as discussed by Ramsey [4], or in a vapor cell, as discussed by Vidal [5].

The cross sections for all three processes of Eq. (7.3) scale as $n^{-3}$. This scaling is easily understood by considering, for example, electron impact excitation of He. If an electron of energy $W_e$ impinges upon a ground state He atom, $\sigma(W_e)$ is the cross section for exciting the He atom from the ground state to all other states, including ionization. Subsequent to excitation the final state of the He atom is an excited He atom or a $He^+$ ion and a free electron. The probability of these outcomes is represented by the differential cross section $d\sigma(W_e)/dW$, where $W$ is the energy of the electron ejected from the He atom. $W < 0$ corresponds to the "ejected" electron remaining in a bound state of He, and $W > 0$ corresponds to ionization of the He. In general, $d\sigma/dW$ is a smooth function of $W$. More important, there is no fundamental difference between $W > 0$ and $W < 0$. The collision process occurs when the electrons are within a short range, $<10$ Å, of the $He^+$ core, and at these distances there is almost no difference between the kinetic energies of a Rydberg electron and a free electron. As a result, the differential cross section $d\sigma/dW$ passes smoothly from $W > 0$ to $W < 0$.

For a fixed value of the incident electron energy $W_e$, we approximate $d\sigma(W_e)/dW$ for $W \approx 0$ by a constant,

$$\sigma'(W_e) = \left. \frac{d\sigma(W_e)}{dW} \right|_{W=0}. \tag{7.4}$$

Below the limit, $W < 0$, the energy $W$ is not a continuous variable, and it is useful to convert $d\sigma(W_e)/dW$ to a cross section per principal quantum number. Explicitly,

$$\sigma(n, W_e) = \frac{d\sigma(W_e)}{dn} = \frac{d\sigma(W_e)}{dW} \cdot \frac{dW}{dn}$$

$$= \frac{\sigma'(W_e)}{n^3}, \tag{7.5}$$

showing the $n^{-3}$ dependence of the cross section for the production of Rydberg states.

The $n$ scaling can also be obtained from the normalization of the Rydberg atom wave functions at small orbital radius. Prior to the collision the He atom is compact, having a radius $\sim 1$ Å. Immediately after the collision the Ryeberg electron is in its large orbit, which has a density at the origin scaling as $n^{-3}$. When the initial state is projected onto the final state in the excitation process, a cross section with an $n^{-3}$ dependence is the result.

Similar arguments apply to charge exchange and photoexcitation, and the basic result is the same; the cross section for the production of Rydberg atoms is the continuation below the limit of the ionization cross section, leading to an $n^{-3}$ dependence of the excitation cross section.

## 7.3 Electron Impact Excitation

To use electron impact excitation of ground state atoms most efficiently, the incident electron energy $W_e$ should be picked so as to maximize the cross section at the ionization limit, i.e., to maximize $\sigma'(W_e)$. In general these energies are from 10 to 100 eV. Typical values of $\sigma'(W_e)$ are $\sim 1$ Å$^2$, and for reference the values of $\sigma'(W_e)$ for the rare gas atoms are given in Table I [6]. For a typical value of $\sigma'(W_e) = 1$ Å, the cross section $\sigma(20, W_e)$ for exciting an $n = 20$ state is $1.25 \times 10^{-4}$ Å$^2$.

An aspect of electron impact ionization which is worth bearing in mind is that the electrons not only excite ground state atoms to Rydberg states, but can also induce transitions between Rydberg states. In fact, the cross sections for

TABLE I. Electron Impact Cross Sections for the Excitation of Rare Gas Rydberg States for the Electron Energy $W_e$ at the Peak of the Cross Section and at 100 eV[a].

| Rare gas | Electron energy $W_e$ (eV) | $\sigma'(W_e)$ (Å)$^2$ |
|---|---|---|
| He | 70 | 0.77 |
| He | 100 | 0.67 |
| Ne | 60 | 0.63 |
| Ne | 100 | 0.61 |
| Ar | 28 | 6.5 |
| Ar | 100 | 1.5 |
| Kr | 20 | 4.0 |
| Kr | 100 | 2.0 |
| Xe | 20 | 10.0 |
| Xe | 100 | 4.6 |

[a] From reference [6].

electrons driving transitions between Rydberg states, typically $\Delta n = 0$, $\Delta \ell$ transitions, are high. They are $\sim 10^6$ Å$^2$ at $n = 20$ and exhibit an $n^4$ scaling [6, 7]. When electron impact excitation is used, both processes

$$e + A_g \rightarrow e + A_{n, \ell} \tag{7.6a}$$

and

$$e + A_{n, \ell} \rightarrow e + A_{n', \ell'} \tag{7.6b}$$

MacAdam and Wing used electron impact excitation to produce He Rydberg atoms to study microwave $\Delta \ell$ transitions [8, 9]. Their apparatus is shown in Figure 1. In a vapor cell filled with He at a pressure of $10^{-3}$ torr, they excited the He with an electron beam with energy 30–50 eV and currents from 10 to 1000 μA. They used such low currents to avoid the production of many $\ell$ states by the process of Eq. (7.6b) and ensure that low $\ell$ Rydberg states were preferentially populated. Preferential population of low $\ell$ states was important, as they monitored the $nd \rightarrow 2p$ emissions to detect resonant microwave transitions from the $nd$ states to higher $\ell$ states.

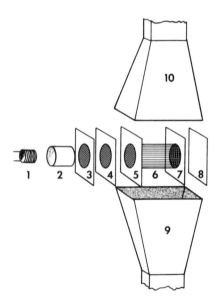

FIG. 1. Exploded view of the resonance module used by MacAdam and Wing: (1) heater; (2) cathode; (3) accelerating grid G1; (4) grid G2; (5) grid B1; (6) the "cage" consisting of wires joining the perimeters of grids B1 and B2; (7) grid B2; (8) collector C; (9) and (10) X-band waveguide horns. The grid supports are nonmagnetic stainless-steel plates, 2.54 cm square. The cage is 2.54 cm long × 1.6 cm diameter. (From reference [10]).

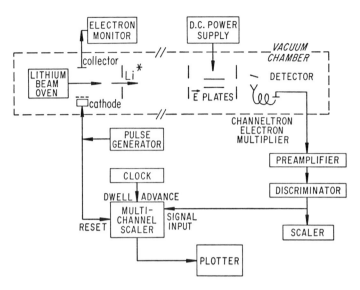

FIG. 2. Atomic beam apparatus for electron-beam excitation of Li atoms to Rydberg states. (From reference [10].)

Kocher and Smith used electron impact excitation to excite Li atoms in a beam to Rydberg states [10]. As shown by Figure 2, they crossed the thermal Li beam with a beam of electrons of energy 10 eV and currents of up to 20 mA. These currents were far in excess of those used by MacAdam and Wing, and, as a result, the Rydberg population was spread over many $\ell$ states, not just the low $\ell$ states. Thus, the radiative lifetimes observed were far longer than the lifetimes of the low $\ell$ states.

On the positive side, electron impact excitation is easy and works with all atoms. On the negative side, it is not at all state-selective and produces ions as well as Rydberg atoms.

## 7.4 Charge Exchange

Charge exchange is the preferred way of producing a fast beam of Rydberg atoms [11]. In essence the method consists of a fast ion beam passing through a charge exchange cell. The charge exchange cross section for the production of a specific $n$ state can be written in a form similar to that of Eq. (7.5). The distribution of product states of charge exchange of ions with incident energy $W_i$ is described by $d\sigma_L(W_i)/dW$, where $W$ is the energy of the Rydberg electron in the neutral product atom and $\sigma_L(W_i)$ is the total electron loss cross section. As in

Eq. (7.3), we can write the charge-exchange cross section for the population of a specific $n$ state as

$$\sigma_{CE}(n, W_i) = \frac{\sigma'_L(W_i)}{n^3},$$ (7)

where

$$\sigma'_L(W_i) = \frac{d\sigma_L(W_i)}{dW}\bigg|_{W=0}.$$

In Table II we give representative values of $\sigma'_L(W_i)$ [12–14]. Only a small fraction of the incident ions which undergo collisions with the gas in the charge exchange cell will be left in any particular Rydberg state. For $n = 10$, the fraction is $<10^{-3}$, and for $n = 30$ it is $<10^{-4}$. If the number density of the charge exchange gas is $N$ and the length of the cell is $L$, in principle the charge exchange cell can be operated with $NL\sigma_L \sim 1$, so that nearly all the ions in the beam undergo charge exchange. In fact it is usually not possible to use such a thick target, because it will scatter atoms in the fast neutral beam into large enough angles to remove them from the beam. Since the cross section for scattering atoms out of the beam is $\sim 1$ Å$^2$, of the same magnitude as the electron-loss cross sections, the cell must be operated at a density length product low enough that the Rydberg atoms produced are not scattered out of the beam. For example, Bayfield et al. [13] used a 20 cm long charge exchange cell containing H at a pressure of $5 \times 10^{-5}$ torr, a pressure low enough to ensure single-collision conditions.

The Rydberg states produced in the cell by charge exchange can be collisionally depopulated by subsequent collisions with the target gas. In the

TABLE II.  Values of the Charge-Exchange Cross Section for Protons on Several Vapor Targets

| Target | Incident energy (keV) | $\sigma'_L(W_i)$ (Å$^2$) |
|--------|----------------------|--------------------------|
| Xe[a]  | 20 | 10 |
| Kr[a]  | 20 | 9 |
| Ar[a]  | 20 | 4 |
| Ar[b]  | 30 | 3.3 |
| He[a]  | 20 | 0.4 |
| He[c]  | 60 | 0.6 |
| H$_2$[c] | 60 | 1.1 |
| N$_2$[c] | 60 | 2.5 |
| CO$_2$[c] | 60 | 3.8 |

[a] See reference [12].
[b] See reference [13].
[c] See reference [14].

charge-exchange collision, it is presumably low $\ell$ states which are populated, by virtue of the overlap of their wave functions with the ground-state wave function. In thermal collisions it has been observed that the cross sections for $\ell$- and $m$-changing collisions are large, with values for Xe reaching $10^5 \text{Å}^2$ at $n = 20$, and minimum values $>20$ Å [15, 16]. These cross sections are far larger than $\sigma_{\text{L}}$, so it is likely that efficient charge exchange is accompanied by collisions which populate high $\ell$ and $m$ states. Thermal energy cross sections for changing $n$ are smaller than the $\ell$-changing cross sections, with typical values of $\sim 10 \text{Å}^2$. Since virtually all $n$ states are populated, some further redistribution over $n$ has no significant effect. In sum, it is reasonable to expect a distribution of $n\ell m$ states from charge exchange, with an $n^{-3}$ scaling of the population in each $n$, approximately the same result as can be expected for electron impact excitation.

In their initial microwave ionization experiment, Bayfield and Koch used a fast beam of H Rydberg atoms [17]. They passed a 10 keV proton beam through a charge-exchange cell filled with Xe to produce the fast beam of Rydberg atoms. They passed the fast beam between two field plates, to which a modulated voltage was applied, to select a band of $n$ states by field ionization. The field required to ionize a state of principal quantum number $n$ scales as $n^{-4}$. If the field is switched between $E_1$ and $E_2$ with $E_2 < E_1$, atoms in states of $n > n_1$ or $n_2$ are ionized by the field. Thus, the difference in signal obtained with $E = E_1$ and $E = E_2$ is due to atoms with $n_1 \leq n \leq n_2$. Using a field switched between 28.5 and 41.0 V/cm, they selected the band of states with $63 \leq n \leq 69$.

An alternative, very selective means of tagging specific Rydberg states was used by Palfrey and Lundeen [18]. They used a $CO_2$ laser to drive He $10\ell \rightarrow n$, $\ell'$ transitions to selectively empty specific He $10\ell$ states which were later refilled by driving $\Delta\ell$ radio-frequency transitions.

## 7.5 Optical Excitation

Optical excitation differs from collisional excitation in a fundamental way: The exciting photon is absorbed by the target atom. As a result, specifying the energy of the absorbed photon specifies the Rydberg state produced. In contrast specifying the energy of an incoming electron does not specify the energy of the Rydberg state produced, because there is no way to control how the energy is shared between the incident electron and the electron which is excited to the Rydberg state. The advent of the tunable laser has enabled us to take advantage of the specificity of optical excitation and has led to the renaissance of interest in Rydberg atoms.

In spite of its difference from collisional excitation, photoexcitation has the similarity of having a cross section with an $n^{-3}$ dependence. This cross section is simply the continuation of the photoionization cross section from above the limit, $W > 0$, to below the limit, $W < 0$. A useful way to describe photo-excitation is to specify $\sigma_{PI}$, the photoexcitation cross section at the ionization limit, $W = 0$. Above the limit, the photoionization cross section is given by $\sigma_{PI}$. Below the limit, the cross section averaged over an integral number of $n$ states equals $\sigma_{PI}$. The cross section $\sigma(n)$ for exciting a resolved Rydberg state is given by

$$\sigma(n) = \frac{\sigma_{PI}}{\Delta W n^3},$$ (7.8)

where $\Delta W$ is the energy resolution of the excitation. The cross section for exciting the Rydberg state is the photoionization cross section times the ratio of the $\Delta n$ energy interval to the experimental resolution. For reference, the photo-ionization cross sections from the ground states of H, alkali, and the alkaline earth atoms are given in Table III [19].

There are several possible methods of optical excitation, which are shown in Figure 3. Atoms can be excited directly from the ground state by single photon excitation, as shown by Figure 3a. This approach is only possible for alkali and alkaline earth atoms, but does require ultraviolet light, as shown by Table III. For all other atoms the required wavelengths are too short for single-photon excitation to the Rydberg states. In this case, excitation by way of a resonant intermediate state, shown by Figure 3b, is often chosen because relatively low-power visible light sources can often be used.

Two-photon excitation via a virtual intermediate state, shown in Figure 3c, can be used for Doppler free excitation. Finally, optical excitation from a state populated by collisions, shown in Figure 3d, is a very powerful technique for selective production of Rydberg states of atoms with large ionization potentials.

TABLE III.   Photoionization Cross Sections at Threshold $(W = 0)^a$

| Atom | Wavelength (Å) | $\sigma_{PI}(10^{-18}\ cm^2)$ |
|------|----------------|-------------------------------|
| H    | 912            | 6.3                           |
| Li   | 2299           | 1.8                           |
| Na   | 2412           | 0.125                         |
| K    | 2856           | 0.007                         |
| Rb   | 2968           | 0.10                          |
| Cs   | 3184           | 0.20                          |
| Mg   | 1621           | 1.18                          |
| Ca   | 2028           | 0.45                          |
| Sr   | 2177           | 3.6                           |

$^a$ From reference [19].

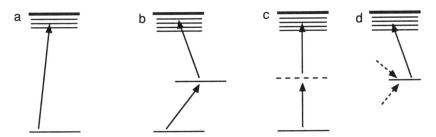

FIG. 3. Methods of optical excitation of Rydberg states: (a) single-photon excitation from the ground state; (b) stepwise two-photon excitation via a resonant intermediate state; (c) two-photon excitation via a virtual intermediate state; (d) single-photon excitation from an excited state which has been produced by a collision process.

An important factor in Eq. (7.8) is the resolution, $\Delta W$, which depends on the experiment. The radiative widths of the levels lead to an intrinsic lower limit. Low $\ell$ Rydberg states of $n \approx 20$ have lifetimes of 5 μs and radiative widths of ~30 kHz, while low-lying intermediate states such as the Na $3p$ state have shorter lifetimes, 16 ns, and correspondingly larger widths, 10 MHz [20]. In addition to the radiative widths are the Doppler widths, which are ~1 GHz for visible transitions. If single-photon excitation of atoms in a vapor cell is used, the Doppler width is unavoidable. On the other hand, if the laser beam crosses a collimated atomic beam at a right angle, the Doppler width vanishes. When two-photon excitation with counterpropagating laser beams is used, the Doppler width also vanishes, even in a vapor cell.

In addition to the foregoing intrinsic widths, there are instrumental widths which must be taken into account. The most common is laser line width. Typical pulsed dye lasers have widths of ~10 GHz, cw dye lasers have widths of ~1 MHz, and cw $CO_2$ lasers have widths of ~50 kHz. Finally, any inhomogeneities in electric or magnetic fields which are present contribute to the resolution.

As a typical example of the use of Eq. (7.8), we consider the most widely used method of optically exciting Rydberg atoms, using a pulsed dye laser pumped by a $N_2$ or Nd:YAG laser. It is straightforward to generate visible laser pulses 5 ns long with 100 μJ energies and 1 cm$^{-1}$ bandwidths. Such a pulse contains ~$3 \times 10^{14}$ photons, and, if collimated into a beam of cross-sectional area $10^{-2}$ cm$^2$, it has an integrated photon flux of $10^{16}$ cm$^{-2}$. At $n = 20$, the $\Delta n$ spacing is ~28 cm$^{-1}$, and with a resolution of 1 cm$^{-1}$, the cross section of Eq. (7.8) is a factor of 28 larger than the photoionization cross section. Using $10^{-18}$ cm$^2$ as the photoionization cross section, we find a probability of 10% for exciting atoms exposed to a photon flux of $10^{16}$ cm$^{-2}$ to the $n = 20$ state.

A typical example of pulsed dye laser excitation is the beam experiment shown in Figure 4 [21]. Li atoms effusing from a resistively heated source (a)

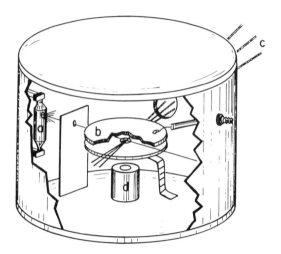

Fig. 4. Apparatus for the study of Rydberg states of alkali metal atoms: (a) the atomic beam source; (b) the electric field plates; (c) the pulsed laser beams; and (d) the electron multiplier. (From reference [21].)

are collimated into a beam which passes between two field plates. Between the field plates the atomic beam is crossed by three pulsed dye laser beams which excite the Li atoms to Rydberg states by the sequence of transitions $2s \rightarrow 2p \rightarrow 3s \rightarrow np$. Shortly after excitation the atoms are selectively ionized by applying a voltage pulse to the upper plate (b) to field-ionize the Rydberg atoms and drive the resulting ions down to the particle detector (d), which is typically an electron multiplier or microchannel plate.

Since the excitation and detection of the Rydberg atoms are not separated spatially, only temporally, it is not the flux of atoms from the source which is important, but rather the density, which is reduced from the density in the source by the factor $A/\ell^2$, where $A$ is the area of the exit hole in the source and $\ell$ is the distance from the source to the interaction region. An upper limit to the source pressure is ~1 torr. Finally, the vacuum requirements for such an apparatus are not stringent. Small-angle scattering of the beam between the source and the interaction region is unimportant, and the collision cross sections of Rydberg atoms with background gas molecules are small enough that pressures of $10^{-6}$ torr are adequate.

Using continuous wave (cw) laser excitation, it is possible to excite atoms with substantially higher efficiency than using pulsed lasers. For example, a single-mode laser of 1 MHz line width has a resolution $3 \times 10^4$ better than the pulsed laser described earlier. Accordingly, for single-photon excitation the resolution-limited cross section can be $3 \times 10^4$ times larger if the excitation is done so as to avoid Doppler broadening. Examples of such an excitation are the experiments of Zollars *et al.* [1] using a frequency-doubled cw dye laser to excite a beam of Rb ground state atoms to the *np* states, and the experiments of Fabre *et al.* [22] in which they used a cw dye laser and diode lasers to excite Na *nf* states.

## 7.6 Collisional–Optical Excitation

Purely optical excitation is possible for metal atoms. For most other atoms the transition from the ground state to any other level is at too short a wavelength to be practical. To produce Rydberg states of such atoms, a combination of collisional and optical excitation is quite effective. A good example is the study of the Rydberg states of Xe by Stebbings *et al.* [23]. As shown in Figure 5, a thermal beam of Xe atoms is excited by electron impact, producing a range of excited states, and a reasonable fraction of the excited atoms accumulates in the metastable state. The production of metastable beams is discussed in detail by Gay [24]. Downstream from electron excitation the atoms in the metastable state are excited to a Rydberg state by pulsed dye laser excitation.

Metastable atoms have also been used as the starting point for laser excitation in cell experiments. Devos *et al.* [25] used metastable He atoms in the stationary afterglow of a pulsed discharge as the starting point for pulsed laser excitation to high-lying *np* states. They observed the time- and wavelength-resolved fluorescence to determine the collision rate constants of the *np* states.

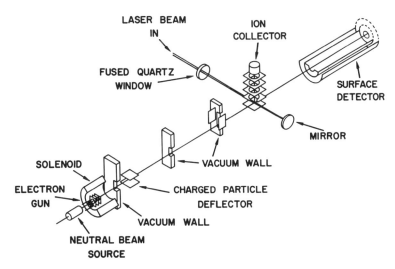

FIG. 5. Schematic diagram of the apparatus using electron impact excitation to the Xe metastable state followed by laser excitation to Rydberg states. (From reference [23].)

Optical excitation of fast beams of H and He atoms has proven to be a very powerful technique [11]. Typically a cw $CO_2$ laser is used for excitation. The $CO_2$ laser has 20 lines 1.3 cm$^{-1}$ apart in each of the 9.6 and 10.6 μm bands. These bands match the frequency of $n \approx 10$ to $n \approx 30$ transitions in an atom. A $CO_2$ laser delivers in excess of 1 watt on most of these lines and has a line width of ~50 kHz. Tuning in discrete steps of 1.3 cm$^{-1}$ is possible by changing lines; Doppler tuning, by changing either the angle at which the laser beam crosses the fast beam or the beam energy, provides continuous tuning. For example, a 10 keV H beam overlapping a counterpropagating 10 μm $CO_2$ laser beam is tuned by 0.2 cm$^{-1}$ for each 100 eV change in beam energy.

Optical excitation of the fast beam is straightforward. The essential notion is to populate the Rydberg levels of the fast beam by charge exchange, remove all atoms in $n > 10$ states by field ionization, and then repopulate the selected level of $n > 20$ using the $CO_2$ laser, as shown by Figure 6 [26, 27]. This approach has two attractions, illustrated by the production of $n = 26$ atoms. First, only one Rydberg state $n = 26$ is present in the final beam. Second, since these $n = 26$ atoms are excited from the $n = 10$ atoms produced by charge exchange, the population is increased by a factor of $(26/10)^3$ over the direct population of the $n = 26$ state by charge exchange, assuming all other collisional process can be ignored. It is worth keeping in mind that there are significant numbers of atoms in states of $n < 10$, and that the bulk of the beam is composed of ground-state atoms, as shown by Figure 6. Since these atoms are fast, they can be ionized in

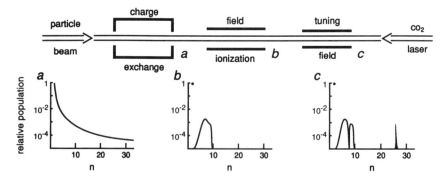

Fɪɢ. 6. Schematic drawing of the method of producing a fast beam of $n \sim 26$ H atoms. A proton beam passes through the charge exchange cell resulting in the $1/n^3$ distribution of Rydberg states at point a, as shown by (a) a plot of the Rydberg state distribution at point a. The beam passes through a field ionization region which ionizes atoms of $n > 10$ and removes all ions from the beam. At point b, all $n > 10$ states are removed, as are the lowest $n$ states, which decay radiatively, as shown by (b). The beam then passes through the tuning field, which brings the $n = 10 \rightarrow n = 26$ transition into resonance, and at point c the $n = 26$ state has been repopulated and is the only state above $n = 10$ which is populated. (Adapted from reference [27].)

collisions with background gas molecules, making the vacuum requirements more stringent than for thermal beams. The approach of Figure 6, or a variant using a second $CO_2$ laser to drive $n = 7$ to $n = 10$ transitions, has become a standard approach in the study of Rydberg atoms using fast beams [27].

The number of Rydberg atoms which can be produced in the approach just outlined is limited to the number of $n = 10$ final states of the charge-exchange process. If ground-state atoms are used as the charge-exchange target, the number of $n = 10$ final states is small. If, however, the target is composed of $n = 10$ atoms, the yield of $n = 10$ atoms in the fast beam is dramatically increased, as recently demonstrated by Deck et al. [28]. They used diode lasers and a color center laser to produce Rb $10f$ atoms as the charge-exchange target for a $S^+$ fast beam. They found that the ratio of the fractional $10h$ population in the resulting S fast beam was increased by a factor of $10^4$ over the ratio obtained with a ground-state charge-exchange target [28].

## References

1. Zollars, B. G., Higgs, C., Lu, F., Walter, C. W., Gray, L. G., Smith, K. A., Dunning, F. B., and Stebbings, R. F. (1985). *Phys. Rev. A* **32**, 3330.
2. Alton, G. (1996). In *Charged Particles*, F. B. Dunning and R. G. Hulet (eds.). Academic Press, New York.
3. Chutjian, A. (1996). In *Atoms and Molecules*, F. B. Dunning and R. G. Hulet (eds.). Academic Press, New York.
4. Ramsey, N. F. (1996). In *Atoms and Molecules*, F. B. Dunning and R. G. Hulet (eds.). Academic Press, New York.
5. Vidal, C. R. (1996). In *Atoms and Molecules*, F. B. Dunning and R. G. Hulet (eds.). Academic Press, New York.
6. Schiavone, J. A., Tarr, S. M., and Freund, R. S. (1979). *Phys. Rev. A* **20**, 71.
7. Foltz, G. W., Beiting, E. J., Jeys, T. H., Smith, K. A., Dunning, F. B., and Stebbings, R. F. (1982). *Phys. Rev.* **A25**, 187.
8. MacAdam, K. B., and Wing, W. H. (1975). *Phys. Rev. A* **12**, 1464.
9. MacAdam, K. B., and Wing, W. H. (1977). *Phys. Rev. A* **13**, 678.
10. Kocher, C. A., and Smith, A. J. (1977). *Phys. Lett.* **61A**, 305.
11. Koch, P. M. (1983). In *Rydberg States of Atoms and Molecules*, R. F. Stebbings and F. B. Dunning (eds.). Cambridge Univ. Press, Cambridge.
12. King, R. F., and Latimer, C. J. (1979). *J. Phys. B* **12**, 1477.
13. Bayfield, J. E., Khayrallah, G. A., and Koch, P. M. (1974). *Phys. Rev. A* **9**, 209.
14. Il'in, R. N., Kikiani, B., Oparin, V. A., Solov'ev, E. S., and Fedorenko, N. V. (1965). *Sov. Phys. JETP* **20**, 835. [(*J. Exptl. Theor. Phys. USSR* **47**, 1235 (1964)].
15. Kachru, R., Gallagher, T. F., Gounand, F., Safinya, K. A., and Sandner, W. (1983). *Phys. Rev. A* **27**, 795.
16. Hugon, M., Gounand, F., Fournier, P. R., and Berlande, J. (1979). *J. Phys. B* **12**, 2707.
17. Bayfield, J. E., and Koch, P. M. (1974). *Phys. Rev. Lett.* **33**, 258.
18. Palfrey, S. L., and Lundeen, S. R. (1984). *Phys. Rev. Lett.* **53**, 1141.
19. Marr, G. V. (1967). *Photoionization Processes in Gases*. Academic Press, New York.

20. Theodosiou, C. E. (1984). *Phys. Rev. A* **30**, 2881.
21. Kleppner, D., Littman, M. G., and Zimmerman, M. L. (1983). In *Rydberg States of Atoms and Molecules*, R. F. Stebbings and F. B. Dunning (eds.). Cambridge University Press, Cambridge.
22. Fabre, C., Kaluzny, Y., Calabrese, R., Jun, L., Goy, P., and Haroche, S. (1984). *J. Phys. B* **17**, 3217.
23. Stebbings, R. F., Latimer, C. J., West, W. P., Dunning, F. B., and Cook, T. B. (1975). *Phys. Rev. A* **12**, 1453.
24. Gay, T. J. (1996). In *Atoms and Molecules*, F. B. Dunning and R. G. Hulet (eds.). Academic Press, New York.
25. Devos, F., Boulmer, J., and Delpech, J. F. (1979). *J. Phys. (Paris)* **40**, 215.
26. Koch, P. M. (1978). *Phys. Rev. Lett.* **41**, 99.
27. Koch, P. M., and Mariani, D. R. (1980). *J. Phys. B* **13**, L645.
28. Deck, F. J., Hessels, E. A., and Lundeen, S. R., Unpublished results.

# 8. LASER COOLING AND TRAPPING OF NEUTRAL ATOMS

## Curtis C. Bradley and Randall G. Hulet

Department of Physics and Rice Quantum Institute
Rice University
Houston, Texas

## 8.1 Introduction

The first proposal for using lasers to deflect atomic beams was made in 1970 [1], followed in 1975 by the suggestion to use laser light to slow and cool atoms [2, 3]. By 1978 the first successful experimental demonstrations of these ideas were realized [4–6]. Since then, the field has advanced at a rapid pace. It is now possible to laser cool atoms to below 1 $\mu$K, to confine as many as $10^{10}$ atoms in traps with densities of greater than $10^{12}$ cm$^{-3}$, and to trap atoms for time periods of up to one hour. Laser cooling techniques are also used to produce slow, monoenergetic atomic beams and to brighten beams by cooling the transverse motion.

In recent years, these techniques have begun to emerge from the specialist's laboratory and are being used in a variety of applications, including cold-atom collisions, time and frequency standards, investigations of quantum degenerate atomic gases, precision measurements of fundamental constants, measurements of parity-violating interactions, and searches for time-reversal–violating permanent electric dipole moments. This chapter is written for those wishing to exploit the properties of laser-cooled atomic gases, but who are not yet expert in the field. We begin with a description of two-level Doppler cooling and sub-Doppler cooling. Techniques for both longitudinal slowing and transverse cooling of atomic beams are presented next. Three types of atom traps are discussed, including pure magnetic, magneto-optical, and dipole traps. The chapter concludes with a brief treatment of evaporative cooling, which has recently been used to attain nano-Kelvin temperatures. We have chosen to focus on a few techniques that we feel are, or will be, the most useful. By necessity, then, large areas of the field are neglected or only briefly mentioned. This chapter is not intended to be an exhaustive review, but rather an introduction to the techniques and their capabilities. We apologize in advance for the many references and contributions that are not included. There are several excellent reviews meant for nonspecialists [7–15], and several special journal issues [16, 17] devoted to laser cooling and trapping. Laser cooling and atom trapping involve many of the

EXPERIMENTAL METHODS IN THE PHYSICAL SCIENCES
Vol. 29B

methods discussed in other chapters of this volume. Relevant chapters are Atomic Beams, Ion Trapping, Laser Stabilization, Diode Lasers, and Frequency Shifting and Modulation.

## 8.2 Laser Cooling

### 8.2.1 Doppler Cooling

The basic idea of laser cooling is that the photons of a laser beam can impart momentum to atoms. The simplest laser cooling, "Doppler cooling," involves just two atomic states, a ground state $|g\rangle$ and an excited state $|e\rangle$, and a laser beam tuned to near the $|g\rangle \leftrightarrow |e\rangle$ resonance frequency. In reality, of course, there are no two-level atoms, but one can be approximated by using a $\sigma^+$ or $\sigma^-$ circularly polarized laser beam tuned to a $J \leftrightarrow J + 1$ transition. In this case, the atoms are optically pumped into the $m_J = \pm J$ ground state sublevel and can only be excited to the $m_J = \pm(J + 1)$ excited state sublevel. By absorbing a photon, the atom acquires the photon momentum $\hbar \mathbf{k}_L$, where $\mathbf{k}_L$ is the wave vector of the laser field. The atom can lose its excitation energy by spontaneous or stimulated emission of a photon. If it is stimulated by the laser beam which originally excited it, the radiated photon rejoins the laser field in the same mode, and there is no net momentum transfer. On the other hand, spontaneous emission may result in a net change of momentum. The change in momentum from sponta- neous decay of a photon with wave vector $\mathbf{k}_s$ is $\hbar \mathbf{k}_s$. But because spontaneous emission is a symmetric process, so that wave vectors $\mathbf{k}_s$ and $-\mathbf{k}_s$ are equally probable, $\langle \mathbf{k}_s \rangle = 0$ and there is no net change in momentum on average in spontaneous decay. So, the round-trip process of stimulated absorption from $|g\rangle$ to $|e\rangle$ followed by spontaneous decay back to $|g\rangle$ results in an average momentum change of $\Delta p = \hbar \mathbf{k}_L$. The rate of spontaneous decay is given by the product of the probability of being in state $|e\rangle$, $\rho_{ee}$, and the decay rate of $|e\rangle$, $\gamma$. Therefore, the average force exerted by the laser beam on the atoms is

$$\langle F \rangle = \left\langle \frac{\Delta p}{\Delta t} \right\rangle = \hbar k_L \rho_{ee} \gamma. \tag{1}$$

Since $\rho_{ee} \leq \frac{1}{2}$ in steady state, $\langle F \rangle \leq \frac{1}{2}\hbar \mathbf{k}_L \gamma$. The surprisingly large maximum acceleration is given by

$$a_{\text{Dop}} = \frac{\hbar k_L \gamma}{2m}, \tag{2}$$

where $m$ is the atomic mass. For example, for the lithium $2s$–$2p$ transition, $\lambda = 671$ nm and $\gamma = (27.1 \text{ ns})^{-1}$, giving $a_{\text{Dop}} = 1.6 \times 10^8$ cm/s$^2$.

Viscous damping of atomic motion can be provided by two or more lasers that intersect symmetrically to form what is often referred to as optical molasses [18, 19]. In molasses, the forces on atoms are directed along the propagation vectors of the laser beams. The ultimate kinetic temperature, $T$, where $k_B T = \langle p^2/2m \rangle$, is determined by a balance of the laser-induced cooling and momentum diffusion. For spontaneously emitted photons, $\langle \mathbf{k}_s \rangle = 0$ and $\langle \mathbf{k}_s^2 \rangle > 0$. Spontaneous emission, therefore gives momentum diffusion for which $\langle p^2/2m \rangle > 0$. This results in laser-cooled atoms undergoing a random walk, or diffusion, out of the molasses region. The diffusion time $\tau_D$ is proportional to $\langle r^2 \rangle$, where $r$ is the displacement due to diffusion. For alkali-metal atoms, $\tau_D$ is approximately 4 seconds for $r = 1$ cm. For certain "misalignments" of the molasses laser beams, it has been observed that $\tau_D$ can actually be increased by more than an order of magnitude [20, 21]. The lower temperature limit of Doppler cooling, $T_{Dop}$, can be estimated from the uncertainty in energy of the spontaneously emitted photon, $\hbar\gamma$. Calculations show that the Doppler cooling limit is given by $k_B T_{Dop} = \frac{1}{2}\hbar\gamma$ [22], which for lithium is 140 μK.

## 8.2.2 Sub-Doppler Cooling

As early as 1988, temperatures less than $T_{Dop}$ were measured in optical molasses [23]. It was soon realized that optical pumping between the degenerate ground-state sublevels due to polarization gradients of certain configurations of laser polarization vectors could occur on time scales much slower than $\gamma^{-1}$ and produce temperatures lower than $T_{Dop}$ [24, 25]. Temperatures equal to several times the "recoil temperature," $T_R$, were measured in molasses. The recoil limit is set by the energy an atom acquires by recoiling from spontaneous emission, so $k_B T_R = (\hbar k_L)^2/m$. For $^7$Li, $T_R = 6$ μK, while for $^{133}$Cs, $T_R = 200$ nK. Temperatures of several times $T_R$ are readily achieved in an optical molasses, when ambient magnetic fields are minimized [19].

Other sub-Doppler cooling techniques have been demonstrated, but have not yet been used in applications. One such method is adiabatic cooling, where atoms confined to the nodes of an optical standing wave are cooled through an adiabatic reduction of the standing wave intensity [26]. This technique has recently been demonstrated in three dimensions [27]. A particularly interesting *subrecoil* technique is velocity-selective coherent population trapping, or VSCPT, in which atoms are optically pumped into a coupled atomic/momentum state that decouples from the laser field for momenta equal to $\pm \hbar k_L$ [28, 29]. This method has also recently been demonstrated in two and three dimensions [30, 31]. For this technique, the width of the resulting momentum distribution is limited only by the interaction time. A different method that produces subrecoil temperatures uses the narrow linewidth of stimulated Raman transitions to provide atoms with very small velocity spread [32–34].

## 8.3 Atomic Beam Cooling

### 8.3.1 Longitudinal Slowing

The most important consideration for slowing an atomic beam with laser radiation is that the effective detuning $\Delta$ of the slowing laser changes as the atoms decelerate. This can easily be seen from Eq. (1), with

$$\rho_{ee} = \frac{\Omega_0^2}{2\Omega_0^2 + 4\Delta^2 + \gamma^2}. \tag{3}$$

$\Omega_0 = -2(\langle g|\mathbf{d}|e\rangle \cdot \mathbf{E})/\hbar \equiv \gamma(I/I_S)^{1/2}$ is the "on-resonance" Rabi frequency, where $\mathbf{d}$ is the atom's electric dipole moment, $\mathbf{E}$ is the laser electric field, $I$ is the laser intensity, and $I_S$ is the saturation intensity. The effective detuning depends on the velocity $\mathbf{v}$ of the atom, the laser beam wave vector $\mathbf{k}$, and the detuning $\Delta_0$ of the laser for an atom at rest. Explicitly, $\Delta = \Delta_0 - \mathbf{k} \cdot \mathbf{v}$, where $\Delta_0 = \omega_L - \omega_0$, $\hbar \omega_L$ is the energy of a laser photon, and $\hbar\omega_0$ is the energy difference between states $|e\rangle$ and $|g\rangle$. The atoms in the beam are slowed by a single laser beam directed against their motion, such that in the frame of the moving atoms the laser frequency appears higher (bluer) than in the lab frame. However, as the atoms scatter photons and slow down, their Doppler shift is reduced until they effectively shift out of resonance and stop decelerating. In order to slow atoms over a larger range of velocity, some compensation for this changing Doppler shift must be made. The two most effective methods for doing this are chirp slowing [6, 35] and Zeeman slowing [36, 37].

For chirp slowing, the laser frequency is linearly chirped in time, so that $\Delta$ remains constant as atoms undergo a constant deceleration, $a$. The frequency is then reset to its initial value and the chirp repeats, slowing a new bunch of atoms. Table I shows some of the quantities relevant to both chirp and Zeeman slowing, for the alkali-metal elements. The time required to stop an atom with initial velocity $v_0$ is $\Delta t = v_0/a$. The required length of travel is

TABLE I. Quantities Relevant to Chirp and Zeeman Slowing for Alkali-Metal Elements

| Atom | $T_0(K)$ | $v_M$ $(10^4 \text{ cm/s})$ | $\Delta v_D$ (GHz) | $\lambda$ (nm) | $m$ (amu) | $\tau$ (ns) | $a_{Dop}$ $(10^6 \text{ cm/s}^2)$ | $\Delta t$ (ms) | $\Delta L$ (cm) |
|------|----------|------|------|------|------|------|------|------|------|
| Li | 900 | 19 | 2.8 | 617 | 7 | 27.1 | 160 | 1.2 | 115 |
| Na | 630 | 8.7 | 1.5 | 589 | 23 | 16.4 | 90 | 0.97 | 43 |
| K | 545 | 6.2 | 0.81 | 766 | 39 | 26 | 26 | 2.4 | 76 |
| Rb | 500 | 4.1 | 0.52 | 780 | 85 | 27 | 11 | 3.6 | 74 |
| Cs | 480 | 3.2 | 0.37 | 852 | 133 | 30.4 | 5.8 | 5.5 | 87 |

$$\Delta L = \upsilon_0 \Delta t - \frac{1}{2} a(\Delta t)^2 = \frac{\upsilon_0^2}{2a}. \tag{4}$$

From Chapter 1 (Eq. 1.32), the median velocity of an atom in the beam is

$$\upsilon_m = 1.30 \sqrt{\frac{2k_B T_0}{m}}, \tag{5}$$

where $T_0$ is the temperature of the atomic beam oven. The values of $\Delta t$ and $\Delta L$ given in Table I assume $\upsilon_0 = \upsilon_m$ ($T_0$ selected for oven pressures of ~0.1 Torr) and $a = a_{Dop}$. The wavelength $\lambda$ is for the principal $S_{1/2} \leftrightarrow P_{3/2}$ transition.

Two methods have been used to generate the required frequency chirp $\Delta\upsilon_D = \upsilon/\lambda$. A traveling-wave electro-optic modulator [35] can be used to produce frequency-modulated sidebands that can be swept in time, or the frequency of a laser diode may be ramped using the injection current [38]. Also, as an improvement over simple single-frequency chirp cooling, it is possible to use multiple laser frequencies in a relay-chirp cooling scheme [39, 40]. Still, there are two significant problems with the chirp slowing technique: (1) It may be technically difficult to sweep the frequency by the amount necessary to cool a significant portion of the velocity distribution, and (2) the atoms do not all arrive at their final velocity at the same longitudinal location. This, combined with the transverse spreading of the slowed atom beam, results in a significantly lower intensity of slow atoms than can be obtained with Zeeman slowing.

Zeeman slowing is, in several respects, a more powerful technique. Instead of the laser frequency being varied, the atoms' transition frequency is changed using a spatially varying magnetic field. For slowing to a specific final velocity, it is best to use the $\sigma^-$ Zeeman slower design [41]. This type of Zeeman slower has its maximum magnetic field with an abrupt field cutoff at the downstream end, so the decelerating atoms are suddenly shifted out of resonance with the counterpropagating laser beam. This results in a minimal spread of slow atom velocities. For such a slower, the desired variation in the magnetic field is given by $\Delta B(z) \sim 1 - (1 - z/z_0)^{1/2}$, where $z$ is the axial displacement along the slower, $z_0$ is given by $\upsilon_0^2/2a$, $\upsilon_0$ is the initial velocity of the fastest atoms to be slowed and $a$ is the deceleration during slowing ($a > 0$). A solenoid having a current density $i(z) \sim i_0[1 - (1 - z/z_0)^{1/4}]$ produces an axial magnetic field with the correct spatial distribution. With this current distribution in mind, a fair approximation (see Fig. 1) of the ideal field is easily generated via stepped layers of windings around the tube of a vacuum nipple [42]. A double-stage Zeeman slower approach, in which the field decreases from a maximum and then rises again, allows slowing from higher initial velocities without resorting to larger solenoid current densities [43].

For either beam slowing technique, it is helpful to note that obtaining maximum slow atom flux does not necessarily require heroic efforts to slow the

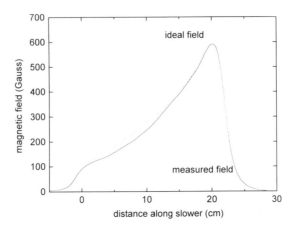

FIG. 1. Comparison of the ideal $\sigma^-$ Zeeman slower magnetic field, designed for decelerating atoms at half the Doppler acceleration, with the measured values for a Zeeman slower used in our laboratory. Atoms with longitudinal velocities near 560 m/s begin to slow at $\sim$0 cm along the slower. The rising magnetic field compensates for the decreasing Doppler shift of the decelerating atoms. The sharp cutoff of magnetic field just beyond $\sim$20 cm helps to minimize the final longitudinal velocity spread of the Zeeman-slowed atoms.

entire thermal velocity distribution. Assuming a conventional atom beam source with the atom-slowing laser beam focused near the beam source aperture, is there an optimal choice of slower length which maximizes the slow atom production? As discussed previously, for a maximum initial velocity $v_0$ to be slowed, the slower length $\Delta L$ scales like $v_0^2$. Consequently, the useful atomic beam solid angle (set by the useable slow-atom–beam diameter) scales as $(\Delta L)^{-2} \sim v_0^{-4}$ (assuming deceleration starts immediately after the source). Since the integrated number of atoms to be slowed from a thermal beam only increases as $v_0^4$ or worse, the resultant flux of slow atoms is roughly independent of $v_0$, or equivalently, the slower length. This line of reasoning shows that a trap can be efficiently loaded directly from the slow atoms already present in a thermal atomic beam, if it is located close to the oven nozzle. Recently, magneto-optical traps have been loaded in this manner, with a small beam block in front of the trapped atoms to prevent trap loss via collisions with the atomic beam [44] . However, when beam slowing is required, a compact design, which is technically easier to build, may work as well as, or even better than, a slower designed to slow a large fraction of the thermal beam. For example, the $\sigma^-$ Zeeman slower illustrated in Fig. 1 is designed to decelerate the slowest 1% of a thermal Li atom beam ($v_0 \leq 560$ m/s), using $a = a_{\mathrm{Dop}}/2$. The slower solenoid produces a 100 G bias along its $\sim$20 cm length and has a field maximum of 590 G. The solenoid consists of several layers of heavily lacquered magnet wire wrapped around a double-wall vacuum nipple. It is energized by three separate current sources (allowing higher currents where needed, without the additional bulk of extra windings so it remains compact) and cooled externally with a fan and "internally" by circulating chilled water (passing inside the nipple's double wall) [42]. On a separate experiment we have installed a $\sim 10$ cm long $\sigma^+$ Zeeman slower which we use to load a Li magneto-optic trap. This slower was conceived and constructed in a couple of days, uses a single 1-amp current source, and is simply air-cooled.

### 8.3.2 Transverse Cooling and Beam Deflection

For many experiments it is necessary to increase the brightness of the slowed atomic beam. The simplest means of doing this is through a reduction of the beam divergence following longitudinal slowing. A 2-D molasses in a plane perpendicular to the atomic beam can significantly reduce the spread of transverse velocities of the slow atoms, such that the beam is effectively collimated.

For loading an atom trap, it is often desirable to separate the slow atoms from the residual fast atomic beam. Directing only the slow atoms toward the trap reduces trap loss caused by collisions with fast atoms. Furthermore, it is helpful to avoid the interaction between the longitudinal slowing laser beam and the trapped atoms. A simple means of doing this is to place the trap just off the atomic beam axis, behind a plate which blocks the fast atomic beam. As mentioned earlier, this approach can be used to effectively load a trap from an unslowed atomic beam [44]. Alternatively, a single laser beam directed transversely to the atom beam axis will deflect the slow atoms. For more precise control and larger angular deflection, the transverse laser beam can be cylindrically focused in the plane of the deflection with the slow atom fraction entering and leaving the laser beam along trajectories approximately perpendicular to the leading and trailing laser beam edges [1, 45]. Using this kind of slow atom beam deflection, alkali atoms with velocities of up to $\sim 10^2$ m/s can easily be deflected through angles of 30° or more. Several groups have also demonstrated methods for collimating and compressing slow atomic beams using laser light combined with inhomogeneous magnetic fields to provide exceptional slow atom beam brightening [46–48]. These brighteners are essentially two-dimensional versions of the magneto-optical trap described later.

## 8.4 Trapping

### 8.4.1 Magnetic Trapping

The first successful neutral atom trap was the quadrupole magnetic trap [49]. Any atom with a magnetic dipole moment $\mu$ will experience a force when located in a gradient of magnetic field. Atoms with their dipole moment aligned with the

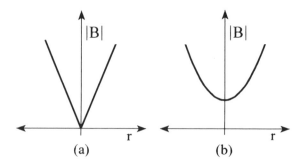

FIG. 2. Contrasting the two magnetic trap configurations. The spatial dependence of field strength is shown for (a) the quadrupole trap and (b) the Ioffe trap. For atoms with a linear Zeeman shift, such as the alkali metals, the trapping potential is proportional to the magnetic field strength $|\mathbf{B}|$.

field direction are attracted to a minimum of the field strength, while those antialigned are repelled. Three-dimensional local minima are easily produced with electromagnets or permanent magnets. Static local field maxima are ruled out by Earnshaw's theorem [50]. The most common magnetic trap configurations are the quadrupole trap and the Ioffe trap [51–53]. Comparing the two types (Fig. 2), the quadrupole trap has a linearly varying field which is more strongly confining than the quadradically varying field of the Ioffe trap, but exhibits a problematic zero-field point at its center. The Ioffe trap offers confinement with a nonzero field minimum.

A quadrupole field results, for example, from two current loops placed in an anti-Helmholtz orientation. For this configuration, the field is zero at a point between the current loops where the field components from each loop (and from other stray fields) exactly cancel. The on-axis field gradient near the trap minimum is given by $1.2\pi NIDR^2(D^2 + R^2)^{-5/2}$ G cm/A, where $R$ is the coil radius, $2D$ is the coil separation, and $NI$ is the total current in each coil [54]. Axial and radial trap depth can be approximately equal for a coil separation of ~1.25 the coil radii, and for this case the gradient is given approximately by $NIR^{-2}$ G cm/A. For 5 cm diameter coils carrying $10^3$ amp-turns of current, the trap magnet field gradients are 160 G/cm, giving a trap depth of 11 mK/cm. Atoms remain confined because as they move about in the trap they adiabatically follow the changing field direction and stay in the same field-repelled spin state (as referred to the quantization direction given by the local magnetic field). Atoms that pass too close to the field zero, however, may undergo a nonadiabatic spin-flip or Majorana transition to the untrapped spin state because of the sudden change in field direction at the trap minimum [49]. The region very near the field zero, then, can act as a sinkhole for trapped atoms, becoming an important loss mechanism for atoms cooled to low kinetic temperatures. Fortunately, several methods have been developed for effectively plugging the hole. One method is to introduce a weak, rotating, transverse magnetic field and thereby dynamically move the hole around in such a way as to prevent the atoms from falling into it (the TOP trap) [55]. A second approach is to repel the trapped atoms from the hole, via the optical dipole force from a blue-detuned laser beam focused on the trapping field minimum [56]. Typically, atoms are loaded into these traps by energizing the trapping coils and forming the trap around a cloud of atoms previously collected and laser-cooled by a combination of techniques including magneto-optic trapping and sub-Doppler molasses cooling. Using these methods, clouds of alkali atoms have been trapped with temperatures ranging from $\sim 20\,\mu K$ to $\sim 1\,mK$ and densities up to $\sim 10^{12}\,cm^{-3}$ [57].

A magnetic trap using a field in the Ioffe configuration provides a nonzero field minimum at the bottom of a harmonic potential. Because of the bias field there is no hole at the bottom. Also, because of the bias field, atoms in the trap can resonantly scatter many photons without being optically pumped to a nontrapped state. This allows for laser Doppler cooling of the atoms in the trap—a means for continuous loading of atoms from an atomic beam or vapor cell. Traps using this field configuration have been produced using conventional current distributions, superconducting coils, and permanent magnets. In our lab, more than $10^{8}$ Li atoms have been confined and laser-cooled to $\sim 200\,\mu K$ in a trap made of six axially magnetized, cylindrical, high-flux NdFeB permanent magnets, positioned and aligned along three mutually orthogonal axes [58], as shown in Fig. 3. Near the center of the trap, the potential experienced by the atoms is harmonic with an oscillation frequency of $\sim 10^{2}\,Hz$. The use of permanent magnets in building atom traps is motivated by the large field gradients that they offer [59, 60] and the desire to have good optical access to the trapped atoms along with overall experimental simplicity.

## 8.4.2 Magneto-Optic Trapping

A very robust trap, first demonstrated in 1987, uses the large dissipative force available from near-resonant laser light, in combination with an inhomogeneous magnetic field, to provide both spatial confinement and damping of atomic motion [61]. The most common configuration for such a magneto-optic trap (MOT) uses three mutually orthogonal pairs of counter propagating laser beams intersecting at the center of a quadrupole magnetic field, with polarizations set as shown in Fig. 4. An alternative arrangement for the laser beams uses a tetrahedral configuration of four beams [62]. In both versions, typical field gradient maxima are near 5 to 20 G/cm, with the field provided by two current-carrying coils placed in an anti-Helmholz configuration. This field distribution produces Zeeman shifts that are proportional to atomic displacement from the trap center. The appropriately polarized trap laser beams, along with these atomic energy

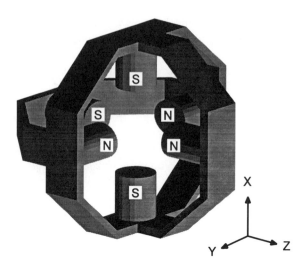

FIG. 3. Diagram showing the construction of our permanent magnet Ioffe trap. The six NdFeB cylindrical trap magnets are held by a magnetic-steel support, which also provides low reluctance paths for the flux to follow between magnets of opposite sign. The letters indicate the inner tip magnetizations of the magnets, N for north and S for south. The magnet tip-to-tip spacing is 4.45 cm.

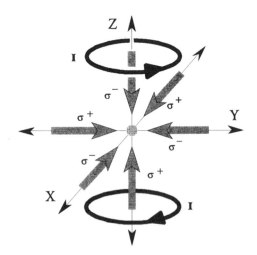

FIG. 4. The six-beam magneto-optic trap. Two coils, each carrying current $I$ in the anti-Helmholtz configuration, provide a quadrupole magnetic field. At the magnetic field center, three pairs of opposing laser beams intersect to provide three-dimensional atom velocity damping. The arrangement of laser beam circular polarizations, given the current directions, are chosen in order to provide spatial confinement.

level shifts, provide a restoring force that keeps the atoms confined near the trap center. To see how this works, it is easiest to consider a simple atom with a $J = 0$, $m_J = 0$ ground state and a $J = 1$, $m_J = 0$, $\pm 1$ excited state, with displacements along one trap axis, say the $z$-axis of Fig. 4. For atoms at positions $z \neq 0$, the $m_J = \pm 1$ excited states have opposite Zeeman shifts, as shown in Fig. 5. This asymmetry produces changes in the detunings for the $\sigma+$ and $\sigma-$ transitions (both are driven via the two opposing laser beams) that consequently lead to an imbalanced optical force for atoms at these positions. By properly arranging the laser polarizations relative to the magnetic field directions, the imbalanced optical forces can be directed inward toward the field center. Also, with the laser frequency detuned below the atomic resonant frequency, the atomic motion undergoes viscous damping. Using this type of trap, many groups have reported trapped atom clouds containing up to $10^8$ atoms at temperatures of $\sim 1\,mK$ and below, with peak densities up to $\sim 10^{11}$ atoms/cm$^{-3}$. With careful balancing and alignment of the laser beams and control of the trap magnetic field, sub-Doppler cooling in a low-density MOT has been demonstrated, with observed

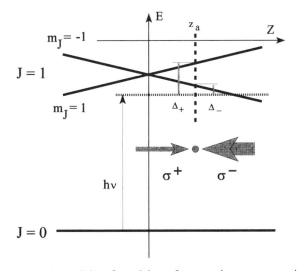

FIG. 5. Diagram showing origin of spatial confinement in magneto-optic traps. We consider a model atom with a $J = 0$ ($m_J = 0$) ground state and $J = 1$ excited state, situated at rest in a quadrupole magnet field, and interacting with opposing laser fields that drive $\Delta m_J = \pm 1$ transitions. The spatial dependence of the Zeeman-shifted levels results in a spatial dependence for the detunings of the two possible transitions. This asymmetry produces an imbalanced optical force that varies with the atom's position. For example, for an atom at $z = z_a > 0$ with the laser frequency $v$ and polarizations as shown, the relative detunings, shown by $\Delta_\pm$, will cause the scattering rate for the $\sigma^-$ transition to be relatively larger, and so the atom will be accelerated toward $z = 0$.

temperatures for trapped Cs as low as 10 μK [63–65]. Furthermore, by blocking the central portion of the hyperfine-repumping light, the central trapped atoms can optically pump into a hyperfine level which is only weakly coupled to the remaining laser beam light. This reduces the trap-losses caused by light scattering and results in increased trap densities. In such a *dark* spontaneous-force optical trap (Dark SPOT), densities of $\sim 10^{12}\,\mathrm{cm}^{-3}$ have been reported [66].

### 8.4.3 Dipole-Force Trapping

The third type of trap utilizes the interaction of the induced atomic electric dipole moment with a gradient in an optical electric field. This type of trap is often referred to as a dipole-force trap. The light force potential for a two-level atom interacting with a sufficiently detuned and/or low-intensity light field is given by [67, 68]

$$U(\mathbf{r}) = \frac{\hbar \gamma^2}{8\Delta} \frac{I(\mathbf{r})}{I_{\mathrm{S}}},$$

where $I_{\mathrm{s}}$ is the intensity required to saturate the two-level atom. Atoms can be confined at a maximum of $I(\mathbf{r})$, as at the focus of a laser beam, with $\Delta$ sufficiently large and negative. The first experimental demonstration of such a trap was in 1986 using Na atoms and a single $\sim 200\,\mathrm{mW}$ laser beam, focused to a 10 μm spot, with the laser frequency tuned to $\sim 0.6\,\mathrm{THz}$ below the D2 resonance frequency of sodium [69]. More recently, Rb has been trapped using a focused laser, detuned up to $\sim 10^2\,\mathrm{nm}$ below the Rb D1 resonance [70]. There are several comparative disadvantages for this type of trap. The numbers of trapped atoms obtained to date are relatively small, typically $\sim 10^3$, largely because of the extremely small trapping volume. Also, because of the heating from scattered light, the lifetime of atoms in pure dipole-force traps is only $\sim 0.2$ seconds.

## 8.5 Evaporative Cooling

While there does not appear to be an ultimate laser cooling limit for a dilute atomic gas, at modest densities the poor optical transmission of the gas sample starts to interfere with the cooling mechanisms discussed. For alkali gases confined to a $\sim 1$ mm region, the on-resonance optical density becomes significant when the average atom density is only $\sim 10^{10}\,\mathrm{cm}^{-3}$. In order to further reduce the temperature of a sample of trapped atoms, while maintaining or increasing its density, a nonlaser technique is needed. Such a technique, evaporative cooling, was first proposed by Hess in 1986 [71] and demonstrated with magnetically trapped atomic hydrogen in 1988 [72], and has since been heavily relied upon to reach beyond laser cooling. By removing the most energetic fraction of the trapped gas and allowing sufficient time for rethermalization of the remaining atoms via elastic collisions, it is possible to achieve dramatic cooling and compression of the gas. The initial method of doing this was to alter the trapping potential in order to provide a sufficiently low barrier. A more effective technique was suggested [8] in 1989 and successfully demonstrated in 1994 in magnetically trapped Na [73] and Rb [74]. In this newer version, atoms are "evaporated" via a forced spin-flip and their subsequent repulsion from the trapping region. In a magnetic trap, the most energetic atoms sample the largest magnetic fields and thereby experience the largest Zeeman shifts. By tuning an RF field to be resonant with atoms at relatively large fields, it is possible to remove a specific high-energy portion of the trapped atoms and leave the low-energy portion intact. After the remaining atoms rethermalize to a lower temperature, the RF frequency can be decreased and the process repeated [75]. Evaporative loss of trapped atoms competes with a separate trap-loss rate, typically from collisions with background gas particles, characterized by $\tau_{TRAP}$. Because of this competition, it is necessary that the rethermalization via elastic collisions be sufficiently rapid for evaporative cooling to dominate and for significant increases in phase-space density to be obtained. An estimate of the average elastic collision rate is given by $R_E = n\sigma_E v$, where $n$ is the average density, $\sigma_E$ is the collisional cross section, and $v$ is the average atom velocity in the trap. For successful evaporation, $R_E > 100/\tau_{TRAP}$ is required [76]. We take as an example evaporative cooling of Li vapor confined in a permanent magnet Ioffe trap. Given the measured triplet elastic cross-section for $^7$Li atoms [77] of $5.0 \times 10^{-13}$ cm$^{-2}$ and an initial density of $\sim 10^{10}$ Li atoms/cm$^3$ at $\sim 200$ µK, for successful evaporation the required $\tau_{TRAP}$ is $\sim 400$ seconds. We obtained this trapped atom lifetime by placing our trap inside a scrupulously cleaned ion-pumped chamber, by sourcing the trap through a small-diameter tube, and by evaporating titanium onto the inside of the ion-pump and trapping-chamber walls. The estimated background gas pressure is $10^{-12}$ torr.

Through the use of evaporative cooling of atoms in a trap, it is possible to increase the phase-space density (given by $d^3r\,d^3p$) of the trapped gas by many orders of magnitude. In the experiments that allowed the first observation of Bose-Einstein condensation in alkali-metal gases [56, 78, 79], evaporative cooling was used to boost the phase-space density by factors of $\sim 10^6$ over what was achieved with laser cooling alone. This breakthrough and other perhaps unanticipated phenomena are now accessible through the use of laser and evaporative cooling, and through the exploitation of the various types of atom traps.

# References

1. Ashkin, A. (1970). *Phys. Rev. Lett.* **25**, 1321.
2. Hansch, T. W., and Schawlow, A. L. (1975), *Opt. Comm.* **13**, 68.

3. Wineland, D. J., and Dehmelt, H. (1975). *Bull. Am. Phys. Soc.* **20**, 637.
4. Wineland, D. J., Drullinger, R. E., and Walls, F. L. (1978). *Phys. Rev. Lett.* **40**, 1639.
5. Neuhauser, W., Hohenstatt, M., Toschek, P., and Dehmelt, H. (1978). *Phys. Rev. Lett.* **41**, 233.
6. Balykin, V. I., Letokhov, V. S., and Mushin, V. I. (1979). *JETP Lett.* **29**, 560.
7. Wineland, D. J., and Itano, W. M. (1987). *Physics Today* **40**, 34.
8. Pritchard, D. E., Helmerson, K., and Martin, A. G. (1989), in *Atomic Physics 11* (S. Haroche, J. C. Gay, and G. Grynberg, eds.), World Scientific, Singapore.
9. Cohen-Tannoudji, C., and Phillips, W. D. (1990). *Physics Today* **43**, 33.
10. Chu, S. (1991). *Science* **253**, 861.
11. Foot, C. J. (1991). *Contemp. Phys.* **32**, 369.
12. Chu, S. (1992). *Sci. Am.* **266**, 70.
13. *Proceedings of the International School of Physics "Enrico Fermi," Course CXVIII, (Varenna 1991): Laser Manipulation of Atoms and Ions* (E. Arimondo, W. D. Phillips, and F. Strumia, eds.), North-Holland, Amsterdam, 1992.
14. Metcalf, H., and van der Straten, P. (1994). *Phys. Reports* **244**, 204.
15. Aspect, A., Kaiser, R, Vansteenkiste, N., and Westbrook, C. I. (1995). *Physica Scripta* **T58**, 69.
16. *J. Opt. Soc. Am. B* **6**, 2020 (1989).
17. *Laser Physics* **4**, 829 (1994).
18. Chu, S., Hollberg, L., Bjorkholm, J. E., Cable, A., and Ashkin, A. (1985). *Phys. Rev. Lett.* **55**, 48.
19. Lett, P. D., Phillips, W. D., Rolston, S. L., Tanner, C. E., Watts, R. N., and Westbrook, C. I. (1989). *J. Opt. Soc. Am. B* **6**, 2084.
20. Chu, S., Prentiss, M. G., Cable, A., and Bjorkholm, J. (1987), in *Laser Spectroscopy VII* (W. Pearson and S. Svanberg, eds.), Springer-Verlag, Berlin.
21. Bagnato, V. S., Bigelow, N. P., Surdutovich, G. I., and Zilio, S. C. (1994). *Opt. Lett.* **19**, 1568.
22. Wineland, D. J., and Itano, W. M. (1979). *Phys. Rev. A* **20**, 1521.
23. Lett, P. D., Watts, R. N., Westbrook, C. I., Phillips, W. D., Gould, P. L., and Metcalf, H. J. (1988). *Phys. Rev. Lett.* **61**, 169.
24. Dalibard, J., and Cohen-Tannoudji, C. (1989). *J. Opt. Soc. Am. B* **6**, 2023.
25. Ungar, P. J., Weiss, D. S., Riis, E., and Chu, S. (1989). *J. Opt. Soc. Am. B* **6**, 2058.
26. Chen, J., Story, J. G., Tollett, J. J., and Hulet, R. G. (1992). *Phys. Rev. Lett.* **69**, 1344.
27. Kastberg, A., Phillips, W. D., Rolston, S. L., Spreeuw, R. J. C., and Jessen, P. S. (1995). *Phys. Rev. Lett.* **74**, 1542; Anderson, B. P., Gustavson, T. L., and Kasevitch, M. A. (1994), postdeadline paper contributed to the Ninth International Quantum Electronics Conference, Anaheim, California.
28. Aspect, A., Arimondo, E., Kaiser, R., Vansteenkiste, N., and Cohen-Tannoudji, C. (1989). *J. Opt. Soc. Am. B* **6**, 2112.
29. Mauri, F., and Arimondo, E. (1991). *Europhys. Lett.* **16**, 717.
30. Lawall, J., Bardou, F., Saubamea, B., Shimizu, K., Leduc, M., Aspect, A., and Cohen-Tannoudji, C. (1994). *Phys. Rev. Lett.* **73**, 1915.
31. Lawall, J., Kulin, S., Saubamea, B., Bigelow, N., Leduc, M., and Cohen-Tannoudji, C. (1995), to be published in *Phys. Rev. Lett.*.
32. Kasevich, M., Weiss, D. S., Riis, E., Moler, K., Kasapi, S., and Chu, S. (1991). *Phys. Rev. Lett.* **66**, 2297.
33. Reichel, J., Morice, O., Tino, G. M., and Salomon, C. (1994). *Europhys. Lett.* **28**, 477.

34. Davidson, N., Lee, H. J., Kasevich, M., and Chu, S. (1994). *Phys. Rev. Lett.* **72**, 3158.
35. Ertmer, W., Blatt, R., Hall, J. L., and Zhu, M. (1985). *Phys. Rev. Lett.* **54**, 996.
36. Phillips, W. D., and Metcalf, H. (1982). *Phys. Rev. Lett.* **48**, 596.
37. Phillips, W. D., Prodan, J. V., and Metcalf, H. J. (1985). *J. Opt. Soc. Am. B* **2**, 1751.
38. Sesko, D., Fan, C. G., and Wieman, C. E. (1988). *J. Opt. Soc. Am. B* **5**, 1225.
39. Salomon, C., and Dalibard, J. (1988). *C.R. Acad. Sci. Paris Ser. II* **306** 1319.
40. Bradley, C. C., Story, J. G., Tollett, J. J., Chen, J., Ritchie, N. W. M., and Hulet, R. G. (1992). *Opt. Lett.* **17**, 349.
41. Barrett, T. E., Dapore-Schwartz, S. W., Ray, M. D., and Lafyatis, G. P. (1991). *Phys. Rev. Lett.* **67**, 3483.
42. Tollett, J. J. (1995). "A Permanent Magnet Trap for Cold Atoms," Ph.D. Thesis, Rice University.
43. Witte, A., Kisters, T., Riehle, F., and Helmcke, J. (1992). *J. Opt. Soc. Am. B* **9**, 1030.
44. Anderson, B. P., and Kasevich, M. A. (1994). *Phys. Rev. A* **50**, R3581.
45. Nellessen, J., Müller, J. H., Sengstock, K., and Ertmer, W. (1989). *J. Opt. Soc. Am. B* **6**, 2149.
46. Nellessen, J., Werner, J., and Ertmer, W. (1990). *Opt. Comm.* **78**, 300.
47. Riis, E., Weiss, D. S., Moler, K. A., and Chu, S. (1990). *Phys. Rev. Lett.* **64**, 1658.
48. Scholz, A., Christ, M., Doll, D., Ludwig, J., and Ertmer, W. (1994). *Opt. Comm.* **111**, 155.
49. Migdall, A. L., Prodan, J. V., Phillips, W. D., Bergeman, T. H., and Metcalf, H. J. (1985). *Phys. Rev. Lett.* **54**, 2596.
50. Wing, W. H. (1984). *Progress in Quantum Electronics* **8**, 181.
51. Gott, Y. V., Ioffe, M. S., and Tel'kovskii, V. G. (1962). *Nucl. Fusion 1962 Suppl.* **Pt. 3**, 1045.
52. Bagnato, V. S., Lafyatis, G. P., Martin, A. G., Raab, E. L., Ahmad-Bitar, R. N., and Pritchard, D. E. (1987). *Phys. Rev. Lett.* **58**, 2194.
53. Hess, H. F., Kochanski, G. P., Doyle, J. M., Masuhara, N., Kleppner, D., and Greytak, T. J. (1987). *Phys. Rev. Lett.* **59**, 672.
54. Bergeman, T., Erez, G., and Metcalf, H. J. (1987). *Phys. Rev. A* **35**, 1535.
55. Petrich, W., Anderson, M. H., Ensher, J., Cornell, R., and Cornell, E. A. (1995). *Phys. Rev. Lett.* **74**, 3352.
56. Davis, K. B., Mewes, M.-O., Andrews, M. R., van Druten, N. J., Durfee, D. S., Kurn, D. M., and Ketterle, W. (1995). *Phys. Rev. Lett.* **75**, 3969.
57. Davis, K. B., Mewes, M.-O., Joffe, M. A., Andrews, M. R., and Ketterle, W. (1995). *Phys. Rev. Lett.* **74**, 5202.
58. Tollett, J. J., Bradley, C. C., Sackett, C. A., and Hulet, R. G. (1995). *Phys. Rev. A* **51**, R22.
59. Tollett, J. J., Bradley, C. C., and Hulet, R. G. (1992). *Bull. Am. Phys. Soc.* **37**, 1126.
60. Frerichs, V., Kaenders, W. G., and Meschede, D. (1992). *Appl. Phys. A* **55**, 242.
61. Raab, E. L., Prentiss, M., Cable, A., Chu, S., and Pritchard, D. E. (1987). *Phys. Rev. Lett.* **59**, 2631.
62. Shimizu, F., Shimizu, K., and Takuma, H. (1991). *Opt. Lett.* **16**, 339.
63. Steane, A. M., Chowdhury, M., and Foot, C. J. (1992). *J. Opt. Soc. Am. B* **9**, 2142.
64. Cooper, C. J., Hillenbrand, G., Rink, J., Townsend, C. G., Zetie, K., and Foot, C. J. (1994). *Europhys. Lett.* **28**, 397.

65. Wallace, C. D., Dinneen, T. P., Tan, K. Y. N., Kumarakrishnan, A., Gould, P. L., and Javanainen, J. (1994). *J. Opt. Soc. Am. B* **11**, 703.
66. Ketterle, W., Davis, K. B., Joffe, M. A., Martin, A., and Pritchard, D. E. (1993). *Phys. Rev. Lett.* **70**, 2253.
67. Gordon, J. P., and Ashkin, A. (1980). *Phys. Rev. A* **21**, 1606.
68. Dalibard, J., and Cohen-Tannoudji, C. (1985). *J. Opt. Soc. Am. B* **2**, 1707.
69. Chu, S., Bjorkholm, J. E., Ashkin, A., and Cable, A. (1986). *Phys. Rev. Lett.* **57**, 314.
70. Miller, J. D., Cline, R. A., and Heinzen, D. J. (1993). *Phys. Rev. A* **47**, R4567.
71. Hess, H. F. (1986). *Phys. Rev. B* **34**, 3476.
72. Masuhara, N., Doyle, J. M., Sandberg, J. C., Kleppner, D., Greytak, T. J., Hess, H. F., and Kochanski, G. P. (1988). *Phys. Rev. Lett.* **61**, 935.
73. Davis, K. B., Mewes, M.-O., Joffe, M. A., and Ketterle, W., (1994), contributed paper to the Fourteenth International Conference on Atomic Physics, University of Colorado, Boulder, Colorado.
74. Petrich, W., Anderson, M. H., Ensher, J. R., and Cornell, E. A., (1994), contributed paper to the Fourteenth International Conference on Atomic Physics, University of Colorado, Boulder, Colorado.
75. Davis, K. B., Mewes, M.-O., and Ketterle, W. (1995). *Appl. Phys. B* **60**, 155.
76. Monroe, C. R., Cornell, E. A., Sackett, C. A., Myatt, C. J., and Wieman, C. E. (1993). *Phys. Rev. Lett.* **70**, 414.
77. Abraham, E. R. I., McAlexander, W. I., Sackett, C. A., and Hulet, R. G. (1995). *Phys. Rev. Lett.* **74**, 1315.
78. Anderson, M. H., Ensher, J. R., Matthews, M. R., Wieman, C. E., and Cornell, E. A. (1995). *Science* **269**, 198.
79. Bradley, C. C., Sackett, C. A., Tollett, J. J., and Hulet, R. G. (1995). *Phys. Rev. Lett.* **75**, 1687.

# 9. OPTICAL STATE-PREPARATION OF ATOMS

## Jabez J. McClelland

Electron Physics Group
National Institute of Standards and Technology
Gaithersburg, Maryland

## 9.1 Introduction

In the study of a broad range of phenomena involving atoms, from collisions to spectroscopy to reaction dynamics, an increasing need has developed for probing ever deeper into the physical processes at work. The realization has evolved over the decades that to quantitatively study atomic interaction phenomena, it is necessary to experimentally resolve as many variables in the interaction as possible. For example, it is not enough to study the average thermal reaction rate of a collision-related process; instead, the process must be studied as a function of relative velocity and scattering angle. Eventually, it has become clear that the internal degrees of freedom of the atom play a crucial role, so these must be experimentally controlled as well. As a result of this need, the use of optical radiation, especially lasers, to control the internal states of atoms through optical pumping has become increasingly popular in atomic physics experiments.

The purpose of this chapter is to provide some guidelines for the experimentalist interested in performing a state-selected experiment using lasers. As this is meant to be a practical guide, an attempt is made to be fairly self-contained, quoting results from the literature as they are needed. Some knowledge is necessarily assumed, however, in particular regarding atomic structure and spectroscopic notation, and the fundamentals of the interaction between atoms and electromagnetic fields. In addition, some familiarity with the nature of quantum coherence will be useful.

Discussion in this chapter is limited to optical pumping in atomic beams, with a strong slant toward collision experiments. A complete review of optical pumping, as it is applied across atomic and molecular physics, would go far beyond the scope of this work. There is a large amount of literature relating to optical pumping in vapor cells, and also cooling and trapping of atoms, that there is simply not enough space to discuss. The reader should be aware of the crossovers with these other fields, though, because much can be learned from parallel developments. In particular, an excellent resource is the review article by Happer [1] on optical pumping in vapor cells.

EXPERIMENTAL METHODS IN THE PHYSICAL SCIENCES
Vol. 29B

As a result of the emphasis on collision work, this chapter focuses mostly on optical pumping process which selectively populate the magnetic sublevels of an atom. This is because of the great amount of interest that exists in probing the channels of a collision which are associated with alignment and orientation of the target. By controlling the magnetic sublevel populations of the target one can, in effect, control orientation and/or alignment, and hence conduct an experiment at the most fundamental level of state selection.

Much of the discussion here is devoted to rate equation calculations and their interpretation, rather than specific experimental arrangements. The reason for this is that the experimental arrangement for a typical optical pumping experiment is quite simple, consisting of an atom beam, a laser (often commercial), and some fluorescence detection. The crucial aspect of doing a good job in optical state-preparation lies in knowing the details of what is happening to the atoms as they interact with the laser. A good understanding of the processes involved and facility with modeling techniques are essential for this. The last section of this chapter is intended as a resource of relevant data on a number of specific atoms, in the hope that this will prove useful for the planning of experiments.


## 9.2  Basic Concepts


The two fundamental facts that make laser optical pumping possible are (a) when an atom is exposed to electromagnetic radiation at a frequency near resonance, transitions are induced between quantum states in the atom, and (b) if an atom is in an excited state, it will decay by spontaneous emission to a lower state. These phenomena lead to transfer of atomic population from one quantum state of the atom to another. By carefully choosing the radiation frequency, intensity and polarization, a significant amount of population can be transferred, and this transfer is known as optical pumping.

The simplest forms of optical pumping are illustrated in Figure 1. Figure 1a shows a model two-state atom, with one ground state and one excited state. The laser field induces transitions between the ground and excited states, and spontaneous emission transfers population from the excited state to the ground state. The net result is that, as long as the laser field is present, a significant population of excited state atoms will exist, i.e., the atoms are "optically pumped" into the excited state. If the laser field is turned off, spontaneous emission causes all the atoms to decay eventually to the ground state. Exactly what fraction of the atoms are in the excited state when the laser is on (an experimentally very important number) is determined by the strength of the laser field, its frequency with respect to the atomic resonance, and the transition probability of the atom.

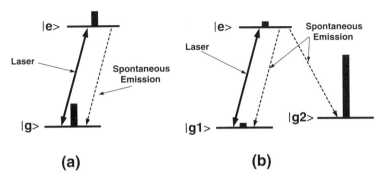

**(a)**                    **(b)**

FIG. 1. Building blocks of optical pumping. (a) A two-level atom. The laser, tuned to resonance, stimulates transitions between the ground state $|g\rangle$ and excited state $|e\rangle$, and spontaneous emission transfers population from the excited to the ground state. In the steady state, a population of excited-state atoms is maintained. (b) A three-level atom. The laser stimulates transitions between a first ground state $|g1\rangle$ and the excited state. Spontaneous emission transfers population back to $|g1\rangle$, and also to another ground state $|g2\rangle$. In the steady state, all atomic population will be optically pumped into $|g2\rangle$.

Figure 1b shows a model three-level atom, in which the laser induces transitions from the ground state to the excited state, and spontaneous emission causes transitions into either of two possible lower states. Atoms can decay back to the original ground state, or they can decay into a new state. If there is no significant transition probability out of the new state, population will continue to accumulate there as long as the laser field is present. Eventually all the atomic population will be transferred to the new state, after which the presence of the laser will have no significant effect.

While Figure 1 illustrates the basic principles of state selection by optical pumping, most atoms have a considerably more complicated level structure than is shown in either Figure 1a or 1b. Generally, an atom will have one or more hyperfine levels in the ground state, each one of these will have degenerate magnetic sublevels, and there will be a variety of states—distinct, degenerate, or quasidegenerate—in the excited-state manifold. Nevertheless, the processes shown in Figure 1 can be thought of as the building blocks of the optical pumping process, and as such they provide a useful framework in which to analyze qualitatively the behavior of a given system.

## 9.3 Calculations of the Optical Pumping Process

In the process of analyzing the feasibility of an experiment, an important step involves making predictions of such things as signal to noise and total count rate expected. Thus, it is important to be able to estimate the population fraction that the optical pumping process can produce in the desired atomic state. Unfortunately, a great number of difficult-to-characterize experimental parameters affect the optical pumping process to some degree; also, a number of nuances in the photon–atom interaction, relating to coherence, might affect the outcome. Because of these complications, it is sometimes difficult to make truly accurate predictions of the results of a particular optical pumping setup. Nevertheless, experience has shown that reasonable estimates can be obtained by following a few guidelines.

Leaving discussion of some of the potential complications until later, I begin with a rate equation approach to calculating the optical pumping process. Rate equation calculations are relatively simple and give accurate results in many situations. They are based on the phenomenological Einstein model, in which each state of the atom is assigned a population $n_i$, and population transfer between the states occurs via stimulated and spontaneous emission.

### 9.3.1 Two-Level Atoms

Assuming a very narrow band laser tuned exactly to the peak of the atomic resonance, the two-level case of Figure 1a has the rate equations [2]

$$\dot{n}_g = QIn_e - QIn_g + \Gamma n_e, \tag{9.1}$$

$$\dot{n}_e = -QIn_e + QIn_g - \Gamma n_e, \tag{9.2}$$

where $n_g$ and $n_e$ are the ground and excited state populations, respectively, $\Gamma$ is the transition rate for spontaneous decay, $I$ is the laser intensity (i.e., the energy per unit area per second), and $Q$ is the stimulated rate per unit intensity. The first two terms in Eqs. (9.1) and (9.2) describe stimulated emission into and out of the ground state, and the third term describes spontaneous emission. Note that if Eqs. (9.1) and (9.2) are added, the result is $\dot{n}_{tot} \equiv \dot{n}_g + \dot{n}_e = 0$, which is as expected, because the total number of atoms does not change.

The coefficient $Q$ is given by

$$Q = \frac{8\pi \left| \boldsymbol{\varepsilon} \cdot \mathbf{d} \right|^2}{c \hbar^2 \Delta\omega_0}, \tag{9.3}$$

where $c$ is the speed of light, $\hbar$ is Planck's constant divided by $2\pi$, $\boldsymbol{\varepsilon}$ is the electromagnetic field polarization vector, $\mathbf{d}$ is the dipole matrix element of the atomic transition, and $\Delta\omega_0$ is the full width at half maximum atomic line width (in radians per second). The derivation of Eq. (9.3) involves using time-dependent perturbation theory to obtain the transition rate for an atom in an electromagnetic field, and then averaging over the atomic line shape. It is a little too lengthy to include here, but it is discussed in reference [2].

Equation (9.3) can be simplified in the case of an atom whose line shape is determined entirely by spontaneous emission. In this case, $\Delta\omega_0 = \Gamma$, and use can be made of the relationship [3] between the magnitude of the dipole matrix element and the spontaneous decay rate $\Gamma$, i.e.,

$$d^2 = \frac{3\hbar c^3}{4\omega_0^3}\Gamma,$$

(9.4)

where $\omega_0$ is the frequency of the atomic transition (in radians per second). With this simplification,

$$Q = \frac{3\lambda^3}{2\pi ch}|\varepsilon\cdot\hat{d}|^2,$$

(9.5)

where $\lambda = 2\pi c/\omega_0$ is the wavelength of the atomic resonance radiation, and $\hat{d}$ is the unit vector along the dipole moment $d$. The quantity $|\varepsilon\cdot\hat{d}|^2$ ranges between zero and one and contains dipole selection rule information associated with the transition. It is unity for a truly two-level atom, and hence can be ignored for the present. It will be important, however, in the later discussion of multilevel atoms.

The steady-state value of $n_e$, obtained by setting the time derivatives to zero in Eqs. (9.1) and (9.2), is

$$n_e = \frac{QI}{2QI + \Gamma}n_{tot}.$$

(9.6)

Note that for $I \to \infty$, $n_e \to \frac{1}{2}n_{tot}$, or at most only half the atoms can be pumped into the excited state. It is common practice to define a saturation intensity $I_{sat}$ such that when $I = I_{sat}$, one-fourth of the atoms are excited (assuming $|\varepsilon\cdot\hat{d}|^2 = 1$). Thus,

$$I_{sat} \equiv \frac{\Gamma}{2Q} = \frac{\pi hc\Gamma}{3\lambda^3}.$$

(9.7)

The rate equations then take on the convenient form

$$\dot{n}_g = \frac{I}{2I_{sat}}\Gamma(n_e - n_g) + \Gamma n_e,$$

(9.8)

$$\dot{n}_e = -\frac{I}{2I_{sat}}\Gamma(n_e - n_g) - \Gamma n_e,$$

(9.9)

This form allows straightforward solution of the equations, with the natural lifetime of the transition as the unit of time. To apply Eqs. (9.8) and (9.9) to a given two-level atomic transition, then, all that is needed is the natural transition probability $\Gamma$ and the wavelength $\lambda$. $\Gamma$, which is equivalent to the Einstein $A$ coefficient, is tabulated for a number of atomic transitions of interest in Table I; values of $\Gamma$ for a great many other atomic transitions can be found in reference [4].

## 9.3.2 Multilevel Atoms

For many situations, extension of the rate equations to multilevel atoms is simply a matter of adding terms to take account of stimulated and spontaneous emission for each additional state. When a complete set of equations is arrived at, the solution can be obtained using standard numerical methods for solving coupled linear differential equations [5].

The type of multilevel problem that is of interest for state preparation of atoms often consists of optical pumping between two or more manifolds of magnetic sublevels. The preferential population of one or more of the magnetic sublevels of a particular state is often the goal of the optical pumping process, because the result is an atomic population that is oriented or aligned in the laboratory. To correctly determine the individual state-to-state stimulated and spontaneous emission rates for this situation, account must be taken of the branching ratios and selection rules for different possible transitions to or from a given level. This information is contained in the quantity $|\boldsymbol{\varepsilon} \cdot \hat{\mathbf{d}}|^2$.

TABLE I.    Optical Pumping Parameters for Several Atoms

| Atom | Transition | $\lambda$ (nm) | $\Gamma/2\pi$ (MHz)[a] | $I_{sat}$ (mW/cm$^2$)[b] |
|------|-----------|---------|---------|---------|
| Alkalis: | | | | |
| Li | $2S_{1/2}-2P_{1/2,\,3/2}$ | 670.8 | 5.8 | 2.5 |
| Na | $3S_{1/2}-3P_{1/2}$ | 589.6 | 10 | 6.4 |
|    | $3S_{1/2}-3P_{3/2}$ | 589.0 | 10 | 6.4 |
| K | $4S_{1/2}-4P_{1/2}$ | 769.9 | 6.1 | 1.7 |
|   | $4S_{1/2}-4P_{3/2}$ | 766.5 | 6.2 | 1.8 |
| Rb | $5S_{1/2}-5P_{1/2}$ | 794.8 | 5.4 | 1.4 |
|    | $5S_{1/2}-5P_{3/2}$ | 780.1 | 5.9 | 1.6 |
| Cs | $6S_{1/2}-6P_{1/2}$ | 894.3 | 4.4 | 0.80 |
|    | $6S_{1/2}-6P_{3/2}$ | 852.1 | 5.2 | 1.1 |
| Alkaline earths: | | | | |
| Ca | $4^1S_0-4P_1$ | 422.6 | 35 | 61 |
| Ba | $6^1S_0-6^1P_1$ | 553.5 | 19 | 15 |
| Metastable rare gases: | | | | |
| He | $2^3S_1-2^3P_{0,1,2}$ | 1083 | 1.6 | 0.16 |
| Ne | $^3P_2-\alpha_9$ | 640.2 | 6.9 | 3.4 |
| Ar | $^3P_2-\alpha_9$ | 811.5 | 5.9 | 1.4 |
| Other atoms: | | | | |
| Cr | $4^7S_3-4^7P^0_2$ | 429.0 | 5.0 | 8.3 |
|    | $4^7S_3-4^7P^0_3$ | 427.5 | 4.9 | 8.2 |
|    | $4^7S_3-4^7P^0_4$ | 425.4 | 5.0 | 8.5 |

[a] The transition probability $\Gamma$ is given divided by $2\pi$, as this corresponds to the natural line width in megahertz observed spectroscopically.
[b] The saturation intensity $I_{sat}$ is calculated from Eq. (9.7).

For an atom with resolvable hyperfine structure, the complete specification of a magnetic substate consists of the orbital angular momentum $L$, the spin $S$, their combination $J$, the nuclear spin $I$, the total angular momentum $F$, and the magnetic quantum number $M$. For a transition between an "excited" level specified by $(L, S, J, I, F, M)$ and a "ground" level specified by $(L', S', J', I', F', M')$,[1] the branching ratio factor is given by [6]

$$|\varepsilon \cdot \hat{\mathbf{d}}|^2 = (2F + 1)(2F' + 1)(2J + 1)(2J' + 1)(2L + 1)$$

$$\times \begin{Bmatrix} L' & J' & S \\ J & L & 1 \end{Bmatrix}^2 \begin{Bmatrix} J' & F' & I \\ F & J & 1 \end{Bmatrix}^2 \begin{pmatrix} F' & F & 1 \\ M' & -M & q \end{pmatrix}^2, \qquad (9.10)$$

where the braces denote a 6-$j$ symbol, the large parentheses denote a 3-$j$ symbol, and $q$ is $\pm 1$ for $\sigma^\pm$ light or 0 for linearly polarized light [7]. If an atom has no hyperfine structure (i.e., $I = 0$), Eq. (9.10) is still valid; one need only set $I = 0$, $F = J$, and $F' = J'$. As complicated as Eq. (9.10) seems, in many cases the 6-$j$ symbols are the same for all transitions and can be ignored for practical purposes, leaving only the 3-$j$ symbol. The 6-$j$ symbols do need to be considered, however, when several different $F$-levels (or $J$-levels) in either the excited or ground state are part of the optical pumping process.

For the stimulated terms, Eq. (9.10) is used in expression (9.5) for the stimulated rate $Q$. For the spontaneous emission terms, the rate for a specific $M$ to $M'$ transition is given by $\Gamma|\varepsilon \cdot \hat{\mathbf{d}}|^2$. It should be noted that all magnetic sublevels within a given hyperfine level have the same total decay rate $\Gamma$; this rate is "split up" into the possible decay channels according to the values of $|\varepsilon \cdot \hat{\mathbf{d}}|^2$. This is reflected in the fact that all the $|\varepsilon \cdot \hat{\mathbf{d}}|^2$ coefficients for spontaneous decay from a given $M$-level add to unity, as predicted by the sum rules for 3-$j$ and 6-$j$ coefficients.

As a specific example of how a multilevel problem is set up, consider the optical pumping of sodium on the $3S_{1/2}(F' = 2) \rightarrow 3P_{3/2}(F = 3)$ transition, ignoring any role that might be played by other hyperfine levels in the atom. This transition has many of the elements contained in the optical pumping of almost any atomic transition, and it has seen a wide range of applications. The level structure is shown in Figure 2. The quantum numbers $(L', S', J', I', F')$ are $(0, \frac{1}{2}, \frac{1}{2}, \frac{3}{2}, 2)$ for the $3S_{1/2}(F' = 2)$ ground state and $(L, S, J, I, F) = (1, \frac{1}{2}, \frac{3}{2}, \frac{3}{2}, 3)$ for the $3P_{3/2}(F = 3)$ excited state. The $F' = 2$ state has five magnetic sublevels, corresponding to values of $M'$ ranging from $-2$ to $+2$, and the $F = 3$ excited state has seven sublevels, corresponding to $M = -3 \cdots +3$. Altogether, then, there are 12 state populations to follow.

In the absence of a magnetic field, all the magnetic sublevels in each hyperfine state are degenerate. Which transitions are induced by the laser, however,

[1] I follow the convention of labeling ground-state levels with primes, and excited-state levels without primes.

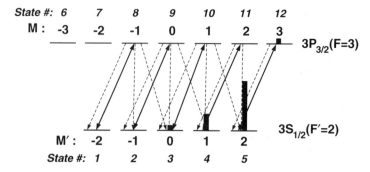

FIG. 2. Optical pumping of the $3S_{1/2}(F' = 2) \rightarrow 3P_{3/2}(F = 3)$ transition in sodium with circularly polarized light. Selection rules limit laser-stimulated transitions to $\Delta M = +1$ for $\sigma^+$ light, and spontaneous emission transfers population with $\Delta M = 0, \pm 1$. The result is a transfer of population to the $M' = +2$ state, which corresponds to a spin-polarized (both electronic and nuclear) ground state, and the $M = +3$ state, which consists of a spin-polarized, orbitally oriented state. The state numbers refer to the numbering scheme in Eqs. (9.11)–(9.20).

depends on optical selection rules and the polarization of the laser. For instance, if $\sigma^+$ (or $\sigma^-$) circularly polarized laser light is used, only transitions with $\Delta M = +1$ (or $-1$) are allowed. If linearly polarized light is used, only $\Delta M = 0$ is allowed. Thus, the number and character of the equations is different, depending on the polarization of the light.

The case of circularly polarized light is of specific interest because it has been used to create populations of spin-polarized ground-state, and also pure angular momentum, spin-polarized excited-state, sodium atoms [8]. For $\sigma^+$ (i.e., left-handed circularly polarized) light [9], transitions to the excited states $|3, -3\rangle$ and $|3, -2\rangle$ are never stimulated, so these states can be eliminated from the rate equations (here the notation $|F, M\rangle$ is used to denote a given magnetic state, with $L$, $S$, $J$, and $I$ dropped for simplicity). Ten equations then describe the remaining states, taking into account stimulated transitions between $|2', -2'\rangle$ and $|3, -1\rangle$; $|2', -1'\rangle$ and $|3, 0\rangle$; etc., as well as all the possible spontaneous transitions. For compactness in the equations, the states can be numbered 1 to 12, starting with $|2', -2'\rangle$ and moving from left to right across the ground and then the excited state, as shown in Figure 2. The 10 equations for $\sigma^+$ optical pumping are then

$$\dot{n}_1 = \frac{I}{2I_{\text{sat}}} \frac{\Gamma}{15} (n_8 - n_1) + \frac{\Gamma}{15} n_8, \tag{9.11}$$

$$\dot{n}_2 = \frac{I}{2I_{\text{sat}}} \frac{\Gamma}{5} (n_9 - n_2) + \frac{8\Gamma}{15} n_8 + \frac{\Gamma}{5} n_9, \tag{9.12}$$

$$\dot{n}_3 = \frac{I}{2I_{sat}} \frac{2\Gamma}{5}(n_{10} - n_3) + \frac{2\Gamma}{5}n_8 + \frac{3\Gamma}{5}n_9 + \frac{2\Gamma}{5}n_{10}, \qquad (9.13)$$

$$\dot{n}_4 = \frac{I}{2I_{sat}} \frac{2\Gamma}{3}(n_{11} - n_4) + \frac{\Gamma}{5}n_9 + \frac{8\Gamma}{15}n_{10} + \frac{2\Gamma}{3}n_{11}, \qquad (9.14)$$

$$\dot{n}_5 = \frac{I}{2I_{sat}}\Gamma(n_{12} - n_5) + \frac{\Gamma}{15}n_{10} + \frac{\Gamma}{3}n_{11} + \Gamma n_{10}, \qquad (9.15)$$

$$\dot{n}_8 = \frac{I}{2I_{sat}} \frac{\Gamma}{15}(n_1 - n_8) - \Gamma n_8, \qquad (9.16)$$

$$\dot{n}_9 = \frac{I}{2I_{sat}} \frac{\Gamma}{5}(n_2 - n_9) - \Gamma n_9, \qquad (9.17)$$

$$\dot{n}_{10} = \frac{I}{2I_{sat}} \frac{2\Gamma}{5}(n_3 - n_{10}) - \Gamma n_{10}, \qquad (9.18)$$

$$\dot{n}_{11} = \frac{I}{2I_{sat}} \frac{2\Gamma}{3}(n_4 - n_{11}) - \Gamma n_{11}, \qquad (9.19)$$

$$\dot{n}_{12} = \frac{I}{2I_{sat}}\Gamma(n_1 - n_8) - \Gamma n_{12}. \qquad (9.20)$$

Given the symmetry of the magnetic sublevels, the equations for $\sigma^-$ optical pumping are identical to Eqs. (9.11)–(9.20) with a simple relabeling of states. The equations for linear polarized excitation will of course be different, but similar in character.

Before solving Eqs. (9.11)–(9.20) numerically, it is useful to observe a few qualitative features. For instance, as always should be the case, $\dot{n}_{tot} \equiv \Sigma_i \dot{n}_i = 0$ (this provides a useful check to make sure the coefficients are correct!). Furthermore, the evolution of the population can be followed by inspection, with the help of Figure 2. It should be clear that as time progresses, population will be transferred to states with more and more positive values of $M$, eventually leading to a steady-state configuration that is reduced to a two-level system with population in only the $|2', 2'\rangle$ and $|3, 3\rangle$ states. This is, in fact, one of the main reasons why optical pumping of this transition in sodium has seen so much application. On one hand, the generation of a two-level atom allows experimental tests of a number of fundamental quantum electrodynamical phenomena. On the other hand, sodium in the $|2', 2'\rangle$ ground state is a completely spin-polarized atom, with nuclear and electron spin both oriented in the laboratory. The orientaion, either parallel or antiparallel to the propagation direction of the laser light, can be selected by pumping with either $\sigma^+$ or $\sigma^-$ light. Similarly, the state $|3, 3\rangle$ is a state with electron spin, nuclear spin, *and* orbital angular momentum oriented in the laboratory.

Using the rate equations, the analysis of circularly polarized excitation in sodium can then be broken down into (a) determining the time it takes to reach the two-level condition (the optical pumping time), which is found by solving the time-dependent rate equations, and (b) determining the steady-state excited $|3, 3\rangle$ population fraction, which can be obtained using Eq. (9.6) for a two-level system. A similar analysis can be applied to linear polarized excitation, but the steady-state excited state population is found by solving the simultaneous equations obtained when all time derivatives are set to zero.

When we compare linear to circularly polarized excitation, an important point must not be overlooked. When the selection rule $\Delta M = 0$ is invoked for linearly polarized excitation, an assumption is made that the axis of quantization is *along the electric vector of the laser field*. The selection rule $\Delta M = \pm 1$ for circularly polarized light, on the other hand, assumes the quantization axis to be *along the direction of light propagation*. Thus, care must be exercised when interpreting the results of an optical pumping calculation in the reference frame of the laboratory.

### 9.3.3 Laser Frequency Dependence and Power Broadening

Until now, all the rate equations that have been discussed assume an infinitely narrow laser frequency tuned exactly to the center of the atomic resonance. It is possible, under some circumstances, to extend the applicability of rate equations to situations in which the laser frequency is still narrow, but detuned from the center of the atomic resonance by an amount $\Delta$. As will be discussed in more detail later, this is possible when there are no other nearby atomic levels, and when the long-time behavior (compared to $1/\Gamma$) is of interest.

While a correct treatment requires the optical Bloch equations, the dependence on $\Delta$ can be introduced into the rate equation approach by arguing that the instantaneous stimulated rate $QI$ is dependent on $\Delta$ because the atomic transition has an instantaneous Lorentzian line shape with a full width at half maximum of $\Gamma$. This argument is borne out by comparison with exact optical Bloch equation results [2]. Thus, an extra factor

$$\mathcal{L} = \frac{\Gamma^2}{\Gamma^2 + 4\Delta^2} \tag{9.21}$$

is included in all the stimulated rate terms. This can be done for a two-level atom, and also for a multilevel atom such as the one shown in Figure 2, where there are only two energies involved.

With the addition of $\Delta$-dependence, the two-level steady-state excited-state population [Eq. (9.6)] becomes

$$n_e = \frac{QI\mathscr{L}}{2QI\mathscr{L} + \Gamma} n_{tot} \tag{9.22}$$

$$= \frac{\Gamma^2(I/I_{sat})}{2[\Gamma^2 + 4\Delta^2 + \Gamma^2(I/I_{sat})]} n_{tot}. \tag{9.23}$$

It is evident from examination of Eq. (9.23) that with increasing intensity, the profile broadens because the center of the line saturates earlier than the wings. The full width at half maximum $\delta\omega_{FWHM}$ of the power broadened profile described by Eq. (9.23) is given by

$$\delta\omega_{FWHM} = \Gamma[1 + (I/I_{sat})]^{1/2}. \tag{9.24}$$

It should be emphasized that Eq. (9.23) describes the *steady-state* frequency dependence of the excited-state population. Because of the intensity dependence in Eq. (9.23), it is tempting to say that the atomic transition is power-broadened. It must be remembered, however, that this is only true in the steady state; the instantaneous (i.e., on a time scale short compared to the stimulated rate) atomic line width is still governed by the unbroadened line shape given in Eq. (9.21).

Power broadening is a very real effect in most practical situations, since saturation intensities are generally a few milliwatts per square centimeter, and a laser with diameter of order 1 mm and power only a few milliwatts can have an intensity of several hundred milliwatts per square centimeter. Thus, the line width observed in a laser-induced fluorescence measurement of a beam of atoms will almost always be power-broadened unless that laser intensity is kept to a minimum.

### 9.3.4 Limitations of Rate Equations

While rate equations provide a practical tool for evaluating a broad range of optical pumping systems, it must be remembered that they represent only an approximate solution. If they are to be used in a given situation, some under-standing of what is left out is essential in order to avoid misleading results. In this section an attempt is made to set out some guidelines for the appropriate use of rate equations. The assumption is that in many instances they will still provide a useful tool for evaluating experiments in state selection, even though in some cases they clearly are inadequate.

The major omission in a rate equation approach is any coherences that might develop between the quantum states of the system. In a fully quantum treatment of a system of atomic levels, the full state of the system is described by a set of complex amplitudes, one for each atomic level in the system. If one allows for a number of posible sets of amplitudes, each with a given probability, the system is properly described by a density matrix. In general, there will be specific phase relationships between the complex amplitudes for each level, and these relative phases represent coherences between the states. The relative phases show up as off-diagonal elements in the density matrix. Since rate equations follow state populations, which are associated with the square magnitudes of the complex amplitudes, or just the diagonal elements of the density matrix, they do not keep track of the relative phases, and hence have no coherence information.

Exact treatment of the coherence between levels in a two-level system can be obtained using the optical Bloch equations. In a multilevel system, such as might be encountered in a state-selection experiment, extensions of the optical Bloch equations can be used, though the problem rapidly becomes a complex one. Discussion of the optical Bloch equations is beyond the scope of this work; for such a discussion and some approaches to the multilevel problem, the reader is encouraged to consult references [2, 10, 11]. Instead, let us concentrate on the following question: Given a specific set of atomic levels and a laser to pump them, will a rate equation approach be adequate for predicting what we want to know about the system? To answer this, the following queries should be useful.

### 9.3.4.1 Are Transient Populations Important?

If the experiment to be carried out relies only on steady-state populations produced by cw (or quasi-cw) laser illumination, rate equations may well be quite adequate. If not, there may be problems. For example, a major, well-known effect that is not seen in a rate equation treatment is Rabi oscillations. These are a manifestation of coherence induced by the laser between the ground and excited states, and consist of transient oscillations in the atomic population between the ground and excited state. Their frequency (given by $\omega_R = \Gamma(I/2I_{sat})^{1/2}$ for a two-level atom [2]) is determined by the stimulated rate, and they can cause as much as a 100% movement in population between the ground and the excited states. They do decay, however, on the time scale of the natural lifetime (or faster in the presence of collisions), and when they have decayed, the state populations often follow the rate equation predictions quite closely. So, if the state distribution of the atom within the first natural lifetime or two after the laser is turned on is important, rate equations should not be used.

It should be noted that in some situations it is not completely trivial to determine if the experiment to be performed is sensitive to transient populations, since the eventual population distribution in the steady state could depend on transient effects. This occurs, for example, in the case of circularly polarized excitation of sodium when one considers the influence of the other hyperfine levels $F' = 1$ and $F = 2$. The $F' = 1$ state lies 1772 MHz below the $F' = 2$ state, and the $F = 2$ state lies 59.6 MHz below the $F = 3$ state. Before the population is completely transferred to the $|2', 2'\rangle \rightarrow |3, 3\rangle$ system, where it is isolated by selection rules from all other atomic levels, there is some probability that the $F = 2$ state will be excited, because it is only a few atomic line widths away from the $F = 3$ state. This opens the pathway for decay into the $F' = 1$ ground state during the transient period before optical pumping is complete. Thus, how many atoms find their way into the $|2', 2'\rangle \rightarrow |3, 3\rangle$ two-level system depends critically on the transient behavior of the state populations, the correct modeling of which requires more than rate equations.

### 9.3.4.2 Do Coherences Matter for the Planned Experiment? Even in the steady state, coherences are induced between all states connected by the laser. These coherences do not appear in a rate equation approach, so if the experiment is somehow sensitive to them, rate equations should not be used. In some cases, the effect of these coherences can be dramatic. For instance, if two different ground-state sublevels are coupled by two lasers to the same excited state, it can happen that there is no steady-state excited state population. This phenomenon, known as coherent population trapping [12], is a result of the coherence that develops between the two ground-state sublevels. It does not show up in a rate equation treatment of this situation. In some situations, however, it can be put to use if population transfer between two ground states is desired without subjecting the atoms to spontaneous emission [13].

One must take care, however, not to be misled by trivial coherences induced by coordinate rotations. Consider, for example, linearly polarized excitation of a $J' = 0 \rightarrow J = 1$ transition. With a quantization axis along the electric vector of the linearly polarized light, the selection rule $\Delta M = 0$ applies, and the system becomes a simple two-level atom. The rate equation approach will work just fine. If one insists, however, on a quantization axis along the direction of the light propagation, then the linearly polarized light must be broken down into $\sigma^+$ and $\sigma^-$ components, which are coherent with each other. The result is a coherent excitation of the $|1, 1\rangle$ and the $|1, -1\rangle$ states. The coherence between these two states, which must be maintained to get the right answer, will be lost in a rate equation approach. Thus, before applying the rate equations, it is important to rotate the coordinate system of a given problem to a system that eliminates as many coherences as possible.

In sum, it can be said that rate equations can be used in situations where the steady-state populations are of interest, and where coherence either does not matter or cannot develop between more than two states (ground and excited) at a time. As a result, a simple situation such as that shown in Figure 2 is quite appropriate for rate equations. However, cases with multiple excited levels within a few natural line widths, or multiple lasers tuned to the same state, should be approached with caution.

## 9.4 Calculations and Experimental Reality

Even if the rate equation approach were exact for all situations, its usefulness (and, indeed, the usefulness of any theory) can still be hampered by a number of experimental factors that can influence the outcome of an optical pumping setup. These factors must be taken into account in planning an experiment, and an evaluation made as to whether they will have a significant influence. If the effect of these factors is unknown or ignored, it is useless to perform any modeling calculations because the predicted results could be dramatically different from reality. The following discussion covers a number of effects that can influence the optical pumping process.

### 9.4.1 Laser Spatial Profile

In a crossed-beam arrangement, where the laser beam intersects the atom beam at 90°, the laser spatial profile has two effects. The profile along the direction of motion of atoms translates into a time-dependent laser intensity for the atoms, because the atoms are travelling with some velocity $v$. This can be taken into account relatively easily by using a time-dependent intensity in the rate equations. It must be remembered, however, that (a) not all atoms travel with the same velocity, so some averaging will need to be done over the velocity distribution of the atoms, and (b) if the time dependent intensity varies too rapidly, there may be important nonadiabatic or transient effects. The profile transverse to the atom beam is important because if it is nonuniform, different parts of the atom beam experience different laser intensities. If the optical pumping process is strongly intensity dependent, a nonuniformly pumped atom beam will result.

### 9.4.2 Doppler Shift Effects

Because of the relatively high velocity of thermal atoms, and the precision with which a laser is tuned to the atomic frequency in an optical pumping experiment, the Doppler shift can be a substantial effect. If an atom is traveling at exactly 90° to the laser beam, there will be no first-order Doppler shift. However, atom beams and laser beams generally have some amount of divergence, so there will always be a range of angles between the atomic trajectories and the laser. The result is that atoms in different parts of the beam will experience different laser frequencies, shifted by an amount

$$\delta\omega = \mathbf{k} \cdot \mathbf{v} = \omega \frac{v}{c} \cos \theta, \qquad (9.25)$$

where $\mathbf{k}$ is the wave vector of the laser light, $\mathbf{v}$ is the velocity of the atom, $\omega = 2\pi c/\lambda$ is the laser frequency, and $\theta$ is the angle between $\mathbf{k}$ and $\mathbf{v}$. As an example, consider a thermal sodium beam pumped on the 589 nm resonance line. For $\theta = 1$ degree (17 mrad), $\delta v \equiv \delta\omega/2\pi = 24$ MHz, a quite significant shift compared to the natural line width of 10 MHz. Of course, not only do different parts of the beam experience different laser frequencies, but there is also the thermal velocity spread in the beam, which results in a range of Doppler shifts for any given position in the beam. Clearly, the Doppler shifts in a given experiment must be estimated and either deemed insignificant or averaged over before optical pumping calculations can be relied on.

### 9.4.3 Atom Deflection Effects

In addition to ordinary Doppler shifts, there can arise more complicated effects resulting from the deflection of the atom beam by light pressure. Because the optical pumping process usually involves a number of spontaneous emission events, there will in general be a net transfer of momentum from the laser to the atom. Thus, as the atom passes through the optical pumping region, the atom is accelerated along the laser propagation direction, resulting in a changing Doppler shift as a function of time. The transverse acceleration of the beam depends on the laser intensity, the detuning, and the instantaneous Doppler shift; it is given by [14]

$$\mathbf{a} = \frac{\Gamma \hbar \mathbf{k}}{2m} \frac{I/I_{\text{sat}}}{1 + I/I_{\text{sat}} + 4[(\Delta - \mathbf{k} \cdot \mathbf{v})/\Gamma]^2}. \qquad (9.26)$$

For sodium with $I \gg I_{\text{sat}}$, $\mathbf{a}$ can be as large as $10^6$ m/s$^2$, which means that the angle that the atom beam makes with respect to the laser direction changes by about 1 mrad per millimeter of travel along the beam. Under some circumstances, the changing Doppler shifts induced by this changing angle have significant effects on the optical pumping process.

### 9.4.4 Incomplete Laser Polarization

In the ideal optical pumping experiment, the polarization of the laser is assumed to be exactly as specified—100% circularly polarized, for example. Because of imperfect polarizers, natural birefringence in windows, and reflection effects which occur in a real experimental situation, this is often not the case. It is thus important to make an estimate of how sensitive an optical pumping scheme is to the "wrong" polarization. Often, the effects are not severe if the amount of "wrong" polarization is small. In a circularly polarized $J \rightarrow J + 1$

optical pumping scheme such as the one in Figure 2, the effect is only a somewhat incomplete transfer of the population to the desired subelevels. The effects, if small, can be estimated with rate equations, but one should be aware of the possible coherences between excited sublevels that would be neglected.

Situations with more levels must be treated with more care, however. The case of sodium is an example of this. The problem arises from the loss mechanism to $F' = 1$ mentioned earlier under transient effects. If the polarization of the pumping laser is not completely circular, the loss mechanism is no longer transient. With imperfect $\sigma^+$ light, there is always some probability that the $|2', 2'\rangle$ state will make a transition to the $|3, 1\rangle$, or worse, the $|2, 1\rangle$ state, from which the pathway is open to the $F' = 1$ state. Because of the nature of rate equations, if a pathway to a population "sink" exists, the only true steady-state solution consists of all population eventually in the "sink." Thus, a laser with only slightly impure polarization can result in a total loss of population from the desired steady-state result.

### 9.4.5 Magnetic Fields

Magnetic fields can have profound effects on the optical pumping process in the situation where one is trying to produce a specific distribution of magnetic sublevels in the atoms. First of all, magnetic fields will Zeeman-shift the energy levels, changing the effective frequency of the laser. For an atom with hyperfine structure in a field that is not too strong, this shift is given by [15]

$$\Delta\omega_Z = \frac{eB}{2m_e} g_F M, \tag{9.27}$$

where $B$ is the magnetic field strength in teslas, $m_e$ is the electron mass, $M$ is the magnetic quantum number, and

$$g_F = g_J \frac{F(F + 1) + J(J + 1) - I(I + 1)}{2F(F + 1)}, \tag{9.28}$$

with

$$g_J = 1 + \frac{J(J + 1) + S(S + 1) - L(L + 1)}{2J(J + 1)}. \tag{9.29}$$

If the atom does not have hyperfine structure, Eq. (9.27) still holds, with $F$ replaced by $J$. For example, the Zeeman shift of the $|3, 3\rangle$ sublevel of the $3P_{3/2}$ level in sodium is $\Delta\omega_Z/2\pi = 2.8 \times 10^4$ MHz/tesla.

More troublesome, however, is the fact that the atomic angular momentum will precess if there is a magnetic field that is not along the quantization axis. This precession is related to the Zeeman effect, and the rotation of the atomic angular momentum vector (in radians per second) is given by the same expression (9.27), multiplied by $\sin \alpha$, where $\alpha$ is the angle between the quantization axis and the magnetic field. The result is that the orientation of the atom, or equivalently, the magnetic sublevel population distribution, will change with time. In the case of the sodium $|2', 2'\rangle$ state in a field perpendicular to the original quantization axis, the orientation changes at $8.9 \times 10^{10}$ radians per second per tesla. At a thermal velocity of 800 m/s, this corresponds to a rotation of 6 degrees after traveling only 1 cm in a $10^{-7}$ tesla (1 milligauss) field.

Besides ruining the atomic orientation, this precession also has an effect similar to incomplete polarization in any situation analogous to circularly polarized pumping of sodium. Because the precession causes a continual redistribution of population toward sublevels with smaller $M$, the opportunity to make transitions to the $F = 2$ sublevels is always there. Again, this opens the loss channel to $F' = 1$, making the complete loss of population a possibility.

From a practical point of view, then, magnetic fields must be well controlled in an optical pumping experiment. Ideally, they should be reduced to the $10^{-7}$ tesla (1 milligauss) level to prevent unwanted precession and Zeeman shifts. In some circumstances, however, a weak "guide" field (i.e., weak enough to cause minimal Zeeman shifts but strong enough to overpower any residual stray fields) can be applied along the desired quantization axis, provided other aspects of the experiment are not sensitive to this field.

### 9.4.6 Radiation Trapping

This effect can lead to problems in experiments where high-density atomic beams are employed [16]. If the density is high enough, the fluorescence emitted by an optically pumped atom can be large enough to affect the optical pumping of neighboring atoms. Since this fluorescence comes from arbitrary directions and has varying polarization, the result is a redistribution of magnetic sublevel populations. As a rule of thumb, atomic densities below about $10^{10}$ atoms-cm$^{-3}$ are generally safe from radiation trapping. A check can be made relatively simply by carrying out the experiment to be done with the optically pumped atoms at a range of atomic densities and observing whether the outcome varies. Theoretical calculations can also be made based on numerical simulations if so desired [17].

## 9.5 Diagnosis of an Optically Pumped Beam

While calculations provide a very useful guide for setting up an experiment, the possible theoretical shortcomings and experimental pitfalls are such that some sort of probe of the state distribution should be carried out before an experimental result can be relied on. Usually some sort of spectroscopic probe can be arranged, expecially since there will generally be laser light available from the optical pumping.

Diagnosis of experiments that involve degenerate magnetic sublevel populations, such as alignment or orientation of an alkali atom, have been approached
by a number of methods. A very straightforward method is to expose the atoms
to a magnetic field strong enough to Zeeman-shift the magnetic sublevels to the
point where they are resolvable spectroscopically. The relative populations can
then be observed by monitoring the relative peak heights in a laser-induced
fluorescence spectrum. This method has been used with sodium [18, 19] and
appears to work quite well. Care must be exercised on a few points, however.
The method could prove problematic if the experiment to be done with the
optically pumped atoms is sensitive to magnetic fields. In principle, it is possible
to arrange for the experiment to be conducted in a field-free region, after which
the atoms enter the magnetic field, but one must always worry about how well
isolated the experimental region is. Furthermore, changes can occur in the
magnetic sublevel population on entering the magnetic field if it is not along the
quantization axis defined for the optical pumping, or if there are strong field
gradients.

Another approach involves measuring the degree of polarization and angular
dependence of the fluorescence in either the optical pumping region itself or in
a probe region. This method has the advantage that no magnetic fields need to be
applied, but has the disadvantage of sometimes relying on some degree of
modeling of the state distribution.

When a given population distribution in an excited-state magnetic sublevel
manifold decays and emits fluorescence, the light intensity and polarization will
have well-defined angular dependencies. Each excited magnetic sublevel contributes to the intensity according to the decay paths open to it, as dictated by the
selection rules $\Delta M = 0, \pm 1$. The contribution from each decay path is determined by the relevant Clebsch–Gordan coefficient. The angular dependence of
the intensity is given by $\sin^2 \theta$ for $\Delta M = 0$ transitions, or $(1 + \cos^2 \theta)/2$ for
$\Delta M = \pm 1$, where $\theta$ is the angle between the quantization axis ($\hat{z}$) and the
direction of observation. The polarization of the light for $\Delta M = 0$ transitions is
linear along $\hat{z}$ for $\theta = \pi/2$. For $\Delta M = \pm 1$, the light propagating along $\theta = 0$ is
circular, $\sigma^{\pm}$, but the light propagating along $\theta = \pi/2$ is linear in the plane
perpendicular to $\hat{z}$ [20].

To probe an excited state, then, the fluorescence from the optical pumping
region can be analyzed and used to infer the relative sublevel populations. In
some cases it may be desirable to characterize the sublevel population distribution in terms of state multipoles. These can provide a physical picture of the
excited-state configuration, which may lead to some insights. Reference [16]
provides more details on this technique as applied to sodium.

A ground-state distribution can be investigated by using a very weak probe of
well-characterized polarization. Using knowledge of the selection rules and
Clebsch–Gordan coefficients, it is possible to predict what excited-state magnetic

sublevel populations would be produced by the probe, given a specific ground state distribution, assuming no saturation or optical pumping.[2] The polarization and angular distribution of the probe fluorescence can then be predicted, and measurements compared with this. This approach is not a direct measurement, since it relies on first modeling the ground state population, and then seeing if the model agrees with experiment, but nevertheless it can be used with success in many instances [21].

In cases where fluorescence probes are not convenient and the ground-state magnetic sublevel population distribution is of interest, another approach, making use of a Stern–Gerlach-type magnetic field, can be utilized [22]. The atom beam is passed through a region of magnetic field $\mathbf{B}$ with a strong gradient, and the spatial profile is measured downstream. The atoms feel a deflection force $\mathbf{F} = (\boldsymbol{\mu} \cdot \boldsymbol{\nabla})\mathbf{B}$ according to their magnetic moment $\boldsymbol{\mu}$, which is related to their magnetic quantum number $M$ by $\boldsymbol{\mu} = \mu_B g_F M \hat{\mathbf{z}}$ [$\mu_B$ is the Bohr magneton $e\hbar/2m_e$, and $g_F$ is the $g$-factor, given by Eq. (9.28)]. With a suitable arrangement of strong fields and long flight times, individual peaks corresponding to each magnetic sublevel can be resolved in the spatial profile of the atom beam. The relative populations can then be easily monitored as a function of optical pumping parameters. Of course, as in the case with Zeeman separation of the sublevels, one must be extremely careful with stray magnetic fields when using this technique.

## 9.6 Specific Atoms

This chapter ends with a discussion of some experimental aspects of specific atomic species that have been or could be optically pumped for a state-selected experiment. The atoms are grouped in categories that have similar properties, with a few general words about each category. Tables I and II contain relevant parameters for the transitions of interest in the species discussed. For atoms not discussed here, energy levels can be found in reference [23], transition probabilities can be found in reference [4], some hyperfine structure information is available in reference [24], and some isotope shifts can be found in reference [25].

### 9.6.1 Alkalis

The combination of a simple electronic structure, strong resonance transitions in the visible or near infrared, and relative ease of forming an atomic beam has made alkali atoms a popular choice for many applications.

---

[2] Optical pumping and even saturation can begin to occur at very low power levels. The experimenter is warned to resist the temptation to turn up the power to get a little more signal in this type of experiment!

For state-selected collision studies, as well as other applications, sodium has been widely used [8, 19, 21, 26–32]. A good description of the fundamentals of sodium optical pumping, as applied to state selection, has been given by Hertel and Stoll [26]. Atomic beams can be formed with oven temperatures in the 300–400°C range, and the strong $D_1$ and $D_2$ lines, at 589.6 and 589.0 nm respectively, lie within the peak range of the laser dye rhodamine 6G. There is only one stable isotope, $^{23}$Na, which has a nuclear spin $I = 3/2$. Thus, the atom has hyperfine structure, and it generally must be resolved in an optical state-preparation experiment. Laser line widths of order 1 MHz are required for

TABLE II.    Hyperfine Splittings (HFS) for Selected Isotopes[a]

| Isotope | Abundance | $I$ | State | HFS [$F$–splitting (MHz)–$F$][b] |
|---|---|---|---|---|
| $^6$Li | 7.4% | 1 | $2S_{1/2}$ | $\frac{1}{2}$–228–$\frac{3}{2}$ |
| | | | $2P_{1/2}$ | $\frac{1}{2}$–26.2–$\frac{3}{2}$ |
| | | | $2P_{3/2}$ | $\frac{5}{2}$–3.2–$\frac{3}{2}$–1.9–$\frac{1}{2}$ |
| $^7$Li | 92.6% | 3/2 | $2S_{1/2}$ | 1–803.5–2 |
| | | | $2P_{1/2}$ | 1–92–2 |
| | | | $2P_{3/2}$ | 3–9.3–2–5.9–1–2.8–0 |
| $^{23}$Na | 100% | 3/2 | $3S_{1/2}$ | 1–1772–2 |
| | | | $3P_{1/2}$ | 1–189–2 |
| | | | $3P_{3/2}$ | 0–16.5–1–35.5–2–59.6–3 |
| $^{39}$K | 94% | 3/2 | $4S_{1/2}$ | 1–462–2 |
| | | | $4P_{1/2}$ | 1–56–2 |
| | | | $4P_{3/2}$ | 0–3.3–1–9.4–2–21–3 |
| $^{85}$Rb | 72.2% | 5/2 | $5S_{1/2}$ | 2–3035.7–3 |
| | | | $5P_{1/2}$ | 2–362.1–3 |
| | | | $5P_{3/2}$ | 1–29.3–2–63.4–3–120.7–4 |
| $^{87}$Rb | 27.8% | 3/2 | $5S_{1/2}$ | 1–6834.7–2 |
| | | | $5P_{1/2}$ | 1–812–2 |
| | | | $5P_{3/2}$ | 0–72.3–1–157.1–2–267.2–3 |
| $^{133}$Cs | 100% | 7/2 | $6S_{1/2}$ | 3–9192.6–4 |
| | | | $6P_{1/2}$ | 3–1168–4 |
| | | | $6P_{3/2}$ | 2–151–3–201–4–251–5 |
| $^{53}$Cr | 9.6% | 3/2 | $^7S_3$ | $\frac{3}{2}$–206.5–$\frac{5}{2}$–289.1–$\frac{7}{2}$–371.7–$\frac{9}{2}$ |
| | | | $^7P_2$ | $\frac{1}{2}$–40–$\frac{3}{2}$–66–$\frac{5}{2}$–93–$\frac{7}{2}$ |
| | | | $^7P_3$ | $\frac{9}{2}$–7–$\frac{7}{2}$–5–$\frac{5}{2}$–4–$\frac{3}{2}$ |
| | | | $^7P_4$ | $\frac{11}{2}$–65–$\frac{9}{2}$–53–$\frac{7}{2}$–41–$\frac{5}{2}$ |

[a] It should be noted that in an atomic beam consisting of a mixed isotopic sample, the hyperfine manifolds of the isotopes will generally be offset from each other by the isotope shift.

[b] $F$-values are displayed with the lowest energy level first.

this, so a frequency-stabilized cw dye laser, pumped by a visible argon ion laser, is necessary.

One aspect of single-frequency optical pumping in sodium that must be kept in mind is the potential for transfer between hyperfine ground states [10]. In some cases this may be the desired optical pumping process, but in others the goal may be to maintain a closed system in which the atoms interact repeatedly with the laser photons. Transfer to the other hyperfine level represents loss to the experiment in this case. While it is possible in principle to tune only to the $F' = 2 \rightarrow F = 3$ transition in the $D_2$ line, so that transitions to $F' = 1$ are forbidden by the $\Delta F = 0$, $\pm 1$ selection rule, in practice the $F = 3$ state is not perfectly resolved from the $F = 2$ state, being separated by only 59.6 MHz. The $F' = 2 \rightarrow F = 2$ transition has a wing which overlaps the $F' = 2 \rightarrow F = 3$ frequency, and this wing can be quite significant, especially if there is power broadening.

An experimental solution to this problem has been to perform the optical pumping with two laser frequencies, separated by 1712 MHz. One laser frequency is tuned to the $F' = 2 \rightarrow F = 3$ transition, and the other is tuned to the $F' = 1 \rightarrow F = 2$ transition, returning atoms that may have been optically pumped into the wrong state. The extra laser frequency can be generated with either an electro-optic or acousto-optic modulator. Electro-optic modulators produce the second laser frequency without spatial separation, and they always make a symmetric pair of sideband frequencies at plus and minus the modulation frequency. This can be put to advantage by modulating at 856 MHz (an easier frequency to work with) and utilizing the two side bands instead of the carrier frequency. Electro-optic modulators tend to be somewhat inefficient, however, so many experimenters have used acousto-optic modulators. These spatially separate the shifted beam, and they only shift in one direction at a time, the direction being chosen by the angle of incidence on the modulator.

The optical pumping of cesium has also been extensively studied, mainly because of its application in atomic clocks [33], and also because the $D_2$ line at 852.1 nm falls within the range of inexpensive diode lasers [34]. The only stable isotope, [133]Cs, has nuclear spin 7/2. The hyperfine splitting is much larger than in sodium, so the problem of loss to the "wrong" ground-state hyperfine level is less important. Nevertheless, optical pumping with two laser frequencies is often advantageous, because more of the atomic population can be accessed. To achieve this, two or more diode lasers can be used (they are, after all, quite inexpensive), or the frequency of the laser can be modulated by varying the injection current [35].

Rubidium has also been the subject of a number of optical pumping studies, a great majority of which have been in a vapor cell. The $D_1$ and $D_2$ lines, at 794.8 and 780.0 nm, respectively, are also accessible to diode lasers, so it is a good candidate for an inexpensive experiment. There are two stable isotopes, [85]Rb

(72.2% abundance, $I = 5/2$) and [87]Rb (27.8% abundance, $I = 7/2$), each of which has well-resolved hyperfine structure. The two isotopes with significant natural abundance make rubidium unsuitable for an experiment in which all the atoms in the beam must be identically state-selected, such as some collision experiments (though it is in principle possible to use more lasers and pump all atoms). Rubidium has, however, seen wide application in experiments where the existence of other isotopes does not interfere with the measurements, such as in the field of cooling and trapping [36].

Lithium, with its two stable isotopes [6]Li and [7]Li, has also been optically pumped [37], though it poses some special problems. The laser wavelength required is 670.8 for both the $D_1$ and $D_2$ lines, which are separated by 10 GHz. This wavelength can be accessed with a single-frequency stabilized dye laser operating with DCM laser dye, or with a visible laser diode. The problems arise because the hyperfine structure is not well resolved, especially in [7]Li. To avoid the loss mechanism to the "wrong" hyperfine ground state, as discussed for sodium, two-frequency optical pumping is a necessity. For state-selection of ground state atoms, this is not a problem, but if a significant population of excited states is desired, one must be aware of the possibility of coherent population trapping. Coherently exciting the same excited state from two ground-state hyperfine levels could greatly reduce excited-state populations [12]. A possible way to circumvent some of these problems is to use isotopically enriched [6]Li, though this can increase the cost of the experiment.

Potassium, with $D$-lines at 769.9 and 766.5 nm, has seen little use as an optical pumping target for state-selective experiments. The wavelengths require a Ti:sapphire laser, or a dye laser with LDS700 laser dye and a krypton ion pump laser. In principle, however, there is no reason why it could not be optically pumped in the same way as the other alkali atoms.

### 9.6.2 Alkaline Earths

While not as popular as the alkalis, some alkaline earths have been optically pumped to provide state-selected targets. The attraction of these atoms lies in the simple $^1S_0$ ground state and $^1P$ excited states. The excited $P$-state can be aligned and/or oriented, and collisions can be studied without the complication of spin polarization.

Barium can be optically pumped on the $6^1S_0 \rightarrow 6^1P_1$ transition (wavelength 553.5 nm) using rhodamine 110 laser dye [38]. There are quite a few naturally occurring isotopes of barium (134, 135, 136, 137, 138), but these are well separated spectroscopically, so they do not pose a significant problem unless the whole atom beam must be state-selected. A possible cause for caution in using barium is the metastable $6^1D_2$ level, to which the $6^1P_1$ level can decay. The branching ratio for the $6^1P_1$ going to the ground state vs. the $6^1D_2$ state is 425 : 1.

This ratio is fairly large, so if the optical pumping process is modeled with rate equations and monitored experimentally (to the extent possible), the loss mechanism can be kept under control.

Calcium has a similar line, the $4^1S_0 \rightarrow 4^1P_1$ transition at 422.6 nm, which can be optically pumped with a dye laser using stilbene 3 laser dye pumped by a UV argon ion laser [39]. An alternative source of radiation for this wavelength is a Ti:sapphire laser doubled in an external buildup cavity [40]. Isotope shifts and hyperfine structure are not a problem for calcium, because it is naturally 97% $^{40}Ca$, which has zero nuclear spin. As with barium, there is an intermediate metastable $4^1D$ state, to which population can decay from the excited state. In this case, though, the branching ratio is $10^5:1$ in favor of the ground state, so population loss can almost always be ignored.

### 9.6.3 Metastable Rare Gas Atoms

While excitation of rare gas atoms from the ground state to the lowest-lying excited states generally requires many electron volts of energy (i.e., VUV photons), they all have metastable states, easily generated in discharge sources, which can be optically pumped in the near infrared. Applications have included state-selected collision studies, polarized electron sources, and cooling and trapping.

Helium has seen wide application in state-selected experiments, both as a collision participant [41] and as a source of polarized electrons [42]. The $2^3S_1$ metastable state (19.8 eV above the ground state) has accessible transitions to the $2^3P_{0,1,2}$ states at 1.083 μm wavelength. This wavelength can be generated with a specially adapted Ti:sapphire laser, pumped by a visible argon ion laser [43], with a laser using a LNA crystal as its gain medium [44], or with recently available laser diodes.

Metastable neon has also been optically pumped [45–48]. The two metastable states are $^3P_2$ and $^3P_0$, at 16.6 and 16.7 eV above the ground state, respectively. These levels can be excited to an array of states of the configuration $1s^22s^22p^53p$, denoted in the Paschen notation by $\alpha_1 \cdots \alpha_{10}$. Of all these excited states, only the $\alpha_9$ state $[2p^5(^2P_{1/2})3p, J = 3]$ can be excited without opening the possibility of cascade down to the ground state and subsequent loss of population. The wavelength for the transition to this from $^3P_2$ is 640.2 nm, accessible with DCM laser dye. Like the $^3S_1 \rightarrow {}^3P_2$ transition in helium, this transition has the right sublevel configuration ($J \rightarrow J + 1$) to allow creation of a spin-polarized, oriented target, as may be desirable for state-selected collision experiments. The natural isotope distribution of neon is 91% $^{20}Ne$ and 9% $^{22}Ne$. Both of these have zero nuclear spin, so there is no hyperfine structure. There is an isotope shift, however, so the $^{22}Ne$ line is separated from the $^{20}Ne$ line by 1630 MHz.

Argon is similar to neon. The $^3P_2$ and $^3P_0$ states are metastable, with energies of 11.5 and 11.7 eV above the ground state, respectively. The transition most useful for optical pumping is from the $^3P_2$ state to the $3p^5(^2P_{3/2})4p(J = 3)$, or $\alpha_9$ state, with wavelength 811.5 nm [49]. This wavelength is obtainable with a dye laser, a Ti:sapphire laser, or a diode laser. Naturally occurring argon is 99.6% $^{40}$Ar, with zero nuclear spin, so hyperfine splitting and isotope shifts are not important in this case.

### 9.6.4 Other Atoms

Besides the alkalis, the alkaline earths, and the metastable rare gases, few other atoms have been optically state-selected in an atomic beam. Chromium has been optically pumped to produce a polarized high-spin target for collision studies with polarized electrons [50]. The $^7S_3 \rightarrow {}^7P_4^0$ transition at 425.43 nm is accessible with a dye laser operating with stilbene 3 laser dye pumped by a UV argon ion laser. The natural isotope abundance contains 84% $^{52}$Cr, 4% $^{50}$Cr, and 2% $^{54}$Cr, all of which have no hyperfine structure, and 10% $^{53}$Cr, which has hyperfine structure with $I = 3/2$. The $^{50}$Cr 425.43 nm line has an isotope shift of about 132 MHz relative to the $^{52}$Cr line; the $^{54}$Cr isotope shift is negligible, however.

# References

1. Happer, W. (1972). *Rev. Mod. Phys.* **44**, 169.
2. Allen, L., and Eberly, J. H. (1987). *Optical Resonance and Two-Level Atoms*. Dover, New York.
3. See, e.g., Woodgate, G. K. (1970). *Elementary Atomic Structure*, Chapter 3. McGraw-Hill, London.
4. Wiese, W. L., Smith, M. W., and Glennon, B. M. (1966). "Atomic Transition Probabilities—Hydrogen through Neon," Vol. I, *NSRDS-NBS* **4**; Wiese, W. L., Smith, M. W., and Miles, B. M. (1969). "Atomic Transition Probabilities—Sodium through Calcium," Vol. II, *NSRDS-NBS* **22**; Martin, G. A., Fuhr, J. R., and Wiese, W. L. (1988). "Atomic Transition Probabilities—Scandium through Manganese," *J. Phys. Chem. Ref. Data* **17**, Suppl. 3; Fuhr, J. R., Martin, G. A., and Wiese, W. L. (1988). "Atomic Transition Probabilities—Iron through Nickel," *J. Phys. Chem. Ref. Data* **17**, Suppl. 4.
5. See, e.g., Press, W. H., Flannery, B. P., Teukolsky, S. A., and Vetterling, W. T. (1986). *Numerical Recipes*. Cambridge University Press, Cambridge.
6. See, e.g., Edmonds, A. R. (1960). *Angular Momentum in Quantum Mechanics*. Princeton University Press, Princeton.
7. The 3-*j* and 6-*j* symbols are tabulated in a number of places, such as Rotenberg, M. (1959). *The 3-j and 6-j Symbols*. Technology Press, Cambridge. Alternatively, they are now available in many symbolic manipulation computation packages.
8. McClelland, J. J., Kelley, M. H., and Celotta, R. J. (1989). *Phys. Rev. A* **40**, 2321.

9. The subject of the handedness of circular polarization can be the cause of a significant amount of confusion, though it is discussed in detail in most textbooks on optics. A good explanation can be found in Crawford, F. S., Jr. (1968). *Waves*, p. 400. McGraw-Hill, New York.

10. McClelland, J. J., and Kelley, M. H. (1985). *Phys. Rev. A* **31**, 3704.

11. Farrel, P. M., MacGillivray, W. R., and Standage, M. C. (1988). *Phys. Rev. A* **37**, 4240.

12. Gray, H. R., Whitley, R. M., and Stroud, C. R., Jr. (1978). *Opt. Lett.* **3**, 218.

13. Bergmann, K. (1991). In *Atomic Physics 12*, J. C. Zorn and R. R. Lewis (eds.), p. 336. American Institute of Physics, New York.

14. See, e.g., Lett, P. D., Phillips, W. D., Rolston, S. L., Tanner, C. E., Watts, R. N., and Westbrook, C. I. (1989). *J. Opt. Soc. Am. B* **6**, 2084.

15. See, e.g., Woodgate, G. K. (1970). *Elementary Atomic Structure*, pp. 183ff. McGraw-Hill, London.

16. Fischer, A., and Hertel, I. V. (1982). *Z. Phys. A* **304**, 103.

17. Kunasz, C. V., and Kunasz, P. B. (1975). *Comp. Phys. Comm.* **10**, 304.

18. Cusma, J. T., and Anderson, L. W. (1983). *Phys. Rev. A* **28**, 1195.

19. Han, X. L., Schinn, G. W., and Gallagher, A. (1988). *Phys. Rev. A* **38**, 535.

20. See, e.g., Jackson, J. D. (1975). *Classical Electrodynamics*, 2nd Edition, Section 9.2. J. Wiley & Sons, New York.

21. Lorentz, S. R., Scholten, R. E., McClelland, J. J., Kelley, M. H., and Celotta, R. J. (1993). *Phys. Rev. A* **47**, 3000.

22. Riddle, T. W., Oneillian, M., Dunning, F. B., and Walters, G. K. (1981). *Rev. Sci. Instrum.* **52**, 797.

23. Moore, C. E. (1971). *Atomic Energy Levels*, Vols. I and II, *NSRDS-NBS* **35**.

24. Radzig, A. A., and Smirnov, B. M. (1985). *Reference Data on Atoms, Molecules and Ions*, pp. 99ff. Springer Verlag, Berlin.

25. King, W. H. (1984). *Isotope Shifts in Atomic Spectra*. Plenum, New York.

26. Hertel, I. V., and Stoll, W. (1978). *Adv. At. Mol. Phys.* **13**, 113.

27. Balykin, V. I. (1980). *Opt. Comm.* **33**, 31.

28. Jaduszliwer, B., Dang, R., Weiss, P., and Bederson, B. (1980). *Phys. Rev. A* **21**, 808.

29. Dreves, W., Kamke, W., Broermann, W., and Fick, D. (1981). *Z. Phys. A* **303**, 203.

30. Hils, D., Jitschin, W., and Kleinpoppen, H. (1981). *Appl. Phys.* **25**, 39.

31. Scholten, R. E., Shen, G. F., and Teubner, P. J. O. (1993). *J. Phys. B* **26**, 987.

32. Sang, R. T., Farrell, P. M., Madison, D. H., MacGillivray, W. R., and Standage, M. C. (1994). *J. Phys. B* **27**, 2711.

33. Vanier, J., and Audoin, C. (1989). *Quantum Physics of Atomic Frequency Standards*. Adam Hilger, Bristol.

34. Wieman, C. E., and Holberg, L. (1991). *Rev. Sci. Instrum.* **62**, 1.

35. Watts, R. N., and Wieman, C. E. (1986). *Opt. Comm.* **57**, 45.

36. See, e.g., Sheehy, B., Shang, S.-Q., Watts, R., Hatamian, S., and Metcalf, H. (1989). *J. Opt. Soc. B* **6**, 2165.

37. Baum, G., Caldwell, C. D., and Schröder, W. (1980). *Appl. Phys.* **21**, 121.

38. Register, D. F., Trajmar, S., Csanak, G., Jensen, S. W., Fineman, M. A., and Poe, R. T. (1983). *Phys. Rev. A* **28**, 151; Zetner, P. W., Trajmar, S., Csanak, G., and Clark, R. F. H. (1989). *Phys. Rev. A* **39**, 6022.

39. Law, M. R., and Teubner, P. J. O. (1993). In *Abstracts of Contributed Papers to the XVIII ICPEAC*, p. 150, T. Andersen, B. Fastrup, F. Folkman, and H. Knudsen (eds.). Aarhus, Denmark.

40. Polzik, E. S., and Kimble, H. J. (1991). *Opt. Lett.* **16,** 1400.
41. Hart, M. W., Hammond, M. S., Dunning, F. B., and Walters, G. K. (1989). *Phys. Rev. B* **39,** 5488.
42. Rutherford, G. H., Ratliff, J. M., Lynn, J. G., Dunning, F. B., and Walters, G. K. (1990). *Rev. Sci. Instrum.* **61,** 1490.
43. Oro, D. M., Lin, Q., Soletsky, P. A., Zhang, X., Dunning, F. B., and Walters, G. K. (1992). *Rev. Sci. Instrum.* **63,** 3519.
44. Schearer, L. D., Leduc, M., Vivien, D., Lejus, A.-M., and Thery, J. (1986). *IEEE J. Quant. Elec.* **QE22,** 713; Lynn, J. G., and Dunning, F. B. (1986). *Rev. Sci. Instrum.* **61,** 1996.
45. Gaily, T. D., Coggiola, M. J., Peterson, J. R., and Gillen, K. T. (1980). *Rev. Sci. Instrum.* **51,** 1168.
46. Hotop, H., Lorenzen, T., and Zartrow, A. (1981). *J. Electr. Spect. and Rel. Phenom.* **23,** 347.
47. Bussert, W. (1986). *Z. Phys. D* **1,** 321.
48. Brand, J. A., Furst, J. E., Gay, T. J., and Schearer, L. D. (1992). *Rev. Sci. Instrum.* **63,** 163.
49. Giberson, K. W., Hammond, M. S., Hart, M. W., Lynn, J. G., and Dunning, F. B. (1985). *Opt. Lett.* **10,** 119.
50. Hanne, G. F., McClelland, J. J., Scholten, R. E., and Celotta, R. J. (1993). *J. Phys. B* **29,** L753.

# 10. METHODS AND APPLICATIONS OF RESONANCE IONIZATION SPECTROSCOPY

## G. Samuel Hurst and James E. Parks

Institute of Resonance Ionization Spectroscopy
The University of Tennessee
Knoxville, Tennessee

## 10.1 Introduction

Resonance ionization spectroscopy (RIS) is a process for the selective ionization of atoms or molecules. The process uses a series of atomic or molecular photoabsorptions, at least one of which is a resonance step, and it terminates when an electron is removed from the atom or molecule. With modern lasers it is possible, in principle, to carry out the RIS process with every atom in the periodic table and a wide variety of molecules. In practice, some of the noble gases and a selected few other atoms are difficult, but can be resonantly ionized with special laser techniques. The power of the method is derived from two factors: the fact that ionization measurements are highly efficient (approaching 100% in many cases), and the use of tunable lasers to selectively ionize nearly all species of one type in a laser beam while not perturbing species of other types.

This fortuitous combination of selectivity and sensitivity (even at the one-atom level) has led to wide use of the RIS method in physics, chemistry, biology, geology, industry, and other fields. Applications of a simple atom counter include physical and chemical studies of small populations of atoms. Advanced systems incorporating mass spectrometers to greatly reduce isotopic interferences provide state-of-the-art analytical capability. These more elaborate systems can be used, for instance, to date lunar materials and meteorites, date old groundwater and icecaps, measure the neutrino output of the sun, detect trace elements in electronic-grade materials, determine DNA structure, and address a wide range of medical and environmental health problems. Further, RIS has been integrated into accelerators for studies of isomers, isotopes, isobars, and other important aspects of nuclear physics.

Because of the basic work to support and demonstrate RIS capabilities and to find important applications of the method, literature related thereto is extensive and comes from research laboratories worldwide. Fortunately, there is a

EXPERIMENTAL METHODS IN THE PHYSICAL SCIENCES
Vol. 29B

well-organized international symposium on RIS, and much of the literature is documented in the proceedings. For example, the latest of these [1] reports on the Seventh International Symposium on Resonance Ionization Spectroscopy and its Applications, and will key the reader into the previous proceedings. Two textbooks, Letokhov [2], and Hurst and Payne [3], describe RIS in detail. The 1993 issue (15th edition) of the *Encyclopaedia Britannica* puts the subject of RIS in perspective with other methods of analysis and measurement [4]. A recent *Physics Today* article by Hurst and Letokhov [5] describes some highlights on the many applications of RIS. An article on resonance ionization mass spectrometry (RIMS) by Payne *et al.* [6] may be considered as an excellent companion to the present article.

The first experiment to be referred to as resonance ionization spectroscopy involved a measurement of the absolute number of excited states produced when a beam of protons from a small accelerator lost part of its energy in a gas cell filled with helium [7]; see also [3]. Prior to the RIS experiment, considerable work had been done to study the interaction (mainly excitation and ionization) of fast-charged particles with the noble gases. This research on the energy pathways following charged particle interaction with the noble gases involved total ionization measurements and time-dependent studies of VUV emission from pure helium and from low levels of impurities in helium. Although these experiments were elaborate and extensive, they left open one very important question. What quantum state was responsible for the additional ionization produced when minute traces of impurity gases were mixed with helium?

The energy pathways studies suggested it was $He(2^1S)$, but there was no direct measurement for a definite proof. The rationale and a typical result for the $He(2^1S)$ experiment are shown in schematic form in Figure 1. These experiments provided the absolute number of these quantum-selected states and their lifetimes as a function of gas composition and pressure. Conclusive proof was thus made that the $He(2^1S)$ state was responsible for the Penning ionization (also known as the Jesse effect) in helium. A key question in these energy pathways studies was the mechanism by which the $He(2^1S)$ state is populated following proton excitation of helium. According to stopping-power theory, the $He(2^1S)$ state should be abundantly populated, rather than $He(2^1S)$, which has low oscillator strength. However, it was suggested from theory (Payne *et al.* [8]) that a surprisingly large two-body collision is operative to convert $He(2^1P)$ to $He(2^1S)$. Thus, a direct experiment to measure the metastable state population was thought to be crucial. These circumstances were the motivation for RIS. For further discussion, see Section 1.6 of reference [3].

Even though the helium experiments were quite definitive, they were not extended to energy pathways studies for other noble gases, primarily because of the strong interest in RIS detection of atoms in their ground states. This interest was fed by the availability of tunable lasers that could be used to selectively

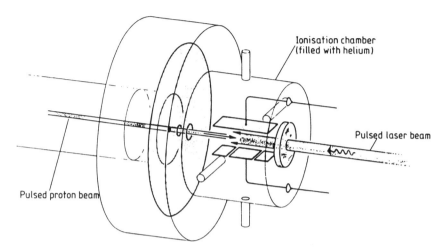

FIG. 1. Schematic representation of the first RIS experiment to measure the absolute population of a selected excited state. (From reference [3].)

ionize nearly every element in nature. Studies of energy pathways in other noble gases are now even more attractive because of advanced techniques of RIS. Track studies can now be done using alpha-particle sources, RIS schemes, and a position-sensitive detector of free electrons. With the optical electron detector, OED (Gibson and Hunter, [9]), the tracks of both the primary ionization and the metastable states could be recorded. Further, the combination of RIS with the OED would make possible the recording of spatial profiles for atoms in their ground states. This should be a very promising method for chemical kinetics studies.

Letokhov's book is an excellent discussion of much of the early work at the Institute of Spectroscopy in the USSR, which concentrated on atoms in their ground states. This work was motivated by the desire for sensitive analytical measurements as well as for isotopic enrichment. Likewise, the early RIS work at Oak Ridge National Laboratory (ORNL) was motivated by the desire for very sensitive detection of atoms. In fact, it was proven at ORNL that single atoms could be counted using proportional counters.

## 10.2 RIS Schemes and Laser Sources

In this section we will give some particulars of the RIS laser schemes developed for the selective and sensitive detection of atoms using RIS. These schemes are useful whether detection is done in a vacuum or in an ionization chamber, a proportional counter, or a Geiger counter, where moderate gas pressures are required.

All RIS schemes involve excitation of a selected atom through one or more resonant transitions followed by a photoionization step (or some other ionization process) of the excited atom. In principle, the number of ways to achieve RIS for all the elements in the periodic chart is large, as illustrated in Figure 2. In practice, the number of RIS schemes for the elements is limited by the availability of commercial tunable lasers in the required wavelength range. The types of schemes can be summarized in a few categories that can be described in a universally accepted notation. The simplest of schemes involve excitation of

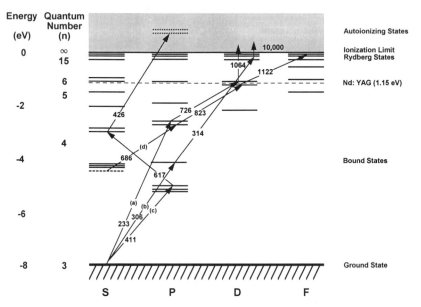

FIG. 2. Schematic atomic energy-level diagram for a "typical" element illustrating some of the more important resonance ionization schemes. The transition wavelengths (in nanometers) and angular momenta are indicated, while the binding energy of the valence electron and its corresponding principal quantum number is tabulated on the left. Utilizing a Nd:YAG laser, scheme (a) enables accessing with high sensitivity and selectivity for most elements, while providing saturated ionization with intense IR photons. The moderately long UV wavelengths of scheme (b) permit very selective and efficient ionization using either copper vapor pumped dye lasers or cw lasers by going to a Rydberg level, followed by ionization with a cw $CO_2$ and inexpensive $N_2$ or excimer lasers with bound–bound transitions to either a Rydberg state within 0.05 eV of the ionization limit, followed by ionization in an electric field of (approximately) 10 kV/cm, or to an autoionizing state, if available. For those elements in which, either by fast atomic collisions, electron bombardment, or recombination, a significant population can be trapped in a long-lived metastable state (d), resonance ionization with tunable semiconductor diode laser excitation, followed by either field ionization or ionization with a $CO_2$ laser, is possible. This last scheme is ideal for noble gas collinear beam RIMS. (From reference [6].)

atoms by the absorption of a single photon, represented by $\omega_1$, followed by photoionization of the excited atom by the absorption of another photon of the same wavelength and energy, $\omega_1$. This simplest scheme is referred to as a one-color scheme and is given the notation $\omega_1 + \omega_1$. Alternatively, this type of scheme is referred to as a $1 + 1$ scheme. It is a one-color, two-photon RIS process in which the first photon is used to excite the atom by a resonant transition and a second photon of the same color is used to photoionize the excited atom. A variation of this type is a $1 + 1'$ or $\omega_1 + \omega_2$ scheme, in which the second photon is of another color, energy, or wavelength.

Very commonly, an atom is excited to a first level and then to a higher level in another allowed resonant transition. This type of process is referred to as a $\omega_1 + \omega_2 + \omega_3$ scheme, where $\omega_3$ may be equal to $\omega_1$ or $\omega_2$, or may be completely different from either of the two. In other shorthand notation, this scheme is labeled as a $1 + 1' + 1''$ scheme, or more simply as a $1 + 1 + 1$ scheme.

When the first excitation step requires more energy than is ordinarily available from a single photon, two-photon excitations can be employed. In the simplest of these cases, two photons of the same color are used to excite an atom to a state allowed by two photon transitions, and then are ionized by the absorption of a third photon of the same wavelength. This process is referred to as a $2\omega_1 + \omega_1$ or a $2 + 1$ RIS scheme. A distinct disadvantage of this type of scheme is that intense laser powers are required for the two-photon absorptions, and non-resonant photoelectron emission can occur as an unwanted source of background noise. This reduces one of the most desirable features of RIS whereby by choosing resonant, single photon transitions, equilibrium between the lower and upper states can be achieved with very little laser power.

The final photoionization step can sometimes involve some interesting variations. For example, an $\omega_1 + \omega_{2AI}$ scheme is a process where the atom is first excited with a single photon of one color and then is photoionized by a second photon properly tuned to the exact wavelength to reach an autoionizing state of the atom. The advantage of such a scheme is that photoionization can be achieved with a much smaller power density. A second variation is a $\omega_1 + \omega_{2R}$ scheme, in which an excited atom is further excited to a high-lying Rydberg state in which the atom can be ionized by collision or with an applied electric field, which is usually synchronously pulsed on at the proper time.

RIS is generally applicable to all the elements of the periodic chart except neon and helium. Figure 3 shows a periodic chart of the elements with typical schemes to achieve the resonance ionization process. The chart also indicates the elements that have been successfully demonstrated to date.

While RIS schemes can be categorized in general for a single figure of the periodic chart, the details of achieving the RIS process are much more intricate and varied, with many alternative means for achieving the same or a more or less

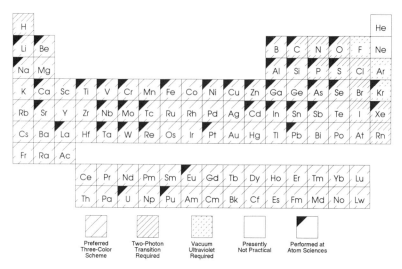

FIG. 3. Periodic chart of the elements, courtesy of Atom Sciences, Inc., showing color schemes required to achieve resonance ionization. Successful ionization of elements by RIS demonstrated to date by Atom Sciences is indicated.

efficient result. Dr. E. B. Saloman has established a data service on resonance ionization spectroscopy at the National Institute of Standards and Technology and has made this information available through a series of publications [10]. The service provides atomic structure and laser information required for resonance ionization and publishes formatted data sheets. To date, summaries for close to a third of all the elements have been published. The data sheets for each element provide its ionization energy and an introduction to the most recent studies, list the element's stable isotopes, isotope shifts, and hyperfine stucture, RIS schemes, atomic energy levels, lifetimes, oscillator strengths, laser schemes, atom sources, and references for both the atomic data and the RIS schemes. Calculations of excited-state photoionization cross-sections by Hartree–Fock techniques are also included, with other instructions on the principles and theory of the technique as it is to be applied to RIS.

Compilations of the various laser schemes that can be successfully used in RIS measurements are also summarized in the data sheets. Most frequently, the laser systems used for RIS schemes are of two types. The first is a YAG pumped tunable dye laser system. The second uses an excimer laser as the primary pumping source. Both systems are pulsed laser sources and are quite versatile in terms of the wavelength ranges that can be covered practically. In each case, the tunable light is generated in a flowing dye laser, and the wavelight ranges are extended by the use of second harmonic light generation and parametric mixing in a crystal. Figure 4 illustrates schematically a YAG laser system that typically

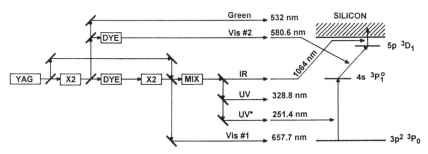

FIG. 4. A single YAG laser system with two dye lasers, frequency doubling and mixing crystals, and dichroic mirrors to selectively separate the different wavelengths represents a typical laser system that can generate two tunable wavelengths of light in the ultraviolet and visible regions to resonantly excite silicon using a three-color scheme. The wavelengths and energy level designations for silicon are shown. Photoionization of excited silicon is achieved by absorption of the fundamental YAG wavelengths.

may be used for the resonance ionization of silicon. First, approximately 1 J of 1064 nm light is generated in the YAG laser during a pulse of about 8 ns. Green light, 532 nm in wavelength and about 350 mJ, is produced in a frequency-doubling crystal by second harmonic generation. This green light is then suitable to be the pumping source of the dye laser. Typically using a $1 + 1 + 1$ scheme for silicon, light of about 251.4 nm is needed for the first excitation step. This light is first generated by using the 532 nm light to excite the dye molecules in a dye laser in order to produce a primary red light of approximately 657.7 nm. Dye laser efficiency in this wavelength range can be as high as 20%, allowing as much as 70 mJ of red light to be produced. The red light is frequency-doubled and then mixed with the fundamental 1064 nm light of the YAG pump to produce 251.4 nm light. The doubling and mixing process is about 3% efficient, so that this light, which is used for the first excitation step, has about 2 mJ of energy in a pulse duration near 8 ns. Because the excitations are chosen to be resonant, the excitation step can be saturated with this small amount of energy during the pulse. The YAG system is fully capable of pumping two dye lasers, so that light for the second excitation step can be produced separately, but simultaneously. Since most of the light of the fundamental wavelengths is not converted in the doubling and mixing processes, this light is still available for other uses. In the case of silicon, a 251.4 nm excitation followed by a 580.6 nm excitation leaves the excited atom very near the ionization continuum, whereby infrared photons of 1064 nm fundamental YAG wavelength are energetic enough to photoionize a silicon atom. This use of the laser system has the advantage that the photoionization process, which has a low cross-section, can be achieved with the more intense fundamental light composed of many photons, each with little energy to photoionize the primary target atoms or other nearby atoms and molecules in or near their ground state. The YAG system advantage is that it can simultaneously generate several different wavelengths of light to carry out a multistep RIS process.

In an excimer laser, the fundamental energy of photons is much higher than with YAG lasers. This fact makies it easier to accomplish some of the RIS schemes. However, there is also a disadvantage: these energetic UV photons, which in many cases can directly ionize surface atoms or molecules, leading to nonselective ionization and unwanted noise from background ions. In practice, the control of unwanted background ions is crucial to the ultimate sensitivities that can be achieved, and the longer fundamental wavelengths of the YAG laser are more attractive for the photoionization steps. Although excimer lasers may have this disadvantage, many researchers choose to use excimers and still obtain outstanding results with their RIS measurements.

# 10.3 Atom Counting

## 10.3.1 General

Nuclear physicists have perfected quite versatile methods of counting single particles. Some methods can be used in a gaseous medium, while others can be used for vacuum applications. With RIS, as shown in the previous section, selective ionization can be carried out for essentially all of the elements. Further, RIS schemes are available that can saturate the ionization process in certain volumes of a laser beam. Putting the sensitive particle detectors together with RIS makes it possible to count atoms in these specified volumes. Of course, the atom to be detected must be a free atom, and its ionization must be done either inside the detector itself or in a volume of space from which electrically charged particles can be swept into the detector. The following examples are chosen to illustrate that atom counting can be carried out in a variety of ways.

## 10.3.2 Cesium Atoms

Because of their favorable RIS scheme, Cs atoms were first used to demonstrate the possibility of counting ordinary atoms in their ground states (Hurst *et al.* [11]) even in the background of enormously larger concentrations of other atoms or molecules. For this demonstration, a proportional counter was used as a single-electron detector. Rutherford and Geiger originated the proportional counter, even before the Geiger–Muller counter was invented, and much later Curran showed the detector could be made sensitive enough to detect nearly all of the single-electron events. A typical counting gas for proportional counters is a mixture of argon with 10% of methane (P-10 gas). One-atom detection was demonstrated by evaporating a small population of Cs atoms from a metallic source into the P-10 gas and directing the RIS beam through the proportional counter. This showed that a single atom of Cs can be detected even when there are $10^{19}$ atoms of Ar and $10^{18}$ molecules of $CH_4$ in the same laser volume. Incidentally, this volume could be very large in the case of Cs. For example, it could be the product of 5 cm$^2$ × $L$, where $L$ is the length of the laser beam in the proportional counter, or easily greater than 100 cm$^3$. Also, of course, it is possible to make the volume quite small where single atoms are detected, if spatial resolution is desired. Since pulsed lasers are commonly used in RIS, good time resolution is easily accomplished.

### 10.3.3 Cesium Atoms in Spontaneous Fission Decays

A much more challenging type of one-atom detection was demonstrated by detecting the Cs atoms that are a part of the heavy-mass peak in the spontaneous decay of Cf-252 by nuclear fission (Kramer *et al.* [12]). Because of the nearly 100 meV of energy ejected into the Cs atom during the binary fission decay, one expects about $3 \times 10^6$ ion pairs in the fission track containing the one cesium atom (expected about 14% of the time in the heavy-mass peaks). The RIS process on the Cs atom can give only one additional electron. Obviously, it was first necessary to sweep these electrons out of the active volume before firing the laser to detect the Cs atom. This was accomplished by using a drift chamber that could communicate with the adjacent proportional counter. Electrons could be drifted out (away from the counter) in less than a microsecond, so the neutral atom could be contained near its original location because of the much slower diffusion process at the gas pressure used in proportional counters. After the large burden of free electrons was cleared out, the RIS laser was fired at the single atom, and the free electron produced was drifted into the proportional counter. This demonstrated that RIS methods can be used to count individual atoms in time coincidence with nuclear decay.

### 10.3.4 Noble Gas Atoms

While the RIS process is simple for an alkali atom, it takes on great difficulty for the noble gases. In spite of the difficulty of generating the required VUV radiation (see Section 10.2), an effective RIS scheme has been worked out for the Kr atom. However, the sensitive volume is quite small in this case, and other developments were required (Hurst *et al.* [13]) to achieve a useful sensitivity. In particular, it was necessary to develop an "atom buncher," which would make it more likely that an inert atom would be in the laser beam at the time of the pulse. The way the RIS process has been augmented with the atom buncher is shown schematically in Figure 5 in context with some special features of atom counting. This method takes advantage of the fact that noble gases are chemically inert and can be kept in a sample volume for an indefinite period. In fact, it is possible to continue firing a laser until all of the atoms have been removed from the sample

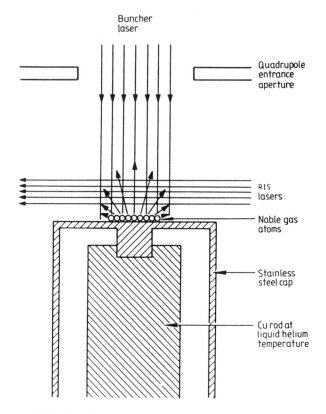

FIG. 5. Concept of the "atom buncher," which can make it far more likely that an atom to be resonantly ionized will be in the RIS beam. Such a concept should be useful for many types of spectroscopic studies with rare atoms or molecules. (From reference [3].)

volume and implanted into a solid, where they are retained during the measurement cycle. However, the time required to do this would be unreasonably long without the atom buncher. In the schematic, we also show a small mass spectrometer to provide isotopic selectivity. In principle, then, each atom of a given isotope can be implanted and counted with an electron detector, which counts the pulses of electrons given off when each accelerated positive ion is implanted. We will give some details later on variations of this detection scheme.

Polar ice-cap dating, solar neutrino experiments, and related applications have motivated scientists to further develop the use of RIS to count Kr-81 atoms. In these applications it is necessary to count samples of a small number of atoms, for example a few hundred, that have a half-life of 210,000 years. Clearly, decay counting is not practical; hence, the potential for direct counting based on RIS is being perused. Initially (see Chapter 10 of reference [3]), mass analysis was accomplished with a quadrupole mass filter and the ions were accelerated to sufficient energy (about 10 keV) that nearly all of them would be implanted into a target. The initial number of atoms in the gaseous sample could be determined by integrating the total counts as the sample was depleted. This mode of data analysis is suggestive of the operation of a Maxwell's demon, in that nearly all of the atoms of a selected isotope could be implanted into the target after a few hours of operation.

In an extension of this work (Thonnard *et al.*), a time-of-flight mass spectrometer was used that has the obvious advantage of producing additional information as the sample is analyzed. Further, depletion of the gaseous sample by implantation was not stressed, because the more conventional method of data analysis, in which the atoms are sampled without implantation, was used. More recent work (see [6]) has shown that fewer than 100 atoms of Kr-81 can be counted even in a much larger background of Kr-82, the abundant neighboring atom. One of the most important developments making this outstanding achievement possible was the construction of a small volume for the instrument enclosure (shown in Figure 6) from materials that would outgas as little Kr as possible. A small volume also reduces the recurrence time for atoms to return to the cold finger once they are blown off with the buncher laser, making possible increased pulsed rates.

The important common feature in all versions of the RIS method for counting Kr atoms is the combination of the laser scheme and the atom buncher. The laser scheme for generating the necessary VUV radiation to initiate the RIS process in Kr is shown in Figure 7. This elaborate scheme, which uses a number of nonlinear steps, was investigated theoretically by M. G. Payne and was first

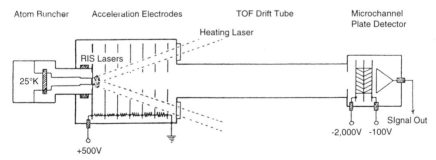

FIG. 6. Schematic of a static RIMS system for use with noble gases (or extremely inert elements if the temperature of the chamber is sufficiently high). Between laser pulses, the sample collects on the "atom buncher," a tiny cold spot. Periodically, a laser pulse heats the atom buncher, releasing the condensed atoms into the beam of a properly time-delayed laser pulse, which then ionizes the selected element. When lasers are available at the required VUV wavelength, nearly all of the selected sample can be ionized, mass-selected, and implanted into the detector within a few minutes. (From reference [6].)

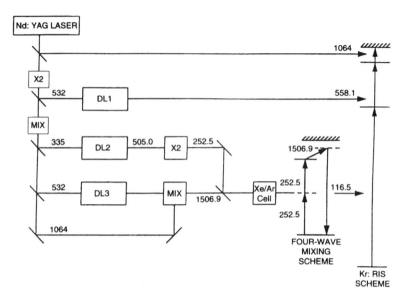

FIG. 7. Energy-level diagram ionization scheme, and laser setup used to detect [81]Kr or [85]Kr at the hundred, or less, atom level. The vacuum-ultraviolet light is generated by four-wave mixing in mixtures of xenon and argon. Most wavelengths between 110 and 200 nm can be generated by this method in sufficient quantity for use in RIMS detection. (From reference [6].)

demonstrated in practice by Kramer *et al.* (1978). Later work by N. Thonnard showed that a combination of the RIS scheme with an atom buncher of the Thonnard design enabled the depletion or pumping of 50% of the Kr-81 atoms from an enclosure volume of 1.8 L in about 2 min. The general references [3] and [6] may be consulted for more details on this topic.

In this section we have described how atoms can be counted with a laser technique known as RIS. In the next section we will discuss the use of RIS in other analytical systems involving ion sputtering and laser ablation. In these systems, groups of atoms can be ionized and detected as a single pulse with the use of suitable detectors. Further, when ultimate sensitivity is required, atom counting can be done with these methods as well.

## 10.4 SIRIS and LARIS

The analysis of solid materials using RIS requires that the atoms be in a free state. To meet this requirement, several methods have been developed to break samples into their constituent atoms, and several of these techniques have become standards in practice. In this section, atomization by energetic ions and by photons will be discussed. These RIS analysis methods are known as

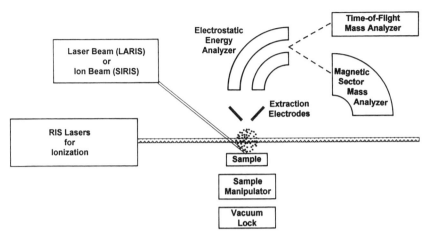

FIG. 8. SIRIS and LARIS instrumentation systems commonly consist of six key sections: (1) an atomization source consisting of an ion beam for SIRIS or a laser beam for LARIS, (2) an RIS laser system, (3) a sample and RIS interaction section, (4) an ion extraction and ion discrimination electrode section, (5) an energy analyzer for energy discrimination of SIMS ions, and (6) a mass spectrometer with ion detector for mass analysis.

sputtered initiated resonance ionization spectroscopy (SIRIS) [14]–[16] and laser ablation resonance ionization spectroscopy (LARIS) [17]. Thermal methods for atomization will be discussed in the section following.

The established technique for the analysis of solids, particularly in the semiconductor industry, is secondary ion mass spectrometry or SIMS. In the SIMS technique, solid samples are analyzed by sputtering atoms and ions from a sample using an energetic ion beam and then analyzing the sputtered ions with a mass spectrometer. In another technique, lasers are used to ablate and ionize solid samples for analysis by mass spectrometry. Often the ions produced are a poor representation of the constituents of the sample, with the ion formation in the sputtering process depending heavily on factors such as surface condition. SIRIS and LARIS avoid these pitfalls by having the ionization process independent of the sputtering process, thus providing more reliable results and a better characterization of the sample.

The concepts for SIRIS and LARIS are very similar and are represented schematically in Figure 8. Each is a laser post-ionization technique, first involving atomization of the sample by sputtering with a pulsed energetic ion beam or by ablation with a high-power pulsed laser beam. A method is usually included to suppress the secondary ions formed in the sputtering process prior to the selectively ionization of the neutral atoms by the resonance ionization process. In order to provide additional selectivity and discrimination of the secondary ions, a mass spectrometer is used for mass analysis and isotopic identification. SIRIS and LARIS instrumentation systems commonly consist of six key sections: (1) an atomization source, (2) a RIS laser system, (3) a sample and RIS interaction section, (4) an ion extraction and ion discrimination electrode section, (5) an energy analyzer for energy discrimination of SIMS ions, and (6) a mass spectrometer with ion detector for mass analysis and energy discrimination. The design characteristics of each section are critical to the instrument's overall effectiveness and to making sensitive, selective measurements that truly characterize the sample. The design and the functional dependencies of these sections will be discussed in the following.

The subject of sputtering by energetic ion beams has long since been thoroughly investigated and reported. For simplified practical applications, the essence of the sputtering process can be summarized succinctly as follows. The yield of sputtered atoms per incident ion depends on the energy of the ion beam, the type of primary ion, the target matrix material, and the angle of incidence between the primary ion and the target. Measured values for sputtering yield are usually quoted for normal incidence between the ion beam and sample plane, with the number of sputtered atoms inversely proportional to the cosine of the angle of incidence, i.e., $1/\cos\theta$, for reasonable values of $\theta$. In other words, the sputtering yield is least for normal incidence ions. For 10 keV argon ions at normal incidence, a sputtering yield of 2 is typical in silicon, and the yield is about 4 at 60°. The distribution for atoms sputtered by an energetic ion beam, as well as those atomized by laser ablation, follow a cosine distribution whereby the number of sputtered atoms per unit solid angle is greatest in the direction normal to the sample surface.

A number of factors affect the detection limits of SIRIS and LARIS instruments for sensitive analysis. These include (1) the sputtering or ablation process, (2) the spatial and temporal overlap of the RIS laser beam and the moving atom cloud, (3) the efficiency of the RIS scheme, (4) the efficiency of extracting the ionized atoms from the multitude of other sputtered atoms, (5) the transmission efficiency of the mass spectrometer, and (6) the detection efficiency of the ion detectors.

Sputtering is usually accomplished with a pulsed ion beam to better provide a good spatial and temporal overlap with the pulsed laser beam. The energy distribution of sputtered atoms has the general form $f(E)$, given by

$$f(E) = \frac{CE}{(E + U)^3} \tag{10.1}$$

Using this energy distribution, its associated angular distribution, the distribution of the primary ion beam pulse, the geometry of the laser beam, and its location relative to the sample, the optimum time for pulsing the RIS laser beam can be calculated from an equation for the fraction of sputtered atoms in the path of the laser beam when it is turned on. Typically, for an ion beam pulse, 10 keV argon ions of 1 μs duration with a cylindrical laser beam of 1 mm diameter centered at 2.5 mm parallel to the sample surface, the optimum time to turn the RIS laser on is about 12 μs. The RIS laser beam pulse is typically of a few nanoseconds duration and essentially probes the atom cloud at an instant of time. The fraction of atoms in the RIS laser probe beam at the optimal time can be typically as much as 20% or better.

In principle, LARIS has two distinct advantages over ion sputtering. First, the number of atoms ablated in a single pulse can be much larger than can be practically achieved with sputtering. It is possible to ablate as many as $10^{17}$ atoms from a sample irradiated with a single pulse of light from a YAG laser system. As attractive as this might seem, a disadvantage is that a plasma is formed in the process that is difficult to control, making rejection of ablated ions very difficult. A second advantage is that the light pulse is short (compared to typical ion beam pulses), thus concentrating the atom cloud in time so that the temporal overlap with the RIS laser beam is more efficient. Only the energy distribution of the atom cloud will disperse the atoms in time and decrease the amount of overlap.

Ideally, an atomization source would yield atoms all in the same atomic state. This is never the case, but sputtering and ablation do leave the atoms in or near their ground states, and often there is a distribution among several states separated by spin differences. Usually, however, only one of these spin states can be accessed with the RIS process. Partitioning of the energy typically may be between two spin states, allowing only half the atoms to be probed. A key feature of the RIS process is that for a given state, there is unity probability that a selected atom in a particular state will be ionized if it is in the laser beam. With state partitioning among the sputtered atoms, RIS efficiencies of 50% have been shown and are a reasonable expectation.

The extraction efficiency for atoms once they have been ionized by the RIS laser depends on the electrode designs and the type of mass spectrometer system used. With magnetic sector instruments, extraction efficiencies of 50% are reasonable. With time-of-flight mass spectrometers, efficiencies approaching 100% are possible. Transmission efficiencies for mass spectrometers after the ions have been extracted are very close to 100%, and detector efficiencies can also be near 100%.

The practical sensitivity of the RIS technique depends heavily on how effective background ions can be rejected. Several background reduction techniques are generally employed to reduce unwanted ions. Background ions come from (1) secondary ions initiated by the primary sputtering process, (2) sputtered ions from the walls and electrode surfaces generated by the secondary ions, and (3) nonlinear and nonresonant ionization generated by the intense laser beams. Background ions can be reduced with clever designs of the extraction electrodes, given the fact that sputtered ions originate from the surface of the sample and

RIS ions are produced at some distance from the sample. Because of this difference in origin for secondary ions and RIS ions, extraction electrodes can be designed to only extract ions from a certain region of space. The electrode design can also create very different electric fields for the two regions, causing the secondary ions to have a much different energy distribution from the RIS ions. This allows energy discrimination in the mass spectrometer to be effective in reducing background ions.

Because sputtering is usually accomplished with a pulsed ion beam, a delay between the ion pulse and the RIS laser pulse can be used to time-discriminate against the secondary ions. Electric potentials on the electrodes can be changed by pulsing them so that secondary ions can be swept out of the way before the RIS lasers are turned on. Electronic gating of the ion detector also can be used to reduce unwanted background ions, provided there is sufficient time difference between the secondary ions and the RIS ions.

SIRIS and LARIS techniques have been shown to have a linear response over a wide range of concentrations. While in principle concentrations of atoms could be computed from the measurements, standard samples are usually used to calibrate the measurements, using standards set by the National Institute of Standards and Technology. One feature of the SIRIS/LARIS techniques is that ionization occurs away from the sample surface, freeing the measurements from the effects of differences in sample composition. This feature allows for calibrations where comparisons are made between different materials. For semi-conductor analysis, ion implant samples are easily fabricated to use as calibration samples. Implant doses can easily be determined, and the distributions of ions with depth are well characterized.

As noted, measurements on semiconductor materials are a prime application for SARIS/LARIS methods [18]. Measurements of both intrinsic and extrinsic dopants in bulk semiconductor materials are of interest to the semiconductor industry and are important in the development of new materials. Quantum well devices such as diode lasers require characterization of multilayed materials fabricated by techniques such as metal vapor deposition and molecular beam epitaxy. Stoichometry of samples near the interfaces of the layered materials is important, and diffusion across the boundaries is of interest. Depth profiles of implanted materials can be measured by switching between DC and pulsed ion beams. A DC ion beam can remove layers of a sample quickly, and the beam can be rastered in the $x$ and $y$ directions to mill out a crater to a depth at which measurements can be made using the pulsed ion beam mode of operation. Normally in a SIRIS measurement with a pulsed ion beam of short duration, no more than an equivalent monolayer of material is sputtered from the sample during a typical five-minute measurement time. Equivalent monolayer of material is used here because in the sputtering process, atoms are ejected from several monolayers down from the surface of a sample.

As an ion beam is rastered over the surface of a sample, electrons are emitted. These electrons can be both a benefit and a problem. The electrons can be collected and used for imaging the sample. Secondary ions that are generated as the ion beam is rastered across the sample surface also can be used to image the sample. This image can be element-selective with the mass spectrometer and can be used to identify the location of concentrations of a given element. With the additional use of resonance ionization, additional selectivity can be achieved in the imaging process. Analysis of geological samples is a typical application where it is important to have imaging capability in the measurement. Imaging by selected element can lead to greater understanding of microscopic images of the samples. As geological samples are most frequently nonconducting materials, the emission of their electrons leads to the buildup of an electrostatic charge on the sample, which in turn creates an electric field that perturbs the extraction of ions and the measurement. An auxiliary electron gun can be used to spray electrons on the sample during the measurement, alleviating this problem and stabilizing the measurements.

As a practical matter, RIS generally does not provide isotopic information, although in principle it could. Instead, the incorporation of the mass spectrometer into the instrument serves to provide this information.

## 10.5 Thermal Atomization Methods

Many RIS applications for the analysis of impurities in bulk materials are done with thermal atomization. These methods are not as elaborate as SIRIS and are appropriate whenever the detailed spatial information (surface raster or depth profile) is not required.

The review article by Payne *et al.* [6] contains an excellent summary of thermal atomization, which is usually implemented to produce continuous atomization.

A number of methods for continuous atomization use thin metal ribbons made of materials such as rhenium or tantalum that can endure temperatures exceeding 2000°C. At these temperatures, many solids are evaporated directly into the atomic state. However, not all of the atoms are neutral, and a sizable fraction of these can be in excited states.

Naturally, for RIS it is desirable that any atomizer produce most of the product atoms in their ground state; otherwise, backgrounds can arise in a number of ways. For this reason, samples are often coated with carbon or some other reducing agent with a low work function.

Another problem with continuous atomization is the loss of sensitivity in terms of the amount of starting material needed for an analysis. This problem, of course, is caused by the fact that most of the atoms pass through the sensitive volume when the pulsed laser beam is off. One effective way to keep these losses at an acceptable level is to use a copper vapor laser that can be pulsed at a rate of about 6000 Hz.

In spite of these difficulties, a number of laboratories have used filaments and ribbons for continuous atomization for RIS applications and have obtained impressive results. In the United States, Los Alamos National Laboratory, Oak Ridge National Laboratory, and the National Institute of Standards and Technology have developed and utilized these methods.

By far the most convenient and versatile method for producing a continuous stream of atoms appears to be the graphite furnace. These have been used for atomic absorption spectroscopy for many years, and thus have a long and useful history behind them. Use of the method in the RIMS system has been pioneered by Bekov and his colleagues at the Institute of Spectroscopy (IOS) at Troitzk (near Moscow) [19, 20]. Figure 9 illustrates the scheme in which the graphite oven is used with RIS.

The Troitzk group concentrated on RIS schemes in which the neutral atom is excited to a Rydberg state, followed by field ionization. Examples of this scheme (which typically utilizes Rydberg states with $n^* \simeq 15$ and field strengths of ca. 10,000 volts/cm, or with $n^* \simeq 30$ requiring much more modest field strengths) are included in Figure 2. A clever trick is used to reduce background ionization. As the inset shows, a weak negative pulse is used to sweep out the ionization background before a stronger positive-going pulse is applied to complete the RIS process.

The IOS method has been used in a number of important ways. One of these was a measurement of the rhodium distribution at the Cretaceous–Tertiary

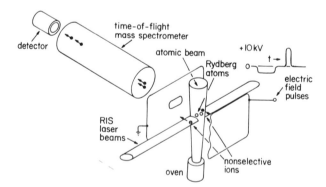

Fig. 9. Schematic representation of a graphite-oven RIMS system. A two- or three-color resonance ionization scheme is used to reach a Rydberg level. The broad, negative electric-field pulse is of insufficient amplitude to ionize atoms of $n^* \sim 15$ levels, but removes any nonselective ions produced by the excitation process, while the narrow 10-kV positive pulse both ionizes and accelerates the selected ions towards the detector. (From reference [6].)

geological boundary. This data strongly supported the hypothesis of Luis Alvarez and his colleagues that a large meteorite was responsible for the extinction, some 65 million years ago, of a great number of species, including the dinosaurs.

At Atom Sciences in Oak Ridge, Tennessee, Bekov and his associates are perfecting the thermal atomization RIS, TARIS, for commercial applications. These include measurement of uranium or plutonium atoms in environmental samples, and of trace elements in biological or medical samples.

## 10.6 On-Line Accelerator Applications

In the preceding examples it was not necessary for the RIS process to be isotopically selective. However, normal spectroscopic lines are slightly affected by nuclear properties. In fact, there are two effects. The first is the general shift due to the mass of the nucleus, known as the *isotope shift*. A more specific effect depends on the magnetic properties of nuclei and is known as *hyperfine structure*. These optical shifts are small and require high resolution in the wavelengths of the RIS lasers. When RIS is provided with isotopic selectivity, the dividend can be large in nuclear physics.

Rare species produced by atomic or nuclear processes in accelerator experiments are studied extensively with RIS. An isotope accelerator delivers ions of a particular isotope into a small target where short-lived nuclei decay. After a brief accumulation time, the target is heated with a plulsed laser to create an atomic beam containing the decay products. These decay products are then subjected to the RIS process followed by time-of-flight analysis of the ions. Analysis of the optical shifts leads to information on magnetic moments of nuclei and the mean square radii of the nuclear charge. Such measurements have been performed on several hundred rare species. These studies continue at various laboratories principally in Europe, the United States, and Japan. For an overview of these important methods, see a recent article by Alkhazov [21].

## References

1. Kluge, H. J., Parks, J. E., and Wendt, K. (1994). "Resonance Ionization Spectroscopy 1994." AIP Press, New York.
2. Letokhov, Vladilen, S. (1987). *Laser Photoionization Spectroscopy*. Academic Press, New York.
3. Hurst, G. S., and Payne, M. G. (1988). *Principles and Applications of Resonance Ionisation Spectroscopy*. IOP Publishing, Bristol.
4. Hurst, G. S. (1993). "Resonance Ionization Spectroscopy," in *Encylopaedia Britannica*. Encyclopaedia Britannica, Inc., Chicago.
5. Hurst, Samuel, G., and Letokhov, Vladilen S. (1994). "Resonance Ionization Spectroscopy," *Physics Today*, Oct. 1994, p. 38.
6. Payne, M. G., Deng, Lu, and Thonnard, N. (1994). *Rev. Sci. Instrum.* **65**(8), 2433.

7. Hurst, G. S., Payne, M. G., Nayfeh, M. H., Judish, J. P., and Wagner, E. B. (1975). *Phys. Rev. Lett.* **35,** 82.

8. Payne, M. G., Klots, C. E., and Hurst, G. S. (1975). *J. Chem. Phys.* **63,** 1422.

9. Gibson, W., and Hunter, S., submitted for publication.

10. Saloman, E. B. *Spectrochim. Acta* **45B**(1/2), 37 (1990); **46B**(3), 319 (1991); **47B**(4), 517 (1992); and **45B,** 1139 (1993).

11. Hurst, G. S., Nayfeh, M. H., and Young, J. P. (1977). *Appl. Phys. Lett.* **30,** 229.

12. Kramer, S. D., Bemis, C. E., Jr., Young, J. P., and Hurst, G. S. (1978). *Opt. Lett.* **3,** 16.

13. Hurst, G. S., Payne, M. G., Kramer, S. D., Chen, C. H., Phillips, R. C., Allman, S. L., Alton, G. D., Dabbs, J. W. T., Willis, R. D., and Lehmann, B. E. (1985). *Rep. Prog. Phys.* **48,** 1333.

14. Parks, J. E., Schmitt, H. W., Hurst, G. S., and Fairbank, W. M. (1983). *Thin Solid Films* **108,** 69.

15. Pellin, M. J., Young, C. E., Calaway, W. F., and Gruen, D. M. (1984). *Surf. Sci.* **144,** 619.

16. Winograd, N., Baxter, J. P., and Kimock, F. M. (1982). *Chem. Phys. Lett.* **88,** 581.

17. Arlinghaus, H. F., Thonnard, N., and Schmitt, H. W. (1989). In *Microbeam Analysis*, P. E. Russell (ed.). San Francisco Press, Inc.

18. Arlinghaus, H. F., Spaar, M. T., Thonnard, N., McMahon, A. W., Tanigaki, T., Shichi, H., and Holloway, P. H. (1993). *J. Vac. Sci. Technol.* **A11,** 2317.

19. Bekov, G. I., and Letokhov, V. S. (1983). *App. Phys. B* **30,** 161.

20. Bekov, G. I., and Letokhov, V. S. (1983). *Trends Anal. Chem.* **2,** 252.

21. Alkhazov, G. D. (1994). "Investigation of Radioactive Nuclei by Resonance Ionization Spectroscopy," in Resonance Ionization Spectroscopy 1994, Kluge *et al.* (eds.). AIP Press, New York.

# 11. DETECTION OF METASTABLE ATOMS AND MOLECULES

## H. Hotop

Fachbereich Physik, Universität Kaiserslautern
Kaiserslautern, Germany

## 11.1 Introduction

This chapter deals with the principles and experimental methods for the detection of metastable (i.e., long-lived, electronically excited) atoms and molecules. The discussion will focus on neutral metastable particles at low kinetic energies ($<1$ eV), for which the key properties governing detection are the excitation energy $E^*$ above the ground state, the ionization energy $I^*$ from the metastable neutral to the ionic ground state, and the natural lifetime $\tau$ of the metastable species. Table I summarizes these properties for selected metastable states of atoms and molecules [1–28]. For the listed examples, $E^*$ spans the range 4.6–20.6 eV, but we note that important metastable states with lower excitation energies exist, e.g., $N(^2D)$, $N(^2P)$, $O(^1D)$, $O(^1S)$, or $O_2(a^1\Delta)$. The ionization energies $I^*$, which are given by the difference $I - E^*$ ($I$ = ionization energy of ground state neutral to ground state ion), fall in the range 3–5 eV for many of the cases listed in Table I, but may approach the ground-state ionization energy $I$ for low-lying metastable levels. Rydberg atoms and molecules, in which one electron is excited to an orbital of high principal quantum number $n$, represent an important class of long-lived particles with energies of only a few milli–electron volts below the ionization threshold. Rydberg states of atoms and molecules have been extensively discussed recently (see, e.g., [29]) and are also covered in other chapters of this volume; we therefore omit this class of long-lived states from the subsequent discussion. We also do not dwell on long-lived autoionizing states [30], which have excitation energies above the ionization threshold and decay into the ionization continuum by electron–electron interactions.

The term "metastable" does not imply a precise minimum value for the lifetime $\tau$. We regard species as metastable if their flight time through some apparatus is shorter than their natural lifetime (typically $\tau \geq 10^{-4}$ s). The relatively long lifetimes of metastable levels are due to the fact that spontaneous decay to lower-lying states by electric dipole emission is forbidden. Because of their long lifetimes, metastable species are important carriers of potential energy in various media such as the atmospheres of planets, gaseous discharges, and

EXPERIMENTAL METHODS IN THE PHYSICAL SCIENCES
Vol. 29B

TABLE I.  Selected Metastable States of Atoms and Molecules

| Atom/ molecule | Metastable state | Excitation energy $E^*$ (eV)† | Ionization energy $I^*$ (eV)† | Lifetime $\tau$ [s] |
|---|---|---|---|---|
| $^1$H | $2s\,^2S_{1/2}$ | 10.1988 | 3.3996 | $0.14$[a] |
| $^4$He | $1s2s\,^3S_1$ | 19.8196[b] | 4.7678[b] | 7870[c]; 9090 ± 30%[d] |
|  | $1s2s\,^1S_0$ | 20.6158[b] | 3.9716[b] | 0.0196[e]; 0.0197(10)[f] |
| $^{16}$O | $2p^3 3s\,^5S_2$ | 9.146 | 4.472 | $185(20) \times 10^{-6}$[g,h] |
| $^{20}$Ne | $2p^5 3s\,^3P_2$ | 16.6191[i] | 4.9454[i] | 24.4[j]; >0.8[k] |
|  | $2p^5 3s\,^3P_0$ | 16.7154[i] | 4.8491[i] | 430[j] |
| $^{40}$Ar | $3p^5 4s\,^3P_2$ | 11.5484[l] | 4.2112[l] | $55.9$[j]; $38^{+8}_{-5}$[m] |
|  | $3p^5 4s\,^3P_0$ | 11.7232[l] | 4.0364[l] | 44.9[j] |
| $^{84}$Kr | $4p^5 5s\,^3P_2$ | 9.9152[n] | 4.0844[n] | $85.1$[j]; $39^{+5}_{-4}$[m] |
|  | $4p^5 5s\,^3P_0$ | 10.5624[n] | 3.4372[m] | 0.488[j] |
| $^{132}$Xe | $5p^5 6s\,^3P_2$ | 8.3153[o] | 3.8145[o] | $149.5$[j]; $42.9(9)$[p] |
|  | $5p^5 6s\,^3P_0$ | 9.4472[o] | 2.6826[o] | $0.078$[j]; $(128^{+122}_{-42}) \times 10^{-3}$[p] |
| Hg | $5d^{10}6s6p'\,^3P_0$ | 4.6674 | 5.7701 | 5.6[q] |
|  | $5d^{10}6s6p\,^3P_2$ | 5.4606 | 4.9769 | ≈0.1[r] |
|  | $5d^9 6s^2 6p'\,^3D_3$ | 8.7945[s] | 1.6430 | ≈0.06[r] |
| H$_2$ | $c^3\Pi_u^-$ | 11.764 | 3.662 | $1.02(5) \times 10^{-3}$[t] |
| N$_2$ | $A^3\Sigma_u^+$ | 6.169 | 9.412 | 1 − 2[u] |
|  | $a^1\Pi_g$ | 8.549 | 7.032 | $120(20) \times 10^{-6}$[h,u] |
|  | $E^3\Sigma_g^+$ | 11.875 | 3.706 | $190(30) \times 10^{-6}$[v] |
| CO | $a^3\Pi$ | 6.010 | 8.004 | $\geq 3 \cdot 10^{-3}$[w,x] |

*Note.* The entries a, c, e, j, q, and x represent theoretical predictions: *a*: [1]; *b*: [2]; *c*: [3]; *d*: [4]; *e*: [5, 6]; *f*: [7]; *g*: [8]; *h*: [9]; *i*: [10]; *j*: [11]; *k*: [12]; *l*: [13]; *m*: [14]; *n*: [15]; *o*: [16]; *p*: [17a] for $^3P_2$, [17b] for $^3P_0$; *q*: effective lifetime for natural isotopic mixture [18]; *r*: [19]; *s*: [20]; *t*: [21]; *u*: [22]; *v*: [23]; *w*: [24]; *x*: [25].
† Energies in electron volts calculated from spectroscopic values with the conversion 1 eV = 8065 · 541 cm$^{-1}$ [26]. In addition to the individual references, the tables of Ch. E. Moore [27] were used for the energies of the atoms. Energies for the molecules were taken from K. P. Huber and G. Herzberg [28]; they correspond to the lowest rovibrational state.

lasers. In such environments the effective lifetime is often limited by inelastic collisions, in which the excitation energy is transferred to other atoms or molecules or to surfaces. In studies of the basic properties of metastable species (e.g., their interaction with photons, electrons, and heavy particles), efficient and selective detection is desirable and often mandatory. If absolute rate constants or cross-sections are to be determined, it may even be necessary to measure the absolute flux or density of the metastable species; even today this is a difficult task and has been accomplished for only a limited range of species, notably the rare gas metastable atoms.

In the following part, we discuss methods for the (relative) detection of metastable species and present selected examples in more detail. In the last part, we describe methods for the quantitative, absolute detection of metastable atoms.

## 11.2 Methods for the Detection of Metastable Species

Schemes for the detection of metastable excited atoms and molecules often exploit their long lifetime and stored potential energy. In Sections 11.2.1 to 11.2.4, we discuss detection methods based on de-excitation of the metastable (normally to the ground state), either by electron emission from solid or gaseous targets with sufficiently low work function or ionization potential, or by spontaneous or collision-induced photon emission. In Section 11.2.5 we describe the schemes based on laser-induced photoexcitation and photoionization; they combine high selectivity and sensitivity. We omit special methods of metastable detection such as ionization of metastable species in collisions with electrons [31–33] or surfaces [34–36] and quenching in electric fields [37, 38].

### 11.2.1 Electron Emission from Surfaces

Metastable atoms or molecules incident on a surface may eject electrons, if their internal energy $E^*$ exceeds the work function $\Phi$ of the surface. This was first established by Webb [39] for metastable Hg* atoms incident on a nickel surface. In most of the subsequent work, gas-covered ("contaminated") metal surfaces were used, often in conjunction with electron multipliers used to detect the ejected electrons. The current of emitted electrons provides a measure of the relative incident metastable flux. Detectors based on (secondary) electron ejection are easy to construct and operate at typical overall efficiencies between about $10^{-4}$ and 0.5. The value of the electron emission coefficient $\gamma$, defined as the average number of electrons emitted per incident metastable particle, depends strongly on the excitation energy $E^*$ and on the nature of the detector surface. Quantification of $\gamma$ is not an easy task and will be discussed in some detail in Section 11.3. For chemically cleaned (i.e., simply cleaned with the normal chemical solvents before insertion in the vacuum system), adsorbate-covered surfaces, *in situ* measurement of $\gamma$ is mandatory for quantitative investigations [40, 41]. True reproducibility can only be obtained at "atomically clean" (i.e., adsorbate-free) solid [42–56] or liquid [57] surfaces. If electron emission from contaminated surfaces is used for the relative detection of metastable species on a routine basis, care must be taken to ensure that the detection efficiency remains constant over extended periods of time; heating of the surface to a sufficiently high temperature may help to achieve this (especially for metastables with low excitation energies) [24, 41, 58–61].

Mechanisms for Electron Emission from Surfaces. The basic ideas underlying our present understanding of electron emission from surfaces by slow ions or excited neutral species ("potential emission") were developed by Hagstrum [42, 51, 62a, b]. Figure 1 illustrates the processes that can lead to electron emission from a surface upon impact of an excited atom. Figures 1a and 1b refer to atomically clean conducting (e.g., metal) surfaces. If a metastable atom M* approaches a surface with a sufficiently high work function $\Phi$, resonance ionization (RI) occurs [indicated by the arrow (1) in Figure 1a], in which the excited electron tunnels into a vacant level above the Fermi surface. This process results in the formation of an ion close to the surface; it is the basic step in surface ionization detectors for atoms with low ionization potentials. If the (effective) recombination energy of the ion to the neutral ground state $E^+$ (which—close to the surface—is typically 1–2 eV lower than the ground state ionization potential $I$ of the isolated neutral) is at least twice the work function $\Phi$, i.e., $E^+ \geq 2\Phi$, Auger neutralization (AN) with electron emission can occur. In this process, an electron from the metal tunnels into the hole state of the ion, with simultaneous excitation of a second electron from the conduction band, which may, if the energy transfer is sufficient, be ejected. The process is shown by the arrows (2) in Figure 1a. Several groups have shown convincingly that the

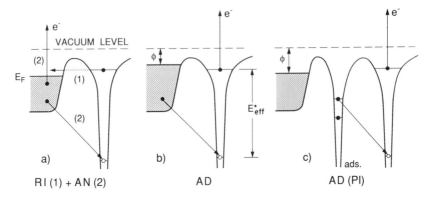

FIG. 1. Illustration of electron-ejecting de-excitation processes that occur upon impact of slow excited atoms or molecules at conducting surfaces. (a) Resonance ionization RI (1), followed by Auger neutralization AN (2); (b) Auger de-excitation AD; (c) Auger de-excitation AD, involving an electron from an adsorbed atom or molecule (i.e., Penning ionization PI of adsorbed species by the excited particle). $E^*_{eff}$ denotes the (distance-dependent) excitation energy near the surface. For surfaces with very low work function, an additional process occurs with substantial probability, namely electron emission due to autodetachment of excited negative ion states which are formed by electron transfer from the surface to the metastable projectile ([64]; see also [56, 77]).

two-step RI + AN electron emission mechanism strongly dominates in the case of de-excitation of metastable $He^*(2^3S, 2^1S)$ atoms [48, 53–57, 63–65] or laser-excited $He^*(2^3P)$ atoms [65] at a clean metal surface of work function $\Phi \gtrsim 4.5$ eV. In the case $E^+ < 2\Phi$, electron emission is prohibited; then ion neutralization with electron excitation to unoccupied states below the vacuum level is likely. Possibly ion escape may occur, but only a few examples for surface ionization of metastable species have been reported to date [34–36].

If no vacant energy levels exist in the solid, that are in resonance with the approaching excited electron (for example, at a low work-function surface or surfaces with a large band gap), RI cannot proceed, and Auger de-excitation (AD) of M* occurs, as illustrated in Figure 1b for the case of AD involving electron exchange [51, 56, 62, 64, 66, 67]. The energetic requirement for AD to occur is $E^*_{\text{eff}} > \Phi$, where $E^*_{\text{eff}}$ represents the effective, distance-dependent excitation energy of M* near the surface. For metastable species with parallel spins of the excited electron and the core hole [e.g., for $He^*(2^3S)$], AD proceeds by electron exchange, as shown in Figure 1b. In other cases, AD by a direct mechanism, involving de-excitation of the excited electron and emission of a metal electron, is possible in addition to electron exchange. For surfaces with very low work functions ($\lesssim 2.2$ eV, e.g., alkalated surfaces), a new important mechanism for electron emission has been identified [64; see also 56], namely the formation of temporary negative ions (i.e., electron resonances attached to excited states of the incoming atom or molecule) with subsequent auto-detachment, yielding sharp peaks in the energy spectrum of the ejected electrons.

At adsorbate-covered surfaces, the probability for RI is often sufficiently reduced that the AD process, illustrated in Figure 1c for the case of electron exchange, dominates [50, 54, 68–71], but mixed behavior (RI + AN)/AD has also been observed; see, e.g. [71]. For thick molecular adsorbates, the AD process is analogous to Penning ionization in the gas phase [50, 54, 68–70].

**Electron Emission Detectors.** An electron emission detector for metastable species consists of a suitable surface and a device to detect the ejected electron current, i.e., an electrometer (for currents $\gtrsim 10^{-12}$ A) or—for increased sensitivity—an electron multiplier in conjunction with an electrometer or a fast counter. In quantitative work, direct measurement of the ejected or of the loss current is preferable. Often, the detector is used in combination with time-of-flight (TOF) techniques in order to discriminate against photons or fast ground-state particles, which may also eject electrons from the detector surface: Moreover, TOF methods can provide information on the lifetime of metastable levels [7–9, 12, 21, 23, 24].

Figure 2a shows an example of an electron emission detector with high sensitivity and an adjustable detection threshold [72, 73]. A beam of metastable atoms or molecules M* ejects electrons upon impact on a detection surface (here,

Mo). Proper choice of the potentials applied to the different electrodes and to the entrance cone of the channel electron multiplier results in efficient detection of the ejected electrons; with an appropriate negative voltage at the retarding grid, one can discriminate against electrons with energies below a certain threshold. This is demonstrated in Figure 2b, which shows effective excitation functions for the formation of metastable $N_2^*$ molecules by electron impact [72]: curve (1) was taken without energy analysis, while spectra (2) and (3) were recorded with a retarding potential difference of $-0.6$ eV and $-1.3$ V, respectively. The onset of spectrum (1) corresponds to the excitation energy of the $N_2^*$ $(B^3\Pi_g)$ state observed via cascading to the $N_2^*$ $(A^3\Sigma_u^+)$ state. In spectrum (3), excitation of the $B$-state is not detected, and the onset of the spectrum corresponds now to the excitation of the $N_2^*$ $(a^1\Pi_g)$ state. The contribution of the $N_2^*$ $(E^3\Sigma_g^+)$ state to this excitation spectrum is strongly enhanced. Subsequently, the energy selectivity of this detector was exploited for the detection of high-lying metastable states in Hg [73]. A state-selective metastable detector of a somewhat different design was recently used in combination with time-of-flight analysis to investigate lifetimes of metastable molecules [74].

The choice, preparation, and operation of the detector surface are critical for the optimal performance of an electron emission detector. Dependable,

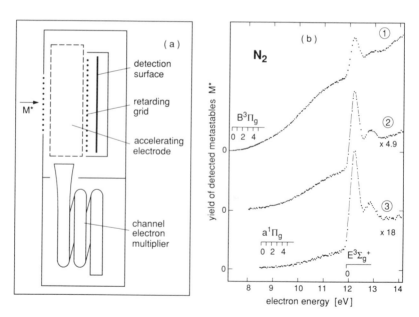

FIG. 2. Sensitive electron emission detector for metastable species M* with an adjustable threshold [72]; (b) electron impact excitation functions for $N_2^*$ molecules as measured with different voltages between the detection surface and the retarding grid [72]; see text.

reproducible operation requires a stable, unchanging surface. This requires that the detector use atomically clean surfaces in ultrahigh vacuum, whose cleanliness and condition is routinely checked by surface analysis techniques. In normal applications, however, metastable particle detectors are operated in a standard high-vacuum environment (background pressure $10^{-7}$–$10^{-6}$ mbar), and therefore the detector surface is contaminated and not well characterized. It is then the task of the experimentalist to find a surface material and mode of operation which yield optimal detection efficiency for the metastable species in question and the best stability over extended periods of time.

Electron emission from atomically clean surfaces has been studied mainly for the metastable rare gas atoms Rg* with the aim of investigating the basic processes. In a few cases electron emission coefficients $\gamma$ were determined [43, 45, 46], yielding values which increase with metastable excitation/ion recombination energy. In Table II we compare these results with values obtained for impact of rare gas ions in their electronic ground state at low kinetic energy [42, 75]. The yields for $Rg^+$ ions depend on the work function $\phi$ of the surface—notably for the heavier rare gases for which the difference $E^+ - 2\phi$ becomes rather small—in a way approximately described by the relation $\gamma = 0.2$

TABLE II. Comparison of Yields for Electron Emission from Atomically Clean Polycrystalline Surfaces by Low-Energy Rare Gas Ions and Thermal Energy Metastable Rare Gas Atoms

| $Rg^+$ | $He^+$ | $Ne^+$ | $Ar^+$ | $Kr^+$ | $Xe^+$ |
|---|---|---|---|---|---|
| $\gamma(Mo)^a$ | 0.300 | 0.254 | 0.122 | 0.069 | 0.022 |
| $\gamma(W)^a$ | 0.290 | 0.213 | 0.095 | 0.050 | 0.013 |
| $\gamma(W)^b$ | 0.260 | 0.245 | 0.090 | 0.049 | 0.013 |
| $\gamma(Au)^c$ | $0.15 \pm 0.01$ | $0.11 \pm 0.006$ | $0.043 \pm 0.005$ | | 0.00025 $\pm 0.00009$ |
| Rg* | $He^*(2^1S, 2^3S)$ | $Ne^*(^3P_{2,0})$ | $Ar^*(^3P_{2,0})$ | | |
| $\gamma(W)^d$ | 0.306(25) | 0.215(20) | | | |
| $\gamma(W)^e$ | 0.280(60) | 0.195(40) | 0.080(16) | | |
| $\gamma(Cd)^f$ | 0.35(6) | 0.22(6) | 0.32(3) | | |

[a] [42]; work function $\Phi$ of Mo/W: 4.27/4.5 eV.
[b] [75a]; $\Phi(W) = 4.5$ eV; ion energies (vertical impact) 100 eV.
[c] [75b]; $\Phi(Au) = 5.1$ eV; numerical values provided by H. Winter. Data are for ion energies (vertical impact) of 100 eV for $He^+$, $Ne^+$, of 120 eV for $Ar^+$, and 103 eV for $Xe^+$.
[d] [43]; yields for $He^*(2^1S)$ and $He^*(2^3S)$ found to be equal within 20%.
[e] [46]; yields for $He^*(2^1S)$ and $He^*(2^3S)$ found to be equal within 10%.
[f] [45, 46]; cadmium evaporated on stainless steel; quoted numbers obtained with a crossed beam method [45]. A gas cell absorption method [46] yielded identical results; see Table 2 in [46]. Yields for $He^*(2^1S)$ and $He^*(2^3S)$ were found to be equal within 10%. The value for Ar* may be too high (F. B. Dunning, private communication).

$(0.8\,E^+ - 2\phi)/E_F$) (where $E_F$ = Fermi energy of metal) [76]. Within the experimental uncertainties, the yields for metastable rare gas atoms Rg* agree with those of the respective $Rg^+$ ion for polycrystalline tungsten, as expected on the basis of the (RI + AN) mechanism for Rg* impact. We note that substantial progress has been recently made in model calculations of the electron emission process, allowing to qualitatively reproduce both measured electron energy spectra and experimental values for the electron emission coefficients $\gamma$ (see [77] and references therein).

Controlled coverage of atomically clean surfaces with adsorbates such as $O_2$, $N_2$, or CO tends to decrease $\gamma$, especially for the heavier rare gases [43, 44, 48], but increases in $\gamma$ have also been observed [45, 46, 70, 71]. For chemically cleaned metal surfaces, the values of $\gamma$ were frequently found to exceed those for the respective atomically clean surfaces, especially for He* and Ne* impact, as discussed by Dunning et al. [45, 46, 78]. Such behavior may be attributed to efficient Penning ionization of a sufficiently thick adsorbate layer (with little coupling to the underlying surface band structure); for He* and Ne*, ionization is the main de-excitation process in collisions with both atoms and molecules. For Ar*, Kr*, and Xe*, on the other hand, nonionizing destruction processes (e.g., excitation transfer, followed by fluorescence or dissociation, or reactive collisions with molecules) are often important or dominant, even in cases, in which the ionization potential of the target is lower than the excitation energy of Rg* [79–83]. In consequence the electron emission coefficient $\gamma$ may decrease for He* and Ne*, yet increase for metastables of lower internal energy (such as Ar*, Kr*, Xe*), when a chemically cleaned surface is partially baked to remove some of the adsorbed molecules, which allows relatively weakly bound electrons in the conduction band to participate in Auger deexcitation. Several groups have reported strong increases in the electron emission efficiency upon heating of detector surfaces [24, 41, 58–61], especially for impact of metastable states with low excitation energy. (For example, an increase by a factor up to 100 was observed for CO $(a^3\Pi)$ incident on a Ta surface heated to 400–600°C [24, 58].

Borst [84] has suggested that the yield $\gamma$ for electron emission from a contaminated, room-temperature Cu–Be–O surface for impact of metastable species is a smoothly varying function of their excitation energy in the range 6–20 eV and can be described by a simple universal curve. Although the excitation energy clearly is an important parameter (see Table III), the use of such a universal curve cannot be recommended; especially at lower excitation energies, the values of $\gamma$ are influenced strongly by the type of contamination, by the temperature of the surface, and by other factors. We note that the $\gamma$-values for Ne*, Ar*, Kr*, and Xe*, used by Borst in constructing his universal curve [84], were actually those for slow rare gas ions, and his value for He* ($\gamma \approx 0.15$) is substantially lower than those measured for several contaminated metal surfaces in more quantitative work, (i.e., $\gamma(He^*) \approx 0.45$–0.9 [40, 45, 46, 85].

Quantitative determinations of $\gamma$ for contaminated conducting surfaces have mainly been carried out with metastable rare gas atoms Rg*, using techniques described in Section 11.3. In Table III we summarize measured values of $\gamma$ [40, 41, 45, 46, 85, 86] for several combinations of metastable rare gas atoms and contaminated surfaces, with the main purpose of indicating the range and spread of the observed $\gamma$ values (see also Table VI in [87]). Several general points are of interest:

TABLE III.  Electron Emission Coefficients $\gamma$ Measured for Metastable Rare Gas Atoms Rg* at Several Contaminated Conducting Surfaces

| | Rg* | | | | | |
|---|---|---|---|---|---|---|
| Detector surface | He*($2^1S$) | He*($2^3S$) | Ne*($^3P_2$) | Ar*($^3P_2$) | Kr*($^3P_2$) | Xe*($^3P_0$) |
| Stainless steel | $0.95(10)^a$ $0.53(8)^b$ $0.6(3)^c$ | $0.95(10)^a$ $0.69(9)^b$ | $0.91(11)^a$ $0.61(8)^b$ $0.30(5)^d$ | $0.97(15)^a$ $0.04$–$0.22^{d,e}$ | $0.01$–$0.05^{d,e}$ | $0.02(1)^c$ |
| Graphite deposited on stainless steel | | | $0.35(5)^d$ | $0.10$–$0.22^{d,e}$ $0.20(2)^f$ | $0.07$–$0.10^{d,e}$ $0.12(2)^f$ | $0.11(2)^f$ |
| Gold (plated) | $0.46(9)^a$ $0.45(2)^g$ | $0.63(7)^a$ $0.55(2)^g$ | $0.52(6)^a$ $0.48(2)^g$ | $0.66(10)^a$ $0.007$–$0.044^d$ $0.24(5)^e$ | $0.003$–$0.008^d$ $0.06(2)^e$ | |
| CuBe | $0.51^b$ | $0.63^b$ | $0.44^b$ $0.40(5)^d$ | $0.06 - 0.25^d$ $0.26^e$ | $0.07^e$ | |
| Molybdenum | | | | $\approx0.02^d$ $\approx0.14^e$ | $\approx0.006^d$ $\approx0.06^e$ | |
| Indium–tin Oxide (ITO) | | | | $0.20(2)^f$ | $\approx0.05^f$ | $0.05(1)^f$ |

[a] [46]; gass cell method; surface at $T = 300$ K. The values for Ar* are considered to be too high (F. B. Dunning, private communication) for reasons discussed in Section 11.3.1.
[b] [40]; improved gas cell method (see Section 11.3.1); surface at $T = 300$ K.
[c] [40]; pulsed-laser photoionization method; surface at $T = 300$ K.
[d] [41]; cw laser photoionization method (see Section 11.3.2); surface at $T = 300$ K. The error in each determination was about 1%, but for Ar* and Kr*, large variations were found between different, apparently identically prepared, detector plates, reflecting varying surface properties.
[e] [41, 61]; cw laser photoionization method; surface at $T = 360$ K. The increase in efficiency was found to be greatest for surfaces with low efficiencies at 300 K.
[f] [86]; cw laser photoionization method; surface at $T = 300$ K. The yields $\gamma$ were observed to vary within the given limits.

- Efficient detection of metastable He* and Ne* atoms occurs at many contaminated conducting surfaces with typical $\gamma$ values of 0.5–0.7 for He* and 0.3–0.5 for Ne* [40, 41, 46, 85]. For He* and Ne*, Penning ionization of absorbed species is energetically allowed and expected to proceed with high probability [40, 41, 46, 69–71].
- For metastable states with lower excitation energies ($E \leq 12$ eV), the yields $\gamma$ can depend strongly on the special properties of the surface and its temperature, as observed for Ar* and Kr* in recent work [41, 61] with yields ranging from 1% to 26% for Ar* ($^3P_2$) and 0.3% to 10% for Kr* ($^3P_2$). Similar behavior is expected for other metastable atoms and molecules with lower excitation energies for which Penning ionization of adsorbates is energetically forbidden or rather unlikely because of other competing de-excitation processes.
- A universal yield curve, representing $\gamma$ as a smooth function which rises monotonously with increasing excitation energy [84], is not supported by the measured yields. Recent work, in which the yields $\gamma$ for metastable Ar* ($4s\ ^3P_2$), Kr* ($5s\ ^3P_2$) and laser-excited Ar* ($4p\ ^3D_3$), Kr* ($5p\ ^3D_3$) states were measured under identical surface conditions, showed conclusively that $\gamma$ does not depend exclusively on excitation energy [61].

The data in Table III refer to metastable species with thermal kinetic energies ($\leq 0.1$ eV). It is reasonable to expect that the electron emission coefficient $\gamma$ will be essentially independent of the particle velocity in the thermal energy range, but direct experimental evidence for such an independence has so far been provided in only few cases [64, 85, 88, 89]. Borst et al. [89], for example, demonstrated that the yield $\gamma$ for metastable O* ($^5S$) atoms on a contaminated Cu–Be–O surface was constant to within 10% for kinetic energies from about 0.03 to 3 eV.

At higher kinetic energies ($\geq 100$ eV), the yields are expected to increase because of kinetic emission of electrons; see, e.g. [90].

Dependable values of $\gamma$ for metastable atoms with low excitation energy ($<9$ eV) and for metastable states of molecules are scarce and estimates— notably for Hg* ($^3P_2$) [60, 91], CO* ($a^3\Pi$) [84, 92, 93] and metastable states of N$_2$ [72, 84]—extend over several decades ($10^{-5}$–$10^{-1}$). Heated metal surfaces (e.g., Ta, Sm) with work functions of 3.5–4.5 eV have been found to be stable detectors for metastable species with excitation energies down to 5.5 eV [e.g., Hg* ($^3P_2$), CO* ($a^3\Pi$), N$_2^*$ ($A^3\Sigma_u$)] [24, 58–60, 92]. As an alternative, alkalated surfaces can be used [19, 56, 77, 92, 94–96] and enable a detection of states with even lower energies [92, 94–96], but operation in ultrahigh vacuum [56, 77, 94] or continuous deposition of alkali vapor [95, 96] is necessary for stable performance. A surface coated with SnO$_2$ has been found to detect both Hg* ($^3P_2$) and the lower-lying Hg* ($^3P_0$) metastable state [97]. The yield for spin–orbit split metastable levels (e.g., for the heavier rare gas atoms and for mercury) will in general be different at contaminated surfaces, especially if the energy difference between the levels in question amounts to a nonnegligible fraction of the excitation energy. Weissmann *et al.* [98] reported efficiency ratios for the two metastable states $^3P_2$ and $^3P_0$ of Ne*, Ar*, and Kr* on stainless steel, finding $\gamma_2/\gamma_0 = 1.00(4)$, 0.96(6), and 0.58(3), respectively.

## 11.2.2 Penning Ionization of Gas Targets

If a metastable atom or molecule M* collides with an atom or a molecule XY, whose ionization potential I(XY) is lower than the excitation energy $E^*$ of M*, Penning ionization (PI),

$$M^* + XY \xrightarrow{\text{PI}} M + XY^+ + e^-, \tag{11.1}$$

can occur with a cross-section that often reaches gas kinetic values (i.e. $\sigma_{PI} \approx 10^{-15}$ cm$^2$) [79–83, 85, 87, 99]. Detection of the Penning ions [100–102] or Penning electrons [98, 99, 103] provides a measure of the incident metastable flux, and metastable to ion–electron pair conversion efficiencies of about $10^{-2}$ can be easily realized with a static gas cell target. PI has been exploited using gas-cell and crossed-beam methods for absolute metastable detection [40, 45, 46, 85] (see Section 11.3).

Penning ionization functions as a high-pass energy filter in that only metastable states with $E^* > I(XY)$ are detected [with an uncertainty governed by the range of collision energies ($\lesssim 0.1$ eV), by the possibility of activation energies and by release of $(M + XY^+)$ binding energy in associative ionization]. Čermák introduced PI as an energy-selective metastable detector [100] and found "a new long-lived state" in $N_2$ [later recognized to be $N_2^*(E^3\Sigma_g^+)$]. PI detection of metastable species can be especially useful in cases where several long-lived states exist that are difficult to separate by time-of-flight techniques or by the limited discrimination offered by retarding field-electron emission detectors [72–74]. Following Čermák, we have exploited this idea, using the fairly simple and efficient apparatus shown in Figure 3a [101, 102]. Metastable atoms or molecules in a molecular beam are created by electron impact and interact with a molecular target beam of suitable ionization potential in the reaction chamber (2 cm from the source region), which functions as the ion source for a 60° magnetic sector field mass spectrometer. The magnetic collimation of the exciting electron beam (typical currents 1–50 μA) in combination with suitable potentials on the blocking grids prevent electrons and ions from passing from the excitation region to the reaction chamber. As an example, Figure 3b shows excitation functions for long-lived states in $N_2$, as obtained for three detector molecules with different ionization potentials between 9.26 eV and 11.42 eV [101]; see also [100]. When using XY = $C_2H_2$, mainly the $N_2^*(E^3\Sigma_g^+)$ state

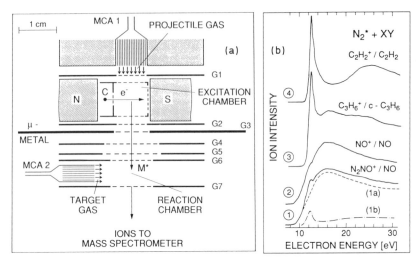

FIG. 3. (a) Apparatus for the production of long-lived excited atoms and molecules M* by electron impact in a collimating magnetic field (N–S) and for the mass spectrometric detection of ions which are formed in the reaction chamber by Penning ionization of a suitable target gas in collisions with the excited species M* [101–103]; (b) excitation functions for the electron impact–induced production of metastable $N_2^*$ molecules (lifetimes >40 μs) as measured by mass spectrometric detection of the listed ions due to Penning or associative ionization processes [101].

(formed either directly or by optical cascade from higher states) contributes to ionization [100, 101].

Note that VUV fluorescence from short-lived excited $N_2$ states is discriminated against in two ways: PI cross sections are normally 10–100 times larger than those for photoionization, and metastable species in the directed projectile beam have a higher probability of reaching the reaction chamber than do fluorescence photons, which are emitted in all directions.

In choosing the threshold energy for detected metastable states, one can also take advantage of the different appearance potentials of fragment ions in dissociative Penning ionization,

$$M^* + XY \rightarrow M + X^+ + Y + e^-,  \tag{11.2}$$

if mass spectrometric ion detection is used [101].

At the expense of sensitivity, state-selective detection of metastable atoms or molecules can be achieved by energy-resolved detection of the electrons formed in the Penning ionization process (11.1) [98, 99, 103].

## 11.2.3 Radiative Decay of Metastable Atoms and Molecules

In principle, metastable species can be detected by monitoring the forbidden optical emission from a sufficiently dense sample. $He^*(2^3S)$ atoms emit

62.556 nm photons in their magnetic dipole decay to the $He(1^1S_0)$ ground state at a rate $A = 1/\tau = 1.4 \times 10^{-4}\ s^{-1}$; in spite of this very low rate, quantitative monitoring of this radiation has allowed experimental measurement of the $He^*(2^3S)$ lifetime [4]. If a beam of metastable species with average velocity $\bar{v}$ is used, a photon detector with an overall efficiency $\eta$ (including solid angle effects) that views the beam over a length $L$ will produce a counting rate of

$$r = \eta \cdot F_0 \left[ 1 - \exp\left( -\frac{L}{\bar{v}\tau} \right) \right],$$ (11.3)

where $F_0$ is the metastable flux at the entrance to the detection region. In a study of electron impact excitation of mercury, Krause et al. [104] were successful in detecting the hyperfine interaction–induced electric dipole decay of metastable $^{199-201}Hg^*(6s\ 6p\ ^3P_{0,2})$ atoms (theoretical lifetimes about 6 s [18]) at 265.66 nm and 227.04 nm, respectively (in this article, we always give vacuum wavelengths). In a measurement of radiative deexcitation of metastable $He^*(2^1S)$ atoms in thermal energy collisions with He, Ar, Kr, and Xe, Dehnbostel et al. [105] monitored the $He^*(2^1S)$ flux by detecting photons from the two-photon decay ($\tau = 19.7$ ms [5–7]) in the absence of target gas.

Measurements of the optical emission spectra of long-lived excited molecules in combination with time-of-flight methods have been essential in characterizing molecular metastable states. The most prominent and important example is the nitrogen molecule, which has several metastable levels [22, 28]. As one selected example, we mention the recent molecular beam work of Neuschäfer et al. [106], who investigated the spectrally resolved photon emission from long-lived excited $N_2^*$ states under collision-free conditions at variable distances (up to 56 cm) from the discharge source region. They determined the vibrational-state dependence of the lifetimes for radiative decay of long-lived $N_2^*$ ($W\ ^3\Delta_u$, $v = 1$–20) molecules to the nearby $N_2^*(B\ ^3\Pi_g, v')$.

With the advent of laser cooling and trapping of atoms, including atoms in metastable levels [14, 17], the detection of forbidden transitions from levels with lifetimes $>1$ s is no longer a problem: Recently, Walhout et al. [17] have precisely determined the lifetime of metastable $Xe^*(^3P_2)$ atoms by measuring the rate of VUV emission due to magnetic quadrupole decay.

## 11.2.4 Collision-Induced Photon Emission

Photon emission following energy transfer from metastable species to suitable solid targets or to gas targets can be used for metastable detection, but because undirected fluorescence is normally detected with a rather low probability (compared to that for electrons or ions), this approach is less well suited for metastable detection if high sensitivity is needed. Energy transfer processes (in combination with dispersed fluorescence detection), however, can yield high selectivity for metastable detection because of the near-resonance needed for energy transfer to take place [59, 79, 107–109]. The basic reactions leading to fluorescence are near-resonant energy transfer from the metastable M* to an atom or a molecule XY,

$$M^* + XY \rightarrow M + XY^*$$

$$\qquad\qquad\qquad \llcorner\!\!\rightarrow \text{photons},$$

(11.4)

and reactive scattering from a target molecule,

$$M^* + XY \rightarrow MX^* + Y$$

$$\qquad\qquad\qquad \llcorner\!\!\rightarrow \text{photons}.$$

(11.5)

Processes (11.4) and (11.5) are important pumping processes in gas lasers such as the He–Ne or the excimer laser [79]. Penning ionization (11.1), (11.2) may also lead to photon emission through the formation of radiating ions or neutral fragments [79, 87, 99, 110–112].

To investigate the use of photon emission as a metastable detection method, Kume *et al.* [113, 114] have studied photon emission from condensed films of organic molecules upon impact of metastable $N_2^*$ molecules. They detected $N_2^* (A^3\Sigma_u^+)$ by phosphorescence from a biacetyl surface with an estimated quantum yield of $10^{-3}$ [113]. Anthracene films, on the other hand, turned out to be quite efficient fluorescence detectors for the singlet states $N_2^* (a\ ^1\Pi_g, a'\ ^1\Sigma_u^-)$ with an estimated quantum yield of $10^{-2}$ [114].

### 11.2.5 Photoexcitation and Photoionization

For samples with sufficient density, conventional photoabsorption techniques can be utilized for the monitoring of metastable species; see, e.g., [80, 115, 116]. Photoexcitation with lasers provides the most selective and also sensitive methods for metastable state detection (see [31–33, 40, 41, 87, 117] and references therein); the principles have been described in detail elsewhere [117–123].

Important detection schemes include the following:

- Laser-induced fluorescence
- Single-photon or resonant two-photon excitation to long-lived Rydberg states in combination with subsequent ionization by electric fields, radiation or collisions
- Single-photon or resonant two-photon ionization

The optimum choice of detection scheme for some particular application will depend on the details of the energy level diagram and on the availability of suitable lasers. Lasers with high duty cycle and narrow bandwidth are desirable.

In some cases, such as the heavier rare gas metastable atoms Rg* ($^3P_2$, $^3P_0$), continuous wave (cw) lasers can be used, allowing efficient detection using all the listed schemes. Recent developments in mode-locked solid-state lasers allow efficient generation of tunable radiation in the range 200 to 1000 nm with picosecond and subpicosecond pulse widths and high repetition rates. Such lasers should prove especially useful in the efficient resonant two-photon ionization detection of low-lying metastable states of atoms and of metastable molecules.

As one specific atomic example, we discuss laser detection schemes for metastable Rg* ($ms\ ^3P_{2,0}$) atoms, with reference to the specific case of Ar* ($4s\ ^3P_{2,0}$) atoms (Figure 4) [41, 108, 124, 125]. They are based on selective excitation of one of the ($4s$–$4p$) resonance transitions, which can be pumped by tunable dye, titanium-sapphire or diode lasers (700–900 nm). The closed transition $4s\ ^3P_2$–$4p\ ^3D_3$ (811.75 nm) is of particular interest because no radiative losses to the $4s\ J = 0$, 1 levels occur; cw laser excitation enables quasistationary population in the $4p\ ^3D_3$ level up to 50% of the initial $4s\ ^3P_2$ population [41]. Therefore, very efficient excitation of Ar** ($ns, nd$) Rydberg levels [126] or photoionization [41, 125] is possible by cw laser two-photon excitation of Ar* ($4s\ ^3P_2$) via the Ar* ($4p\ ^3D_3$) intermediate level. Using an intracavity argon ion laser (458 nm) for the ionizing step (photoionization cross section of Ar* ($4p$)

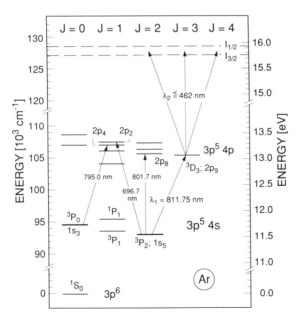

FIG. 4. Energy level diagram of the Ar atom, illustrating the scheme for efficient two-step cw laser photoionization of metastable rare gas atoms Rg* ($ms^3P_2$) via the intermediate Rg* ($mp^3D_3$) level.

$8 \times 10^{-18}$ cm$^2$ [127, 128]), up to 10% of the Ar*$(4s\,^3P_2)$ atoms in a collimated beam were ionized. This efficient cw two-photon ionization scheme was used for absolute calibration of an electron emission detector for metastable Ne*, Ar*, Kr* atoms [41]; see Section 11.3.2.

Excitation to one of the $4p$, $J = 1, 2$ levels results in fluorescence to $4s$ states, which is a measure of the metastable atom density (or flux). Radiative losses to the $4s$, $J = 1$ levels are followed by decay to the ground state with emission of VUV photons. Detection of this VUV fluorescence following selective laser excitation can be used to monitor metastable rare gas atom populations with solar-blind electron multipliers [17, 129], thereby avoiding problems due to laser stray light. Using Doppler-shifted laser excitation spectroscopy with fluorescence detection [118], the velocity distribution of the metastable atoms can be measured. Excitation to $mp$, $J = 1(2)$ also allows efficient state selection of mixed metastable rare gas beams Rg*$(ms\,^3P_2,\,^3P_0)$ [87, 98, 124, 129–133]. We note that single-photon excitation to Rydberg or continuum states is not an optimal choice for sensitive detection of metastable Ne*, Ar*, and Kr* atoms; UV lasers have to be used, and the photoexcitation cross-sections near threshold are low [134].

Laser-induced fluorescence or ionization of metastable molecules has been increasingly studied over the last years [135–145]. A problem with regard to selective and quantitative detection, however, is the fact that closed transitions (such as the Ar*$(4s\,^3P_2{-}4p\,^3D_3)$ transition in Figure 4) do not exist for molecules. Correspondingly, resonant two-photon ionization schemes for molecules normally involve some losses due to fluorescence (or other decay channels such as predissociation) from the intermediate level. As a consequence quantitative detection of metastable molecules with a photoionization depletion method (see Section 11.3.2) requires an efficient single-photon ionization process, carefully chosen to avoid nonionizing paths. UV lasers with sufficient intensity as well as high duty cycle are needed for such a detection scheme. As an alternative to ionization, laser-induced fluorescence using a calibrated photon detector can be used for state-selective depletion of metastable molecules and quantification of the metastable flux. This method can, of course, also be applied to metastable atoms (see [32, 129]).

## 11.3 Absolute Detection of Metastable Atoms

Here we discuss methods for quantitative detection of metastable species. The general procedure consists of converting a known number of metastable atoms into absolutely detectable products (e.g., ion–electron pairs) and observing the corresponding decrease in detector signal. To be somewhat more specific (see Figure 5), let us assume that a metastable atom beam M* produces a signal $I_s$ at a suitable detector (e.g., by electron emission from a surface). On the way to the

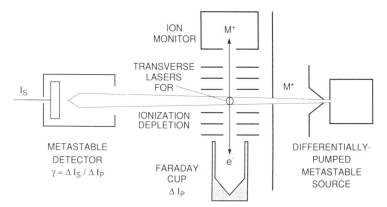

FIG. 5. Semischematic vertical cut through the experimental setup used for the absolute detection of metastable rare gas atoms (M* = Rg*) and for the *in situ* determination of the electron emission coefficient $\gamma$ for normal impact of Rg* atoms on conducting surfaces using a two-step cw laser ionization–depletion method [41].

detector, metastables can be converted to electron–ion pairs (e.g., by laser photoionization [40, 41] or by Penning ionization [40, 43, 45, 46, 85]) resulting in an ion or electron current $\Delta I_p$. The depletion of the metastable flux leads to a reduction $\Delta I_s$ of the detector current. Under the assumption that each depletion process leads to one electron–ion pair and that losses of the metastable flux on the way from the depletion region to the detector can be neglected, the ratio $\Delta I_s/\Delta I_p$ represents the efficiency of the detector for the depleted metastable state. For an electron emission detector, we thereby directly determine the yield $\gamma$ for electron emission: $\gamma = \Delta I_s/\Delta I_p$, as noted in Figure 5. If contributions to the detector current other than those due to the metastable state of interest can be neglected or are known from additional experiments, knowledge of $\gamma$ allows absolute determination of the incident metastable flux.

## 11.3.1 The Gas Cell and the Crossed-Beam Method

Several researchers have developed gas-cell techniques for calibration of surface electron emission coefficients [40, 43, 45, 46, 85] in which the depletion of the metastable beam is due to Penning ionization of a suitable gas and the cell is designed to allow separate measurements of the electron current ejected from a surface and of the product Penning ions. As an alternative to the gas cell, a crossed atomic or molecular beam of low ionization potential (alkali atoms, NO, $C_6H_6$) has also been used for controlled ionization depletion of metastable He*, Ne*, and Ar* beams [45]. Although inherently more precise than the gas-cell technique, the crossed-beam method is subject to error if the metastable atoms are elastically scattered through large angles by the crossed beam particles [45]; the gas-cell method is insensitive to such large-angle scattering [40, 46].

Here we briefly describe the measurement procedure employed with the gas cell [40], shown schematically in Figure 6. The cell consists of a cylinder split longitudinally into two equal sections (4 and 5), which enclose four 0.76 mm grid wires (4a, b; 5a, b) running the length of the cell, a back plate (6), a front plate (3), a monitor grid (2), and a collimating aperture (1). Each component part of the cell is electrically isolated from all others. The cell permits several different, independent procedures to be used to determine the electron emission coefficient $\gamma$ of the back plate surface, one of which we outline here. It involves the measurement of six currents, two for the case of absent target gas, four with the target gas present and Penning ionization occurring (primed quantities).

In the absence of target gas, the current $J$ leaving the monitor grid due to electron emission by the incoming metastable flux $F$ is ($e$ = electron charge)

$$J = ceF, \tag{11.1}$$

where $c$ is a constant. The current $I_-$ leaving the back plate due to electron emission by metastable impact is

$$I_- = \gamma eF. \tag{11.2}$$

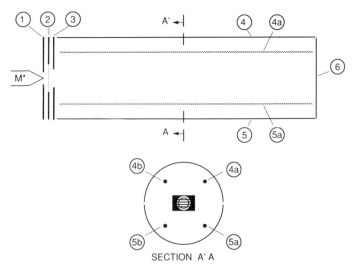

FIG. 6. Schematic drawing of improved gas cell [40] for the determination of electron emission coefficients for metastable rare gas atoms (M* = Rg*) striking a metal surface (6). For details, see text.

If gas is introduced into the cell, the metastable flux entering the cell is reduced to a new value $F'$ as a result of collisions with the gas streaming from the cell, leading to a new monitor grid current $J'$:

$$J' = ceF'. \tag{11.B1}$$

An amount $\Delta F'$ of the metastables entering the cell is deexcited in gas-phase collisions, yielding Penning ions and electrons. If the grid wires are biased negative with respect to the remainder of the cell, a positive current $I'_+$ results because of collection of Penning ions:

$$I'_+ = (1 + \delta) e\Delta F', \tag{11.B2}$$

where $\delta$ is the electron emission coefficient for the ions incident on the grid wires. Geometric considerations indicate the electron emission from the grid wires due to impact of elastically scattered metastables produces a negligible contribution to $I'_+$.

If one half-cylinder is biased negative with respect to all other electrodes, a positive current $R'_+$ will be measured arising from both collection of Penning ions and from electron emission by the fraction $f$ of the elastically scattered metastables striking it:

$$R'_+ = \gamma f e (F' - \Delta F') + e(1 + \delta)\Delta F'. \tag{11.B3}$$

If the back plate, the other half-cylinder, and the front plate are connected and biased negative, a positive current $S'_+$ will be measured, again arising from both the collection of Penning ions and the ejection of electrons:

$$S'_+ = \gamma(1 - f)e(F' - \Delta F') + e(1 + \delta)\Delta F'. \tag{11.B4}$$

Solving for $\gamma$ by eliminating $F$, $F'$, $\Delta F'$, $f$, and $c$, we obtain

$$\gamma = (1 + \delta)[(J'/J)I_- + 2I'_+ - R'_+ - S'_+]/I'_+. \tag{11.AB}$$

The various measured quantities in Eq. (11.AB) have systematic errors of 0.5–1.5%, resulting in errors for $\gamma$ of 1% to 4% [40]. Assuming a value $\delta = 0.02 \pm 0.02$, the RMS sum of all uncertainties in the measured values of $\gamma$ is quoted as $\pm 11\%$ [40].

Reproducible and consistent results for $\gamma$ were obtained with different target gases for $He^*(2^3S)$, $He^*(2^1S)$, and $Ne^*(^3P_{2,0})$ incident on contaminated stainless steel and Cu–Be surfaces with overall uncertainties of about $\pm 16\%$ (see Table III). In contrast, widely different values of $\gamma$ were obtained for $Ar^*(^3P_{2,0})$ and $Kr^*(^3P_{2,0})$ when different molecular target gases were used. The authors attributed these findings to the substantial role of de-excitation processes other than ionization (e.g., reactive scattering or energy transfer followed by dissociation or photon emission) in the interaction of Ar* and Kr* with molecules [40]. As a consequence of metastable destruction through nonionizing processes, too-large values of $\gamma$ result. The gas-cell method therefore is limited to metastable/target gas combinations, for which Penning ionization is known to be the only relevant destruction process.

## 11.3.2 The Laser Photoionization Method

The most direct and dependable method for the absolute detection of metastable species is based on efficient depletion by one- or two-photon ionization. Thus far, this method has been discussed and applied for metastable rare gas atoms [40, 41, 61, 86]. Initial experiments used pulsed dye lasers of low duty cycle in combination with gated detection of the output from two equivalent electron multipliers, which served to measure the electron current due to laser one-photon ionization of He*$(2^1S)$ and Xe*$(^3P_0)$ and the depletion hole in the electron current ejected by the metastables at a contaminated stainless steel surface [40], respectively. Although the results for the electron emission coefficients $\gamma$ had a relatively large uncertainty of $\pm 50\%$ (mainly associated with statistical errors), this work clearly demonstrated the potential of the method. More recently, cw lasers were used for efficient resonant two-photon ionization of metastable Ne*$(^3P_2)$, Ar*$(^3P_2)$, and Kr*$(^3P_2)$ atoms via the intermediate $^3D_3$ level (see Figure 4), permitting accurate determinations of $\gamma$ [41]. The apparatus employed is schematically shown in Figure 5. The $^3P_2$ fraction in a collimated, mixed metastable rare-gas beam Rg*$(^3P_{2,0})$ from a differentially pumped discharge source is transversely excited by a single-mode cw laser to the $^3D_3$ level, from where ionization occurs within the cavity of an argon ion laser (30–60 W at 458/476 nm) [41]. For Ar* and Kr*, the resulting depletions were typically 5–10%, and the photoelectron current $\Delta I_p$, quantitatively extracted and measured with a Faraday cup, amounted to 40–100 pA. The metastable beam was monitored with a rotatable electron emission detector, incorporating several contaminated surfaces, which could be moved into the metastable beam without breaking the vacuum ($10^{-7}$ mbar). Laser ionization depletion results in a decrease $\Delta I_s$ in the detector current, and the electron emission coefficient $\gamma_2$ for Rg*$(^3P_2)$ is directly and simply determined by $\gamma_2 = \Delta I_s / \Delta I_p$. Apart from Rg*$(^3P_2)$ atoms, metastable Rg*$(^3P_0)$ atoms and VUV photons are also present in the beam and contribute to the monitor current $I_s$. These contributions, however, can be reliably determined [41, 98], and the absolute Rg*$(^3P_2)$ flux $F_2$ is then obtained from the $^3P_2$-fraction $I_s(^3P_2)$ of the monitor current and the accurately measured yield $\gamma_2$ by $F_2 = I_s(^3P_2)/e\gamma_2$. Individual values of $\gamma_2$ could be determined with low uncertainties around 1%, enabling the detection of short-time variations of $\gamma$ due to changes of the surface layer associated with temperature variations and reactive chemistry induced by metastable impact [41]. For Ar* and Kr*, the $\gamma$-values were found to vary strongly (by a factor of 30) with the surface material at room temperature ($T = 300$ K), while mild heating to $T = 360$ K led to a much narrower spread in $\gamma$ at levels of 0.14–0.26 for Ar*$(^3P_2)$ and 0.05–0.10 for Kr*$(^3P_2)$. In subsequent work, the role of the excitation energy was investigated by directly comparing the yields $\gamma_2$ for $^3P_2$ and $\gamma_3$ for laser-excited $^3D_3$ in Ar* and Kr* [61]. Transverse laser-excitation of the metastable $^3P_2$ component took place in front of the detector surface. The ratio $\gamma_3/\gamma_2$ was found to be in the range 1–1.5 for $\gamma_2 \gtrsim 0.1$, but rose to values of about 4 for $\gamma_2 < 0.1$ with few exceptions. It was also found that the yields for Ar*$(^3P_2)$ and Kr*$(^3D_3)$ with almost identical excitation energies around 11.5 eV differed by factors of 1.1–2.9 in favor of Ar*$(^3P_2)$, in contradiction to the hypothesis of a $\gamma$-curve which only depends on $E^*$ [84].

From the preceding discussion, it is clear that *in situ* measurement of the yield $\gamma$ is mandatory in quantitative work. The laser ionization depletion method awaits further applications to accurately characterize electron emission from atomically clean surfaces by various metastable species and to provide benchmark Penning ionization cross-sections for suitable target atoms and molecules.

## Acknowledgments

Herewith I acknowledge F. B. Dunning, H. Helm, V. Kempter, T. Kondow, R. M. Martin, N. J. Mason, Ch. Ottinger, M. S. Walhout, G. K. Walters, H. Winter, A. Witte, A. Wucher, and M. Zubek for providing reprints, preprints, or comments in connection with this article. I am grateful to M. S. Walhout for providing a copy of his Ph.D. thesis and for discussions of his experimental results for the lifetimes of metastable Xe* atoms. I thank F. B. Dunning and V. Kempter for their comments on the manuscript, and I gratefully acknowledge my co-authors of references [41], [61], [86], and [128] for their contributions to our work on the absolute detection of metastable rare gas atoms.

## References

1. Breit, G., and Teller, E. (1940). *Astrophys. J.* **91**, 215.
2. Martin, W. C. (1973). *J. Phys. Chem. Ref. Data* **2**, 257.
3. Drake, G. W. F. (1971). *Phys. Rev.* **A3**, 908.
4. Woodworth, J. R., and Moos, H. W. (1975). *Phys. Rev.* **A12**, 2455.
5. Drake, G. W. F., Victor, G. A., and Dalgarno, A. (1969). *Phys. Rev.* **180**, 25.
6. Jacobs, V. (1971). *Phys. Rev.* **A4**, 939.
7. van Dyck, R. S., Jr., Johnson, C. E., and Shugart, H. A. (1971). *Phys. Rev.* **A4**, 1327.
8. Johnson, C. E. (1972). *Phys. Rev.* **A5**, 2688.
9. Mason, N. J. (1990). *Meas. Sci. Technol.* **1**, 596.
10. Kaufman, V., and Minnhagen, L. (1972). *J. Opt. Soc. Am.* **62**, 92.
11. Small-Warren, N. E., and Chow Chiu, L.-Y. (1975). *Phys. Rev.* **A11**, 1777.
12. van Dyck, R. S., Jr., Johnson, C. E., and Shugart, H. A. (1972). *Phys. Rev.* **A5**, 991.
13. Minnhagen, L. (1973). *J. Opt. Soc. Am.* **63**, 1185.
14. Katori, H., and Shimizu, F. (1993). *Phys. Rev. Lett.* **70**, 3545.
15. Sugar, J., and Musgrove, A. (1991). *J. Phys. Chem. Ref. Data* **20**, 859.

16. Yoshino, K., and Freeman, D. E. (1985). *J. Opt. Soc. Am.* **B2**, 1268.
17a. Walhout, M., Witte, A., and Rolston, S. L. (1994). *Phys. Rev. Lett.* **72**, 2843.
17b. Walhout, M., Sterr, U., Witte, A., and Rolston, S. L. (1995). *Opt. Lett.* **20**, 1192.
18. Garstang, R. H. (1962). *J. Opt. Soc. Am.* **52**, 845; (1967). *Astrophys. J.* **148**, 579.
19. McDermitt, M. N., and Lichten, W. L. (1960). *Phys. Rev.* **119**, 134.
20. Learner, R. C. M., and Morris, J. (1971). *J. Phys.* **B4**, 1236.
21. Johnson, C. E. (1972). *Phys. Rev.* **A5**, 1026.
22. Lofthus, A., and Krupenie, P. H. (1977). *J. Phys. Chem. Ref. Data* **6**, 113.
23. Borst, W. L., and Zipf, E. C. (1971). *Phys. Rev.* **A3**, 979.
24. Johnson, C. E., and van Dyck, R. S., Jr. (1972). *J. Chem. Phys.* **56**, 1506.
25. James, T. C. (1971). *J. Chem. Phys.* **55**, 4118.
26. Cohen, E. R., and Taylor, B. N. (1988). *J. Phys. Chem. Ref. Data* **17**, 1795.
27. Moore, Ch. E. (1971). "Atomic Energy Levels," *Nat. Stand. Ref. Data Ser., Nat. Bur. Stand. (USA)* **35**.
28. Huber, K. P., and Herzberg, G. (1979). *Constants of Diatomic Molecules.* Vol. IV of *Molecular Spectra and Molecular Structure.* Van Nostrand Reinhold Company, New York.
29. Stebbings, R. F., and Dunning, F. B. (1983). *Rydberg States of Atoms and Molecules.* Cambridge Univ. Press, New York.
30. Feldman, P., and Novick, R. (1967). *Phys. Rev.* **160**, 143.
31. Fabrikant, I. I., Shpenik, O. B., Snegursky, A. V., and Zavilopulo, A. N. (1988). *Phys. Rep.* **159**, 1.
32. Lin, C. C., and Anderson, L. W. (1992). *Adv. At. Mol. Opt. Phys.* **29**, 1.
33. Trajmar, S., and Nickel, J. C. (1993). *Adv. At. Mol. Opt. Phys.* **30**, 45.
34. Bühl, A. (1933). *Helv. Phys. Acta* **6**, 231.
35. Varney, H. N. (1967). *Phys. Rev.* **157**, 116; (1968). *Phys. Rev.* **175**, 98.
36. Roussel, J., and Boiziau, C. (1977). *J. de Phys.* **38**, 757.
37. Lamb, W. E., Jr. (1951). *Rep. Progr. Phys.* **XIV**, 19.
38. Petrasso, R., and Ramsey, A. T. (1972). *Phys. Rev.* **A5**, 79.
39. Webb, H. W. (1924). *Phys. Rev.* **24**, 113.
40. Dunning, F. B., Rundel, R. D., and Stebbings, R. F. (1975). *Rev. Sci. Instrum.* **46**, 697.
41. Schohl, S., Klar, D., Kraft, T., Meijer, H. A. J., Ruf, M.-W., Schmitz, U., Smith, S. J., and Hotop, H. (1991). *Z. Phys.* **D21**, 25.
42. Hagstrum, H. D. (1954). *Phys. Rev.* **96**, 325; (1956). *Phys. Rev.* **104**, 672; (1956). *Phys. Rev.* **104**, 1516.
43. MacLennan, D. A. (1966). *Phys. Rev.* **148**, 218.
44. Delchar, T. A., MacLennan, D. A., and Landers, A. M. (1969). *J. Chem. Phys.* **50**, 1779.
45. Dunning, F. B., Smith, A. C. H., and Stebbings, R. F. (1971). *J. Phys.* **B4**, 1683.
46. Dunning, F. B., and Smith, A. C. H. (1971). *J. Phys.* **B4**, 1696.
47. Shibata, T., Hirooka, T., and Kuchitsu, K. (1975). *Chem. Phys. Lett.* **30**, 241.
48. Boiziau, C., Dose, V., and Roussel, J. (1976). *Surf. Sci.* **61**, 412.
49. Johnson, P. D., and Delchar, T. A. (1978). *Surf. Sci.* **77**, 400.
50. Conrad, H., Ertl, G., Küppers, J., Wang, S. W., Gérard, K., and Haberland, H. (1979). *Phys. Rev. Lett.* **42**, 1082.
51. Hagstrum, H. D. (1979). *Phys. Rev. Lett.* **43**, 1050.
52. Munakata, T., Ohno, K., and Harada, Y. (1980). *J. Chem. Phys.* **72**, 2880.
53. Sesselmann, W., Conrad, H., Ertl, G., Küppers, J., Woratschek, B., and Haberland, H. (1983). *Phys. Rev. Lett.* **50**, 446.

54. Bozso, F., Yates, J. T., Jr., Arias, J., Metiu, H., and Martin, R. M. (1983). *J. Chem. Phys.* **78**, 4256.
55. Hart, M. W., Hammond, M. S., Dunning, F. B., and Walters, G. K. (1989). *Phys. Rev.* **B39**, 5488.
56. Maus-Friedrichs, W., Dieckhoff, S., Wehrhahn, M., and Kempter, V. (1991). *Surf. Sci.* **253**, 137; Brenten, H., Müller, H., and Kempter, V. (1992). *Z. Phys.* **D22**, 563.
57. Keller, W., Morgner, H., and Müller, W. A. (1986). *Mol. Phys.* **58**, 1039.
58. Freund, R. S. (1971). *J. Chem. Phys.* **55**, 3569.
59. Lee, W., and Martin, R. M. (1976). *J. Chem. Phys.* **64**, 678.
60. Zubek, M., and King, G. C. (1982). *J. Phys.* **E15**, 511.
61. Schohl, S., Meijer, H. A. J., Ruf, M.-W., and Hotop, H. (1992). *Meas. Sci. Technol.* **3**, 544.
62a. Hagstrum, H. D. (1954). *Phys. Rev.* **96**, 336.
62b. Hagstrum, H. D. (1978). In *Electron and Ion Spectroscopy of Solids*, L. Fiermans, J. Vennik, and W. Dekeyser (eds.), pp. 273–323. Plenum, New York.
63. Sesselmann, W., Woratschek, B., Küppers, J., Ertl, G., and Haberland, H. (1987). *Phys. Rev.* **B35**, 1547.
64. Hemmen, R., and Conrad, H. (1991). *Phys. Rev. Lett.* **67**, 1314.
65. Oró, D. M., Lin, Q., Soletzky, P. A., Zhang, X., Dunning, F. B., and Walters, G. K. (1992). *Phys. Rev.* **B46**, 9893.
66. Lee, J., Hanrahan, C., Arias, J., Bozso, F., Martin, R. M., and Metiu, H. (1985). *Phys. Rev. Lett.* **54**, 1440.
67. Woratschek, B., Sesselmann, W., Küppers, J., Ertl, G., and Haberland, H. (1987). *Surf. Sci.* **180**, 187.
68. Sesselmann, W., Woratschek, B., Ertl, G., Küppers, J., and Haberland, H. (1984). *Surf. Sci.* **146**, 17.
69. Harada, Y. (1985). *Surf. Sci.* **158**, 455.
70. Suzuki, T., Suzuki, K., Kondow, T., and Kuchitsu, K. (1988). *J. Phys. Chem.* **92**, 3953.
71. Oró, D. M., Soletzky, P. A., Zhang, X., Dunning, F. B., and Walters, G. K. (1994). *Phys. Rev.* **A49**, 4703.
72. Zubek, M. (1986). *J. Phys.* **E19**, 463; *Proc. 13th Symp. on Physics of Ionized Gases, Sibenik 1986. Abstracts of Contributed Papers*, M. V. Kurepa (ed.), p. 23, Univ. Beograd, Yugoslavia.
73. Zubek, M., and King, G. C. (1987). *J. Phys.* **B20**, 1135.
74. Mason, N. J., and Newell, W. R. (1991). *Meas. Sci. Technol.* **2**, 568.
75a. Varga, P., and Winter, H. (1978). *Phys. Rev.* **A18**, 2453.
75b. Lakits, G., Aumayr, F., Heim, M., and Winter, H. (1990). *Phys. Rev.* **A42**, 5780; Heim, M. (1990). Diplomarbeit, Institut für Allgemeine Physik, TU Wien.
76. Kishinevsky, L. M. (1973). *Radiat. Eff.* **19**, 23; Baragiola, R. A., Alonso, E. V., Ferron, J., and Oliva-Florio, A. (1979). *Surf. Sci.* **90**, 240.
77. Müller, H., Hausmann, R., Brenten, H., Niehaus, A., and Kempter, V. (1993). *Z. Phys.* **D28**, 109.
78. Allison, W., Dunning, F. B., and Smith, A. C. H. (1972). *J. Phys.* **B5**, 1175.
79. Golde, M. F. (1977). In *Gas Kinetics and Energy Transfer*, Vol. 2, P. G. Ashmore and R. J. Donovan (eds.), p. 123f. The Chemical Society, London.
80. Velazco, J. E., Kolts, J. H., and Setser, D. W. (1978). *J. Chem. Phys.* **69**, 4357.
81. Golde, M. F., Ho, Y.-S., and Ogura, H. (1982). *J. Chem. Phys.* **76**, 3535.
82. Jones, M. T., Dreiling, T. D., Setser, D. W., and McDonald, R. N. (1985). *J. Phys. Chem.* **89**, 4501.

83. Chen, X., and Setser, D. W. (1991). *J. Phys. Chem.* **95**, 8473.
84. Borst, W. L. (1971). *Rev. Sci. Instrum.* **42**, 1543.
85. Woodard, M. R., Sharp, R. C., Seely, M., and Muschlitz, E. E., Jr. (1978). *J. Chem. Phys.* **69**, 2978.
86. Kau, R., Leber, E., Reicherts, M., Schramm, A., Ruf, M.-W., and Hotop, H. (Unpublished results).
87. Siska, P. E. (1993). *Rev. Mod. Phys.* **65**, 337.
88. Brutschy, B., and Haberland, H. (1979). *Phys. Rev.* **A19**, 2232.
89. Borst, W. L., Nowak, G., and Fricke, J. (1978). *Phys. Rev.* **A17**, 838.
90. Alvarino, J. M., Hepp, C., Kreiensen, M., Staudenmayer, B., Vecchiocattivi, F., and Kempter, V. (1984). *J. Chem. Phys.* **80**, 765.
91. Borst, W. L. (1969). *Phys. Rev.* **181**, 257.
92. Freund, R. S., and Klemperer, W. (1967). *J. Chem. Phys.* **47**, 2897.
93. Price, J. M., Ludviksson, A., Nooney, M., Xu, M., Martin, R. M., and Wodtke, A. M. (1992). *J. Chem. Phys.* **96**, 1854.
94. Davidson, T. A., Fluendy, M. A., and Lawley, K. P. (1973). *Far. Disc. Chem. Soc.* **55**, 158.
95. Smyth, K. C., Schiavone, J. A., and Freund, R. S. (1974). *J. Chem. Phys.* **61**, 1789.
96. Hemminger, J. C., Wicke, B. G., and Klemperer, W. (1976). *J. Chem. Phys.* **65**, 2798.
97. Koch, L., Heindorff, T., and Reichert, E. (1984). *Z. Phys.* **A316**, 127.
98. Weissmann, G., Ganz, J., Siegel, A., Waibel, H., and Hotop, H. (1984). *Opt. Commun.* **49**, 335.
99. Yencha, A. J. (1984). In *Electron Spectroscopy: Theory, Techniques and Applications*, C. R. Brundle and A. D. Baker (eds.), Vol. 5, p. 197. Academic Press, London.
100. Cermák, V. (1966). *J. Chem. Phys.* **44**, 1318.
101. Hotop, H. (1967). Diplomarbeit, Universität Freiburg.
102. Hotop, H., and Niehaus, A. (1968). *Z. Phys.* **215**, 395.
103. Cermák, V. (1966). *J. Chem. Phys.* **44**, 3774; (1966). *J. Chem. Phys.* **44**, 3781.
104. Krause, H. F., Johnson, S. G., and Datz, S. (1977). *Phys. Rev.* **A15**, 611.
105. Dehnbostel, C., Feltgen, R., and Hoffmann, G. (1990). *Phys. Rev.* **A42**, 5389.
106. Neuschäfer, D., Ottinger, Ch., and Sharma, A. (1987). *Chem. Phys.* **117**, 133.
107. Piper, L. G., Setser, D. W., and Clyne, M. A. A. (1975). *J. Chem. Phys.* **63**, 5018.
108. Derouard, J., Nguyen, T. D., and Sadeghi, N. (1980). *J. Chem. Phys.* **72**, 6698.
109. Krümpelmann, T., and Ottinger, Ch. (1987). *Chem. Phys. Lett.* **140**, 142; (1988). *J. Chem. Phys.* **88**, 5245.
110. Schearer, L. D. (1969). *Phys. Rev. Lett.* **22**, 629.
111. Richardson, W. C., and Setser, D. W. (1973). *J. Chem. Phys.* **58**, 1809.
112. Obase, H., Tsuji, M., and Nishimura, Y. (1987). *J. Chem. Phys.* **87**, 2695.
113. Kume, H., Kondow, T., and Kuchitsu, K. (1986). *J. Chem. Phys.* **84**, 4031.
114. Kume, H., Kondow, T., and Kuchitsu, K. (1986). *J. Phys. Chem.* **90**, 5146.
115. Mityuera, A. A., and Penkin, N. P. (1975). *Opt. Spectrosc.* **38**, 229.
116. Mityureva, A. A., and Smirnov, V. V. (1994). *J. Phys.* **B27**, 1869.
117. Hurst, G. S., Payne, M. G., Kramer, S. D., and Young, J. P. (1979). *Rev. Mod. Phys.* **51**, 767.
118. Kinsey, J. L. (1977). *Ann. Rev. Phys. Chem.* **28**, 349.
119. Demtröder, W. (1982). *Laser Spectroscopy*, Springer Series in Chemical Physics, Vol. 5. Springer-Verlag, Berlin.

120. Letokhov, V. S. (1987). *Laser Photoionization Spectroscopy*. Academic Press, New York.

121. Hurst, G. S., and Payne, M. G. (1988). *Principles and Applications of Resonance Ionization Spectroscopy*, Adam Hilger, Bristol and Philadelphia.

122. Scoles, G. (ed.), *Atomic and Molecular Beam Methods*, Vol. I (1988) and Vol. II (1990). Oxford University Press, New York and Oxford.

123. Shore, B. W. (1990). *The Theory of Coherent Atomic Excitation*, Vols. 1 and 2. Wiley and Sons, New York.

124. Chang, R. S., and Setser, D. W. (1978). *J. Chem. Phys.* **69**, 3885.

125. Hotop, H., Klar, D., and Schohl, S. (1992). *Resonance Ionization Spectroscopy 1992*, Proc. 6th Int. Symp. on Resonance Ionization Spectroscopy and its Applications, Santa Fe, New Mexico, USA, 24–29 May 1992, C. M. Miller and J. E. Parks (eds.), pp. 45–52. *Institute of Physics Conference Series No. 128*. IOP Publ., Bristol and Philadelphia.

126. Kraft, T., Ruf, M.-W., and Hotop, H. (1989). *Z. Phys.* **D14**, 179.

127. Chang, T. N., and Kim, Y. S. (1982). *Phys. Rev.* **A26**, 2728.

128. Kau, R., Leber, E., Schramm, A., Klar, D., Baier, S., and Hotop, H. (1995). In *Resonance Ionization Spectroscopy 1994*, Proc. 7th Int. Symp. on Resonance Ionization Spectroscopy and its Applications, 3–8 July, 1994, Bernkastel-Kues, Germany, pp. 146–149. AIP Press, New York.

129. Verheijen, M. J., and Beijerinck, H. C. W. (1986). *Chem. Phys.* **102**, 255.

130. Dunning, F. B., Cook, T. B., West, W. P., and Stebbings, R. F. (1975). *Rev. Sci. Instrum.* **46**, 1072.

131. Gaily, T. D., Coggiola, M. J., Peterson, J. R., and Gillen, K. T. (1980). *Rev. Sci. Instrum.* **51**, 1168.

132. Hotop, H., Lorenzen, J., and Zastrow, A. (1981). *J. Electron Spectrosc. Relat. Phenom.* **23**, 347.

133. Böhle, W., Geisen, H., Krümpelmann, T., and Ottinger, Ch. (1989). *Chem. Phys.* **133**, 313.

134. Stebbings, R. F., Dunning, F. B., and Rundel, R. D. (1975). In *Atomic Physics 4*, G. zu Putlitz, E. W. Weber, A. Winnacker (eds.), pp. 713–730. Plenum Press, New York.

135. Lichten, W., Wik, T., and Miller, T. A. (1979). *J. Chem. Phys.* **71**, 2441.

136. Eyler, E. E., and Pipkin, F. M. (1982). *J. Chem. Phys.* **77**, 5315.

137. Knight, R. D., and Wang, L.-G. (1985). *Phys. Rev. Lett.* **55**, 1571.

138. Kachru, R., and Helm, H. (1985). *Phys. Rev. Lett.* **55**, 1575.

139. Sharpe, S. W., and Johnson, P. M. (1986). *J. Chem. Phys.* **85**, 4943.

140. Geisen, H., Neuschäfer, D., and Ottinger, Ch. (1987). *Z. Phys.* **D4**, 263.

141. Helm, H., and Cosby, P. C. (1988). *Phys. Rev.* **A38**, 115.

142. Lindsay, M. D., Kam, A. W., Lawall, J. R., Zhao, P., Pipkin, F. M., and Eyler, E. E. (1990). *Phys. Rev.* **A41**, 4974.

143. Kam, A. W., and Pipkin, F. M. (1991). *Phys. Rev.* **A43**, 3279.

144. Ottinger, Ch., and Rox, T. (1991). *Phys. Lett.* **A161**, 135.

145. Ottinger, Ch., and Vilesov, A. F. (1993). *Chem. Phys. Lett.* **211**, 175.

# 12. EXCITED LEVEL LIFETIME MEASUREMENTS

## T. R. O'Brian

Radiometric Physics Division
National Institute of Standards and Technology
Gaithersburg, Maryland

## J. E. Lawler

Department of Physics
University of Wisconsin–Madison
Madison, Wisconsin

## 12.1 Introduction

### 12.1.1 Applications for Radiative Transition Probabilities

The purpose of studying excited level lifetimes is to determine the absolute radiative transition probabilities (or oscillator strengths) needed for quantitative spectroscopy and tests of fundamental quantum mechanical principles. The transition probability translates the intensity of an atomic or molecular absorption or emission line into the population of a particular species in the initial level of the transition. From the relative abundances of the various elements, compounds, and ionization stages, much information about the chemistry and dynamics of the source can be inferred. Observation of elemental abundances and effective temperatures in stellar chromospheres and photospheres—obtained from spectra analyzed using known transition probabilities—provide the only probe of nucleosynthetic and other processes within the stars [1, 2]. Elemental abundances in the stars and the interstellar medium also reveal the evolutionary record of the galaxy [3, 4]. Laboratory plasmas can be investigated using various methods of classical and laser-based spectroscopy that rely on converting measured emission or absorption features into level populations using transition probabilities [5, 6]. Spectroscopic study of plasmas used for fusion research, chemical vapor deposition, semiconductor etching, etc., is relatively nonperturbing to the plasma, is highly species-specific, and provides potentially high spatial and temporal resolution of chemical and physical processes [6]. Transition probabilities also provide basic information about atomic and molecular wavefunctions. While the energy levels of an atom or molecule—known to great precision from the frequency of emitted or absorbed radiation—are principally determined by the part of the wavefuntion closest to the nucleus, the transition

217

EXPERIMENTAL METHODS IN THE PHYSICAL SCIENCES
Vol. 29B

probabilities are more sensitive to the large $r$ behavior and are thus potentially more sensitive probes of the details of the wavefunctions [7]. Although transition probabilities can be accurately calculated for simple systems (one- or two-electron atoms, Rydberg levels, etc.), calculations for multielectron systems generally cannot compete in accuracy with the best modern measurements [8].

## 12.1.2 Radiative Transition Probabilities

An atom or molecule in an excited level $u$, isolated from significant perturbations due to external electromagnetic fields, radiation, or collisions, has a constant probability per unit time $A_{ul}$ of making a spontaneous transition to a lower level $l$ through the emission of electric dipole ($E1$) radiation:

$$A_{ul} = \frac{16\pi^3 v^3}{3g_u \varepsilon_0 hc^3} \sum_{u,l} |\langle \psi_u | e\vec{r} | \psi_l \rangle|^2, \tag{12.1}$$

where $g_u$ is the upper-level degeneracy, $v$ the transition frequency, $e\vec{r}$ the electric dipole operator, and $\psi_u$ ($\psi_l$) the upper (lower) state wave function. The summation is over the magnetic sublevels of the upper and lower levels, and the other constants have their usual meanings. Similar expressions describe the transition probabilities for higher-order, and much weaker, radiative processes (magnetic dipole, electric quadrupole, etc.), and the absorption and stimulated emission cross sections can be determined from $A_{ul}$ [9]. The use of oscillator strengths or $f$ values is often preferred (particularly in astrophysical applications), where the emission (absorption) oscillator strength $f_{ul}$ ($f_{lu}$) is given by

$$g_l f_{lu} = g_u \frac{\varepsilon_0 mc^3}{2\pi e^2 v^2} A_{ul} = -g_u f_{ul} \tag{12.2}$$

for upper (lower) level degeneracy $g_u$ ($g_l$). In general, a particular upper level may be coupled to many unoccupied lower levels. The mean radiative lifetime of the upper level, $\tau_u$, is simply the inverse of the total probability per unit time that the excited level will radiatively decay to any one of the lower levels:

$$\tau_u^{-1} = A_u = \sum_l A_{ul}. \tag{12.3}$$

To determine transition probability—the specific probability that the excited level will decay to a particular lower level—the relative probability of decay through each channel must also be determined. This relative probability is the branching fraction $\phi_{ul}$ for a particular transition:

$$\phi_{ul} = A_{ul} / \sum_l A_{ul} = \tau_u A_{ul}. \tag{12.4}$$

Direct measurement of $A_{ul}$ through emission spectroscopy requires knowledge of the excited-level population and the absolute efficiency of detection of emitted radiation, and each quantity is generally quite difficult to accurately measure. In practice it is usually easier and more accurate to determine $A_{ul}$ indirectly by separately measuring the excited-level radiative lifetime and the branching fraction for each of the appreciable decay channels out of the level.

Radiative lifetimes are now routinely measured with ~5% total uncertainty, and better than 1% uncertainty using special techniques. Emission branching fractions are typically determined to ~5% total uncertainty for all but the weakest decay channels. It is thus common to measure absolute transition probabilities to ≤10% uncertainty. While such precision seems modest compared to other spectroscopic parameters such as transition frequencies (routinely measured to better than part-per-million uncertainty), it is sufficient for many current demands [10]. Tests of fundamental principles generally require lifetimes or transition probabilities with ≤1% uncertainty, pushing the limits of existing methods [11]. Future applications will undoubtedly require transition probabilities of increasing accuracy and precision. The major problem in astrophysical applications is not the lack of precision in the existing data—most measurements over the past few years meet the ~10% precision criterion—but the relative sparsity of measured transition probabilities, particularly in the vacuum ultraviolet and for species in higher ionization stages [3, 12].

## 12.2 Measurement of Radiative Lifetimes

The most widely used and most accurate method for determining lifetimes is direct observation of the spontaneously emitted radiation intensity from a sample of atoms excited to a particular level. The population $N_u(t)$ in a particular excited level decays with the constant probability per unit time $A_u = \Sigma A_{ul}$ (assuming no external perturbations), so

$$\frac{dN_u(t)}{dt} = -A_u N_u(t), \tag{12.5}$$

with the simple solution

$$N_u(t) = N_u(0)\exp(-A_u t) = N_u(0)\exp(-t/\tau_u). \tag{12.6}$$

The exponential decay of the excited level population is mirrored in the detected intensity of the transition $I(t)$ (photons/second),

$$I(t) = I(0)\exp(-t/\tau), \tag{12.7}$$

with radiative lifetime $\tau$ and $I(0)$ the intensity at the (arbitrary) start of the observation period. (For convenience $\tau$ will be understood to represent the upper level lifetime unless otherwise stated.) It is not necessary to know the absolute number of excited atoms at any time, nor related parameters such as the temperature or population distribution, nor the efficiency of detection of the radiation, so long as the number of atoms under observation and the detection efficiency remain constant during the observation period.

Typical effective observation periods in direct lifetime measurements are on the order of two to five lifetimes, roughly corresponding to the reduction in excited level population to 0.1 to 0.01 of the initial value. Shorter observation periods may have insufficient dynamic range for accurate lifetime determination and may not reveal multiexponential decays due to blends or cascades; observation beyond about five lifetimes usually is impractical because of low signal-to-noise ratios.

If the excited atoms do not move appreciably during the observation period, the intensity $I(t)$ of the spontaneously emitted radiation can simply be recorded as a function of time elapsed since the end of the excitation process, $I(t) = I(0) \exp(-t/\tau)$. Time-resolved laser-induced fluorescence on a slow atomic beam is an example of this approach. Alternatively, if the atoms are moving with a nearly constant, common velocity $V$, as in a well-collimated monoenergetic beam, the excitation process can be limited to a particular region in the trajectory of the atoms. After the atoms pass through the excitation region, the intensity of spontaneously emitted radiation can be monitored as a function of the downstream position $z$ of the moving atoms, $I(z) = I(0) \exp(-z/V\tau)$. For constant-velocity atoms, the downstream position is of course an indirect measure of the time elapsed since the excitation. Beam foil or fast beam laser lifetime measurements typify this method.

## 12.3 Time-Resolved Laser-Induced Fluorescence Lifetime Measurements

A commonly used method for measuring excited level lifetimes with good accuracy and precision (typically ~5% total uncertainty) is time-resolved laser induced fluorescence (LIF) on a sample of atoms in a slow beam or in a cell. In this approach, neutral or singly ionized atoms of almost any element are generated through sputtering in an electrical discharge source. A tunable short-pulse laser selectively populates a particular upper atomic level. The spontaneously emitted radiation is detected as a function of time elapsed since the excitation pulse. The discharge beam sources can be operated to produce optically thin, essentially collisionless atomic beams with significant populations in relatively high-lying metastable levels. Lifetime measurements can be made in the beam environment without significant systematic error due to collisional quenching of excited levels or to radiation trapping. The metastable levels provide a platform for excitation of many high-lying levels without the need to use very short laser wavelengths that are more difficult to generate. This general method of time-resolved LIF on discharge-generated species is most useful for neutral and singly-ionized atoms and for lifetimes shorter than about 10 μs.

Many of the concepts and techniques used for time-resolved LIF measurements on atoms from discharge sources are common to other methods of lifetime measurement, and this section is correspondingly more detailed.

## 12.3.1 Hollow Cathode Discharge and Sputtering Cell Sources

The hollow cathode discharge (HCD) source can be used to produce slow beams of neutral and singly ionized atoms of almost any element [13]. In the typical HCD beam source, a cylindrical cathode tube, sealed at one end except for a small hole, is constructed of or lined with the element to be studied. The hollow cathode is coupled to a glass and metal anode assembly filled with buffer gas (typically a noble gas). A DC electrical discharge is generated between the cathode and an anode. Ionized buffer gas atoms sputter the cathode material, and sputtered atoms are excited and ionized through collisions with electrons and buffer gas atoms and ions in the discharge. The small hole connects the HCD source with an evacuated chamber. The buffer gas flows through the hole, sweeping along the sputtered atoms and ions. The exit speed of the buffer gas and sputtered atoms depends on the pressure gradient and orifice size, but HCD beam sources typically operate in a near-effusive manner [14]. As the atoms leave the discharge, excited-level populations spontaneously radiatively decay, collapsing into the ground and metastable levels within millimeters of exiting the orifice.

Detailed descriptions of typical HCD sources can be found in references [13] and [14]. The authors have used a simple HCD source to generate atomic and ionic beams in more than 30 elements so far, including highly refractory metals such as tungsten [15] and titanium [16], and nonmetals such as silicon [17], boron [18], and carbon. Argon or neon are typically used as buffer gases at source pressures of 0.1–1 torr. Discharge currents of 10 mA–10 A at about 1 kV anode potential are typical. Under these conditions the sputtered atom intensity in the beam is on the order of $10^{14}$ atoms/(s sr). The beam contains rich populations of atoms and ions in the ground and metastable levels. The total atomic and ionic intensity, and the populations of the higher-lying metastable levels, can be increased by pulsing the discharge at much higher currents for a few microseconds—the LIF signal can be optimized by controlling the timing of the laser and HCD current pulses. In addition to the sputtered species, the HCD source also can be used to study transitions in the buffer gas atoms and ions, again taking advantage of the significant percentage of metastable populations in the beam [19].

An alternative to the HCD beam source is the sputtering cell source [20, 21]. The cell is sealed with a static fill of buffer gas, typically on the order of 0.1 to 5 torr. The target material forms the cathode, as a wire a few centimeters long or as a bulk pellet supported by small electrodes. Pulsed discharge currents of 1–100 mA at ~10 Hz and a few milliseconds duration are typically used. As in the beam LIF experiments, the timing between the discharge current pulse and the laser excitation pulse is controlled to optimize the LIF signal—firing the laser several milliseconds after the current pulse ends reduces the strong background from the discharge. The sputtering cell generates higher densities of free atoms and metastables than the HCD beam source, typically $\sim 10^{11}\,cm^{-3}$, and thus potentially greater LIF signals. However, the high sputtered-atom and buffer-gas densities significantly increase the chance of systematic error due to radiation trapping and/or collisional quenching of excited levels. One particular potential advantage of the cell is for the measurement of lifetimes in radioactive or highly toxic species such as Tc or Cd. The sputtered material is contained in the cell and can be safely disposed of without contaminating an entire vacuum system.

Figure 1 shows a typical HCD atomic beam source and laser system used for making LIF lifetime measurements.

## 12.3.2 Laser Systems for LIF Lifetime Measurements

Some of the principal considerations in choosing a laser system are spectral tunability and spectral bandwidth, laser power, pulse duration, repetition rate, shot-to-shot temporal and power reproducibility of the laser pulses, and general ease and cost of using the laser system.

Most radiative lifetimes (for non-Rydberg levels) involving electric dipole transitions in neutral and singly ionized atoms are on the order of 1 ns to 10 µs, and most upper levels are accessible (when used with a metastable source) with excitation by radiation in the range $\sim 100$–$1000$ nm. Ideally, the laser pulse duration should be shorter than the lifetime to be measured. For an electric dipole allowed transition, and typical excitation linewidths, the typical saturation photon flux density is on the order of $10^{18}$ to $10^{24}$ photons/(sec cm$^2$), or about 1 W/cm$^2$ to 1 MW/cm$^2$ in the visible range. In a slow beam or cell experiment, LIF data with a reasonable signal-to-noise ratio can be recorded well below saturation, so tunable laser intensities on the order of 10–100 kW/cm$^2$ are usually sufficient for all but the weakest transitions. Typical spectral linewidths in beams from HCD sources are on the order of 0.1 cm$^{-1}$—efficient excitation and usually sufficient resolution result from comparable laser spectral bandwidth. These parameters set the requirements for the laser system.

Most of these requirements are met with pulsed dye laser systems readily available commercially or easily constructed [22]. Available dyes provide continuous spectral coverage in the range from about 350 nm to 900 nm with reasonable efficiency of conversion of the pump laser to narrow band tunable radiation—typically 5–10% conversion efficiency near the extremes of the tuning range, and 25% or better conversion for some dye–pump laser combinations in the middle of the tuning range. Pulse durations of $\sim 3$–$5$ ns are

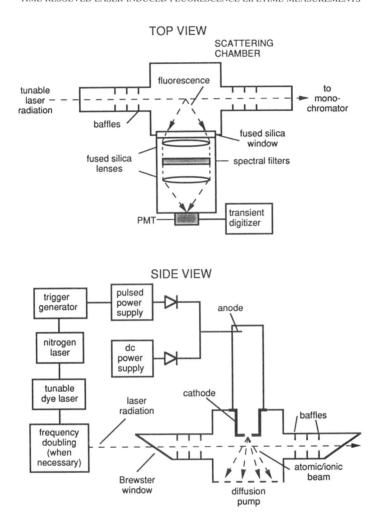

FIG. 1. Schematic diagram of the tunable laser system, hollow cathode discharge beam source, and LIF detection system used by J. E. Lawler's group for LIF lifetime measurements on neutral and singly ionized atoms. See, for example, references [8, 13–18].

common, partly determined by the pump laser pulse duration. Spectral band-widths are typically 0.05 to 0.5 cm$^{-1}$, well-suited for most LIF measurements.

Common pump lasers include a frequency doubled or tripled Q-switched Nd : YAG laser (at 532 nm or 355 nm respectively), an excimer laser operating on XeCl (308 nm), or a molecular nitrogen laser (337 nm). Use of UV fixed-frequency excitation provides the broadest range of dye fundamental wavelengths at reasonable efficiency, since most laser dyes have significant UV absorption bands [23]. The three types of pump laser just listed typically have pulse durations of ~5–10 ns (FWHM) at repetition rates of ~10–200 Hz.

Subnanosecond pulses can be produced by a mode-locked laser, typically producing trains of pulses on the order of ≤100 ps long at ~100 MHz repetition rates [24]. Additional pulse compression techniques can be used to produce subpicosecond dye laser pulses. Ultrafast laser pulses are commonly used in molecular studies where vibrational relaxation times are often much shorter than atomic radiative lifetimes [25].

### 12.3.3 Detection and Recording of LIF Decay Curves

There are two principal methods to record time-resolved LIF decay curves. In the transient recording approach, the fluorescence decay as a function of time is directly measured with a PMT and fast transient recorder. The PMT produces an analog signal proportional to the number of photons detected as a function of time, so finite electronic bandwidth and nonlinearities in the detector and recorder limit the fidelity of short lifetime measurements. However, this type of system is simple and can have a very high effective data collection rate, and is useful for measuring lifetimes longer than about 2 ns with a few percent uncertainty [26]. In the second technique, time-correlated single-photon counting (TC-SPC), a PMT is used only to start and stop an intrinsically fast and linear timer, substantially reducing the electronic bandwidth and linearity requirements on the PMT [27]. This technique is inherently more accurate than transient recording and subpercent lifetime uncertainties are possible. Because of the larger effective electronic bandwidth, accurate measurement of subnanosecond lifetimes is also possible. However, the data collection rate is several orders of magnitude slower than the transient recording technique, limiting the practicality of TC-SPC unless a high repetition-rate laser is used. TC-SPC is most useful when high precision or measurement of very short lifetimes is required.

Transient Detection.   The best PMTs for transient detection LIF lifetime measurements have broad spectral and electronic bandwidths and good linearity of output current over a wide range of incident photon flux. The spectral bandwidth is determined by the photocathode material work function and the window or envelope material. The effective electronic bandwidth is determined by the PMT configuration and the design of the dynode biasing and output coupling base [28]. The side-on squirrel cage type PMT commonly used in LIF studies is rugged and can be constructed for spectral response from ~120 nm to ~1000 nm. In a properly designed base, impulse response widths of ~2 ns with minimal ringing and good linearity over several decades of output current are possible [29].

Commonly used transient recorders are transient digitizers and boxcar averagers. A digitizer collects the entire LIF decay curve for each laser shot; the boxcar slowly scans a short sampling gate across a delay range initiated by a periodic event (typically the firing of the laser). A transient digitizer has the potential for much higher effective data collection rates, typically requiring about one-tenth the time of the boxcar to attain the same effective signal-to-noise ratio and number of detected photons. The boxcar, requiring periods on the order of a minute to record a single decay curve, may record distorted decay curves because of laser-power or source-flux drifts during the observation period. Drifts do not distort the decay curve recorded by the digitizer, since an entire LIF decay is recorded for each trigger event. Figure 2 shows a typical LIF decay curve recorded with a transient digitizer.

**Time-Correlated Single-Photon Counting Detection.**   In the TC-SPC technique, a trigger event such as the dye laser pulse starts a fast timer, and detection of a fluorescence photon by a separate detector stops the timer. The analog signal from the timer is digitized into a number of channels corresponding to the elapsed time since the start event. The start–time–stop cycle is repeated many times (multiple laser shots), and the number of LIF photons detected in each time channel is recorded, generating a time-resolved fluorescence decay curve. To prevent distortions in the decay curve, the detected LIF intensity is typically held sufficiently low so that no more than $\sim 0.1$ photon per laser shot is detected. Corrections to the decay curve must be made considering the statistics of detected "stop" signals—the latter part of the decay curve is suppressed by a multiplicative factor equal to the probability that a photon has not stopped the timer (approximately 0.9 if about 0.1 photons per laser shot are detected). This technique is potentially more accurate for LIF lifetime studies than the transient recording method. The timer (usually a time-to-amplitude converter, TAC) generates a very linear and reproducible time base even for nanosecond-order events. The start and stop detectors are used only as event detectors, and are thus freed from stringent demands of high electronic bandwidth and linearity. Other systematic errors, such as those discussed later in Section 12.3.5, must still be considered in TC-SPC.

The TC-SPC method is particularly well suited to measuring lifetimes at high laser repetition rates such as from a mode-locked laser [30]. The time to generate reasonable signal-to-noise for analysis of a single decay curve—ideally $\geq 10^4$ detected photons—is excessive at typical pulsed laser repetition, rates of $\sim 50$ Hz. TC-SPC systems cannot in general exploit the full $\sim 100$ MHz repetition rate typical of a mode-locked laser—the reset time (primarily the time to discharge the TAC) generally limits the "start" rate to about 100 kHz. The effective data collection rate can be optimized by making the LIF photon the "start" signal and the laser pulse the "stop" signal. Correlation between the detected LIF photon and its generating laser pulse can be maintained with an

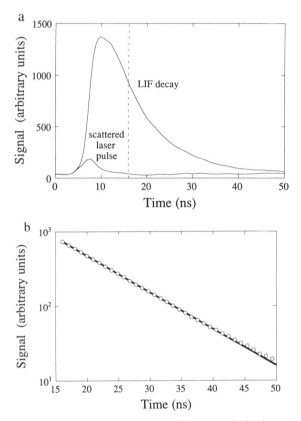

FIG. 2. (a) Sample LIF decay curve for the $3d^2$ ($^1D$)$4s4p$ ($^1P^0$)$y$ $^1F_3$ level neutral Ti, recorded with a transient digitizer with 1024-element temporal resolution and with 390.48 nm laser excitation, using a system similar to that shown in Figure 1. The signal from scattered laser radiation (tuned off the transition) is also shown. Each trace is the average of 640 separate LIF waveforms collected in less than one minute. The dotted line indicates the beginning of the portion of the decay curve, after termination of the laser excitation pulse, used for lifetime measurements. (b) Plot of the LIF signal from (a) (minus the background from the laser pulse) after the laser pulse termination. For clarity, only a few of the LIF data points are shown (open circles). The solid line shows the weighted least-squares single exponential fit. The lifetime of this level is $8.8 \pm 0.4$ ns [16].

appropriate delay line in the "stop" (laser pulse) channel. In this "inverted" configuration, the TAC can operate nearer its maximum repetition rate since only "successful" laser shots (those that apparently produce an LIF photon) are recorded, limiting loss of counts due to TAC reset time [31]. Figure 3 shows LIF decay curves recorded using TC-SPC.

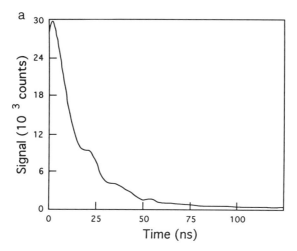

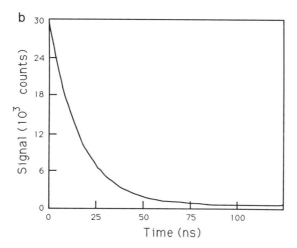

FIG. 3. LIF decay curves of the $3p\,^2P_{3/2}$ level of Na recorded using time-correlated single-photon counting in an atomic beam excited by a mode-locked dye laser at 589.0 nm (approximately 6 ps pulse duration at about 5 MHz repetition rate). (a) Decay curve recorded with linear polarization of the laser and a parallel-oriented linear polarizer in the detection channel. Hyperfine quantum beats distort the single-exponential decay curve. (b) The detection channel polarizer is oriented at the "magic angle" of 54.7° relative to the linearly polarized laser, eliminating distortion of the decay curve by the hyperfine beats. The lifetime of this level was measured to be $16.36 \pm 0.10$ ns, indicating the potential precision of the time-correlated single-photon counting technique. Adapted with permission from J. Carlsson, *Z. Phys. D* **9**, 147 (1988). © 1988 Springer-Verlag, Heidelberg.

## 12.3.4 General Configuration of LIF Lifetime Experiments

In the typical time-resolved LIF experiment using an HCD source, the laser propagation axis, atomic beam axis (for the HCD beam source), and fluorescence collection axis are mutually orthogonal, helping to reduce detection of scattered laser light. In the transient recording scheme, collection optics image the fluorescence onto the PMT photocathode, usually through one or more spectral filters. In the TC-SPC technique, a monochromator is typically used, giving greater spectral resolution and helping to limit the detected signal to 0.1 photon or less per laser shot.

In the transient detection method, it is common to detect as many as $10^4$ LIF photons per laser shot for the stronger transitions, although accurate lifetimes can be determined at signal levels on the order of 10 detected photons per shot. Averaging $\sim 1000$ laser shots usually gives a sufficient signal-to-noise ratio for even the weakest transitions. A typical transient digitizer has an effective waveform collection rate of about 10–1000 Hz, so with a laser repetition rate of $\geq 10$ Hz, at least 1000 separate LIF decays can be recorded and averaged on the order of one minute. Substantially longer times are required for recording a decay curve using TC-SPC.

The initial portion of the LIF decay curve, containing the laser excitation pulse convolved with the spontaneous decay, is usually discarded (see Figure 2). The laser pulse can in principle be recorded and deconvolved from the LIF decay curve [21, 32], but this procedure is subject to error due to saturation of the transition during the high-intensity portion of the excitation pulse. It is preferable to measure shorter lifetimes with ultrafast excitation or an abruptly terminating laser pulse, rather than rely on deconvolution.

The background signal, primarily due to scattered laser radiation and an essentially DC signal from the discharge source light, is recorded by tuning the laser off any transition. The background is subtracted from the LIF signal and a single exponential fit made to determine the radiative lifetime. Because the signal-to-noise ratio decreases along with the exponential decay of the LIF signal, properly weighted fits are more likely to reproduce the actual population decay lifetime [33]. Examination of the residual deviations of the data from the fit can reveal small systematic deviations from single exponential decay.

The use of laser excitation to selectively populate a single upper level of interest generally eliminates systematic error due to repopulation of the target level from higher-lying levels (cascade repopulation). Distortion of the upper-level decay curve due radiative decay to lower levels which then radiate can usually be eliminated with spectral filtering to block the cascade radiation. Whenever possible, it is preferable to record fluorescence decay at two or more different detection wavelengths or spectral bands to minimize the chance of unrecognized cascade contamination.

## 12.3.5 Systematic Errors

Although random errors in beam or cell transient LIF lifetime measurements are frequently smaller than 1%, in practice it is difficult to attain better than 3–5% total uncertainty (including both random and systematic error) in most experiments using the techniques discussed above. Subpercent uncertainties are possible in TC-SPC measurements, particularly when mode-locked lasers are used, but the greater complexity of these techniques has so far limited them to a relatively few studies [30, 34, 35]. Some of the potential systematic errors in an LIF lifetime measurement have already been discussed, such as electronic bandwidth limitations and nonlinearities in the detector and recorder system. Additional sources of error include flight-out-of-view effects, fluorescence decay curve distortion from collisions and radiation trapping, effects of laser polarization, and magnetic field effects.

Flight-out-of-view effects refer to error in the measured lifetime due to motion of excited atoms relative to the detection system. If a significant fraction of the excited atoms leaves the detection region before radiating, the earlier-occurring decays will be effectively oversampled, and the measured lifetime will be shortened. In addition, the effective quantum efficiency varies across the photocathode of a PMT, distorting the recorded decay curve if the image of the atoms moves significantly during the observation period [36]. To partially compensate for these effects, the fluorescence collection optics can be configured to overfill the photocathode, decreasing the motional effects, but with substantial reduction in the signal-to-noise ratio [37]. If a pulsed HCD beam source is used, crude selection of slower atoms is possible by varying the delay between the HCD current pulse and the laser excitation pulse, again with reduced $S/N$ [37]. In most LIF beam or cell experiments, motional effects are significant for lifetimes on the order of 0.1–1 μs and longer, and the upper limit for accurate lifetime measurements is on the order of 1–10 μs.

Radiation trapping, the absorption and re-emission of the spontaneous LIF radiation lengthens the effective lifetime—the photons essentially incur a delay as they "diffuse" out of the excitation volume [21]. The effective lifetime $\tau_{\mathrm{eff}}$ is

$$\tau_{\mathrm{eff}}^{-1} = \tau^{-1}(1 - P_{\mathrm{rad}}), \qquad (12.9)$$

where $\tau$ is the unperturbed lifetime and $P_{\mathrm{rad}}$ the probability that an emitted photon will be absorbed by another atom before reaching the detector. When radiation trapping is a relatively small perturbation, $P_{\mathrm{rad}}$ is a linear function of the density of absorbing atoms. Radiation trapping of sputtered atom transitions is negligible in atomic beams from HCD sources. Resonance transitions in the much higher-density buffer gas often exhibit radiation trapping. In sputtering cells, the higher density of sputtered atoms leads to significant radiation trapping in a small percentage of the transitions [21]. Compensation for radiation trapping is pos-

sible by varying the atom density and thus $P_{rad}$ (by changing the current to vary
the sputtered atom density or the pressure for buffer gas density variation), and
extrapolating $\tau_{eff}^{-1}$ to zero density.

Quenching of laser-excited levels by collisions with buffer gas atoms or ions
generally shortens the apparent lifetime. The effective lifetime due to first-order
collisional quenching is

$$\tau_{eff}^{-1} = \tau^{-1} + n\sigma V, \tag{12.10}$$

where $n$ is the buffer gas density, $\sigma$ the cross-section for quenching collisions
between the buffer gas and the particular excited level, and $V$ the relative average
speed of the buffer gas and sputtered atoms [19, 21]. This approximation is valid
when collisional quenching is a relatively small effect. In the HCD beam, the
buffer gas density is sufficiently low (typically $\leq 10^{-4}$ torr) that collisional
quenching is usually insignificant, although recent work suggests that under
some conditions lifetimes measured in beams from HCD sources may be affected
by collisions between electrons and ions—from both the buffer gas and the
sputtered species—and the neutral excited atoms under study [19]. In the
sputtering cell at typical buffer gas pressures of $\sim 1$ torr and sputtered atom
densities of $\sim 10^{11}$ cm$^{-3}$, the mean time between collisions is $\sim 100$ ns, and
collisional quenching can appreciably shorten the apparent lifetimes in this
temporal range. Recording lifetimes as a function of buffer gas pressure and
extrapolating $\tau_{eff}^{-1}$ to $n = 0$ (Stern–Volmer plot) [38, 33] usually compensates for
quenching effects.

Compensation for radiation trapping or collisional quenching by linear extrapolation to zero density is reliable only when the change in the apparent lifetime
is relatively small. Nonlinearities in Stern–Volmer plots are observed, usually
due to coupling of populations of nearly degenerate levels with significantly
different life times. Cross sections for excitation transfer between nearly degenerate levels of the same spin can be quite large ($\geq 10^{-14}$ cm$^2$). Nonlinearities
may be most pronounced in the region of lowest density where data is sparse
because of low signal-to-noise ratios.

In cell experiments, additional potential systematic error results from alignment effects of the excited atoms. At the end of the short pulse of linearly
polarized laser radiation typically used in LIF experiments, the excited atoms are
aligned (the magnetic sublevel population distribution is nonthermal). Since the
spontaneously emitted radiation is in general anisotropic, collisions or radiation
trapping that disalign the atoms result in fluorescence intensity variations
independent of the decay of the population. The effect of this disalignment can
be eliminated to first order by proper choice of polarization of the excitation and
detected radiation, such as linearly polarizing the laser at the "magic angle" with
respect to the detection system [21]. These effects are usually insignificant in
collisionless, optically thin atomic beams.

Additional alignment effects include Zeeman [39] and hyperfine quantum beats [40]. The magnetic states of the excited atom, coherently excited by the short linearly polarized pulse of the laser, precess in the presence of an external magnetic field (Zeeman) or due to a nonzero nuclear magnetic moment (hyperfine). Because of the anisotropic emission distribution, the precession produces a lighthouse-like sweeping of the emitted radiation across the detector, modulating the fluorescence decay curve. Lifetime determinations may be subject to error when the precession period is on the order of the observation period. Averaging over many precession periods gives accurate lifetimes—short lifetimes are measured in a near-zero residual magnetic field and long lifetimes in a high field to smear out Zeeman beats [13]. Hyperfine beat distortion is eliminated with "magic angle" detection [19, 21] (see Figure 3).

### 12.3.6 Examples of LIF Lifetime Measurements

Time-resolved LIF on HCD-generated beams has been used to make extensive sets of lifetime measurements in many neutral and singly ionized atoms. For example, O'Brian et al. [8] combined 186 level lifetimes in Fe I with measured branching fractions to generate absolute transition probabilities for almost 1200 transitions in the spectral range 225–2700 nm. The solar abundance of iron was recently revised by 35% (resolving a long-standing controversy) because of improved Fe II lifetime measurements made by Hannaford et al. in a sputtering cell [41], and because of improved Fe I lifetimes measured by Bard et al. with a HCD source [42]. Bergeson and Lawler measured Zn II and Cr II lifetimes to provide spectroscopic data for investigation of the relative distribution of condensed and gaseous phases in the interstellar medium [43]. Experiments by Carlsson et al. demonstrate the potential to make 1% or better time-resolved LIF lifetime measurements using pulses from a mode-locked laser and TC-SPC [30. 34].

## 12.4 Beam Foil Lifetime Measurements

Beam foil spectroscopy was one of the earliest methods used to make direct lifetime measurements on ions [44]. In this technique, a well-collimated monoenergetic ion beam passes through a very thin foil which acts as a broadband impulsive excitation source. The incident ions emerge distributed throughout a wide range of ionization stages and a broad range of excited levels. The excited ions radiatively decay as they continue downstream at constant velocity. The decay of the radiation emitted by an excited-level population is monitored as a function of the downstream distance from the foil, which is, of course, a function of time since the foil excitation. The radiative lifetime is thus recovered from the spatially resolved decay of the excited-level emission. The ion beam is optically thin and collisionless, eliminating errors due to radiation trapping or collisional quenching. Spatially resolved beam foil spectroscopy is currently the best method for measuring lifetimes in higher ionization stages ($\geq +2$) and for very short wavelengths (deep VUV and soft X-ray).

## 12.4.1 Techniques of Beam Foil Measurements

A beam foil lifetime experiment consists of an ion source, an accelerator, the foil, a radiation detection system, and a beam monitoring device. The foil and beam must be in a high vacuum.

The RF discharge is a commonly used ion source. The selected element can be made into a volatile compound (e.g., $SiH_4$, CO, $UF_6$) or simply placed as a bulk sample into the discharge chamber. Alternatively, the element can be volatilized in a furnace and ionized by electron beam bombardment. Hollow cathode discharges and many other sources have been used [45].

The most commonly used types of accelerators are single-ended or tandem Van de Graaff accelerators, linear accelerators, and electromagnetic isotope separators. For lifetime experiments, it is advantageous to be able to generate a broad range of beam energies (e.g., 100 keV to 10 MeV) to allow study of lifetimes over a wide temporal range as well as to produce as many different ionization stages as possible. Typical useful beam currents are on the order of 10 na to 10 μA beam a few millimeters in diameter. The ion number density downstream of the foil is usually on the order of $10^4$–$10^6$ cm$^{-3}$.

Carbon foils, 5–50 nm thick, are most commonly used. The low $Z$ value of carbon diminishes Rutherford scattering and energy loss by the incident ions. Carbon foils are also relatively easy to produce and have a relatively long lifetime against degradation from the ion beam. Incident ions with energies $\sim 0.1$ MeV/amu traverse a 50 nm foil in $\sim 10$ fs and emerge into a high-vacuum chamber. The energy loss and angular scattering in the foil can be modeled rather accurately [46]. The emergent ions have an ionization and excitation distribution comparable to a thermal source at $10^5$ to $10^6$ K [47], allowing study of several ionization stages and a wide range of excited levels in each stage. Foils degrade under ion beam bombardment, changing shape and thickness and eventually failing by developing holes. Usually multiple foils are mounted on a wheel so that when one foil fails another can be rotated into place with minimal inter-ruption of the experiment.

Drifts in the ion beam current affect the emission intensity. The beam current is typically monitored by collecting the ions in a Faraday cup and normalizing the recorded signal to the current. Normalizing the detected radiation to the intensity of an emission line monitored at a fixed downstream position provides a more sensitive correction of source drifts. The velocity of the ions emerging from the foil may be determined with a downstream electrostatic analyzer, by observing the Doppler shift of a selected transition, or by calibrating the accelerator potential and modeling the energy loss in the foil.

The typical scheme for detecting the emitted radiation is a monochromator and photon-counting PMT or channeltron detector. A simple normal-incidence concave grating monochromator provides reasonable throughput for wavelengths as short as ~50 nm; grazing incidence monochromators are used in the range ~5–100 nm. A fast (high-throughput) monochromator is usually preferred over one with higher resolution because of the typically low signal levels and the desire to observe short-wavelength transitions. Restricting the field of view to a small portion of the ion beam provides better effective temporal resolution. PMT spectral response can be extended deep into the VUV by coating the window with sodium salicylate.

In the usual beam foil lifetime experiment, the detection system is held fixed and the foil is translated along the ion beam axis. The monochromator field of view typically limits spatial resolution to ~0.1 mm . An effective observation period of 3–5 lifetimes corresponds to translations of several millimeters to several centimeters. To partially compensate for systematic changes in beam velocity or excited level population due to beam-induced foil changes, the foil is alternately translated in opposite directions and the decay curves averaged. The practical range for beam foil lifetime measurements is about 0.1–100 ns. With high-energy beams, and modeling of foil vignetting and limitations of mono-chromator spatial resolution, lifetimes on the order of 10 ps can be measured with moderate uncertainty [48].

## 12.4.2 Systematic Errors

The most significant potential errors in beam foil measurements result from the nonspecific excitation. The relatively low spectral resolution of the mono-chromator (necessitated by the low signal levels and large Doppler widths) and the broad range of excitation over multiple ionization stages makes blending common. The potential for blending often limits beam foil observations to relatively strong transitions where any weak blends introduce insignificant error. The greater problem from nonselective excitation is cascade repopulation of the excited level under observation. In general the radiative decay of any level will thus be multiexponential.

Although some early beam foil lifetime measurements were unreliable, the arbitrarily normalized decay curves (ANDC) method of analyzing multiexponen-tial decays provides more accurate lifetimes [49]. In the ANDC method, the decay curves of the cascade channels must be determined, but the relative intensities are considered as adjustable parameters in a fit to the observed decay curve of the selected level, eliminating the difficult task of spectral response calibration in the VUV and soft X-ray region. The analysis process has been automated, and the effects of failing to identify all appreciable cascades discussed [50]. The ANDC method has significantly improved the reliability of beam foil lifetime measurements. It must be noted, however, that the accuracy of any fitting process depends on the signal-to-noise ratio of the primary decay curve—the ANDC method cannot assure accurate lifetimes from data with a low signal-to-noise ratio.

There are additional sources of systematic error in beam foil measurements that should be considered for evaluating the accuracy and precision of the results. Zeeman and hyperfine quantum beats can occur, since the foil effectively acts as a short pulse excitation source because of the short transit time ($\sim 10$ fs). The different magnetic and hyperfine sublevels are thus excited with well-defined relative phases. Subtle polarization and alignment effects must also be considered [51]. Usually the error from the cascade correction dominates any errors due to quantum beats or alignment effects. Very short lifetime measurements ($<1$ ns) are complicated by the limited spatial resolution of the monochromator ($\sim 0.1$ mm) required for adequate signal. Since most of the decay in such short-lived levels occurs within a millimeter or so of the foil, shadowing effects and changes in the foil shape and thickness under ion bombardment can also introduce error [48]. For lifetimes in the 0.1–100 ns range, uncertainties of $\sim 10\%$ are typical, with larger uncertainties for shorter lifetimes and for measurements with low signal-to-noise ratios.

### 12.4.3 Examples of Beam Foil Lifetime Measurements

The beam foil spectroscopy literature is extensive. Reviews and conference proceedings edited by Sellin and Pegg [52], Cocke [53] and Bashkin [54] summarize earlier work. Beideck et al. [55] reported lifetime measurements in Hg III, relevant to studies of Hg isotopic anomalies in the chemically peculiar star $\chi$ Lupi [56], typifying more recent beam foil studies (see Figure 4). Engstrom and Bengtsson [48] measured lifetimes as short as 6 ps in F VII and discussed the effects of shadowing, foil shape changes, and limited monochromator resolution in making such short lifetime measurements. Maniak et al. [57] measured lifetimes in sodium-like Si IV that should be amenable to accurate ab initio calculations, but persistent small discrepancies between experimental and calculated results are found in this and similar systems.

## 12.5 Fast Beam Laser Experiments for Lifetime Measurements

The fast beam laser method is the best technique for making high-precision lifetime measurements ($\leq 1\%$ uncertainty). A well-collimated monoenergetic ion beam produced in an accelerator is selectively excited with a laser. The decay of the LIF of the excited level is recorded as a function of downstream position and the lifetime inferred from the known beam velocity. Selective excitation eliminates cascade repopulation, and the low-density beam is essentially collisionless and optically thin. The combination of high beam velocities and the inherent precision of mechanical translations gives better effective temporal resolution and linearity than are easily attainable in time-resolved LIF measurements, making higher-precision measurements possible.

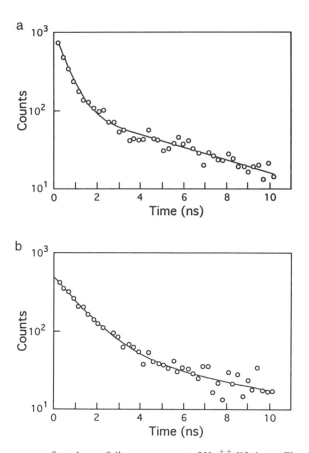

FIG. 4. Decay curves from beam foil measurement of $Hg^{++}$ lifetimes. The data are well fitted by a double exponential decay in each case, and the more rigorous ANDC method was not used. (a) Decay of the $5d^{10}\,^1S_0 - 5d^9 6p\,^1P_1$ transition at 79.017 nm with measured lifetime of $0.52 \pm 0.15$ ns. (b) Decay of the $5d^{10}\,^1S_0 - 5d^9 6p\,^3P_1$ transition at 84.311 nm with measured lifetime of $1.20 \pm 0.15$ ns. Adapted with permission from D. J. Beideck et al., Phys. Rev. A **47**, 884 (1993). © 1993 The American Physical Society.

## 12.5.1 General Techniques

A fast beam laser experiment is essentially a beam foil experiment with the foil replaced by a laser as the excitation source [58]. The ion sources and accelerators used in fast beam laser experiments are the same as those used in beam foil experiments. The beams are typically limited to the singly and doubly ionized species produced in most ion sources, although higher ionization stages could be produced after acceleration. Metastable ion sources can be used [59], and neutral atomic beams may be produced from ion beams through post-acceleration charge-exchange collisions in gases or vapors [60].

Both continuous wave (cw) and pulsed lasers are used as excitation sources. The temporal pulse lengths of a few nanoseconds and maximum repetition rates of $\sim 1$ kHz for pulsed pump lasers impose a rather severe duty cycle mismatch (relative to the continuous ion beam flux). However, the much greater instantaneous laser intensity possible in a pulsed system may generate time-averaged LIF signals comparable to those from cw excitation, particularly when the excitation transition probability is small. Both cw and pulsed lasers may be tuned to the required excitation wavelength either by Doppler shifting relative to the ion beam (determined by the laboratory frame angle between the laser and ion beams) or by use of a tunable laser.

Since the detection system in a fast beam laser experiment typically relies on photon counting, the laser intensity is kept sufficiently low to prevent pile-up effects in the detector. Thus, a laser intensity of $\sim 10$ W/cm$^2$ is a benchmark for fast beam studies of stronger transitions. This intensity is easily achieved with commercial cw dye or Ti : Sapphire lasers pumped by multiwatt ion lasers, or with diode lasers [61]. For weaker transitions, or for excitation wavelengths requiring frequency up-conversion (such as second harmonic generation), pulsed dye lasers are used. A common scheme is to use an excimer pumped dye laser permitting repetition rates up to $\sim 200$ Hz at high peak powers. The ion beam can be passed through the dye laser cavity for greater power.

The detection system consists of a photon-counting PMT with some spectral selection device and often fiber-optic coupling of the LIF to the detector. The selective excitation eliminates much of the background except the scattered laser light, and usually a simple spectral filter is sufficient. The laser excitation region is held fixed and the detection system is translated parallel or antiparallel to the beam. Peak LIF count rates of $\sim 1$–$100$ kHz are typical with cw excitation. Signal-to-noise considerations generally prevent finer spatial resolution than $\sim 0.1$ mm. Ion beam velocity, available translation range, and signal-to-noise ratios usually limit lifetime measurements to $\leq 100$ ns.

In pulsed LIF experiments, the need to avoid pile-up limits peak detection rates, because photon counting is used. Higher data collection rates are possible with gated charge-integration—the detection electronics and PMT are configured to permit accurate counting of up to ~10 coincident photons per laser shot [62]. The dynamic range of detection is also increased because the lowest effective count rate in either single-photon or gated integration techniques is set by the background and/or dark count rates.

For greatest precision the LIF signal is normalized to fluctuations in both ion beam current and laser power. Ion beam current is monitored directly (e.g. with a Faraday cup) and/or through observation of relative LIF intensity at a fixed position in the beam.

## 12.5.2 Systematic Errors and Ultimate Precision in Fast Beam Laser Measurements

Total uncertainties of a few tenths of one percent are possible. The advantage of highly precise lifetime measurements is somewhat diluted by the more limited precision (typically ~5–10%) currently available in most branching fraction measurements (see Section 12.7) necessary to convert the lifetimes to the transition probabilities needed in quantitative spectroscopy. However, highly precise lifetimes can be used to test fundamental principles [7, 60]. In most fast beam experiments, the major source of systematic error is the uncertainty in the ion velocity, typically in the 0.1–1% range. Beam velocities are measured through Doppler shifts of well-known transitions and/or by calibration of the accelerator potential. For subpercent total uncertainty, the contribution from all the other systematic errors must not exceed a few tenths of one percent, imposing formidable requirements on the system. Gosselin et al. [63] discuss some of the effects of small and subtle uncertainties on the overall precision.

## 12.5.3 Examples

The first fast beam laser lifetime experiment was a 1% measurement of the $6\,^2P_{3/2}$ lifetime in Ba II [58]. The review article by Andra [64] summarizes early fast beam laser studies. Although most of the earlier measurements were made with cw excitation, pulsed excitation is now more common, providing sufficient intensity for exciting weaker transitions and producing UV radiation through SHG. For example, Ansbacher et al. [62] made ≤1% precision measurements of Mg II lifetimes using pulsed two-step excitation (near 280 nm) realized by differential Doppler tuning of the retroreflected laser. Guo et al. [59] used SHG (~230–260 nm) and a metastable ion beam to measure lifetimes in low-lying quartet levels of Fe II. Tanner et al. [60] made a subpercent precision measurement of the Cs I $6p\,^2P_{3/2}$ lifetime using charge exchange to generate the neutral beam and a diode laser to excite the transition (see Figure 5). This measurement was precise enough to test the accuracy of calculations relevant to atomic parity-nonconservation experiments. Dumont, Baudinet-Robinet, and Garnir and

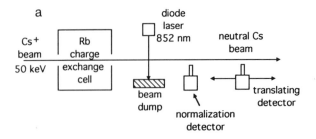

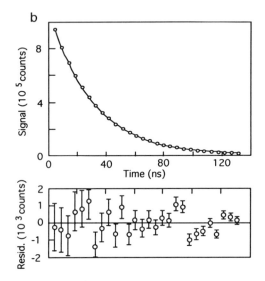

FIG. 5. (a) Schematic diagram of the fast beam laser precision measurement of the $6p\,^2P_{3/2}$ level lifetime in $^{133}$Cs. Note the charge-exchange cell for production of a neutral atomic beam and the use of diode laser excitation, unusual features in a fast beam laser experiment. The "normalization detector" monitors the beam flux (through the LIF signal), and the "translating detector" is moved along the beam to record the spatial dependence of the LIF decay. (b) Typical LIF decay curve with single exponential fit (solid line) and the residuals (fit minus data) shown. Adapted with permission from C. E. Tanner *et al.*, *Phys. Rev. Lett.* **69**, 2765 (1992). © 1992 The American Physical Society.

co-workers have developed a hybrid beam foil laser technique for measuring radiative lifetimes [65]. The beam foil process generates ions in high excitation levels, and cascade-free lifetimes are referred from changes in the populations of two levels coupled by tunable laser excitation. This relatively new method has been used for only a few lifetime measurements thus far, but should help reduce systematic error in high ionization stage lifetimes due to cascade repopulation (see Figure 6).

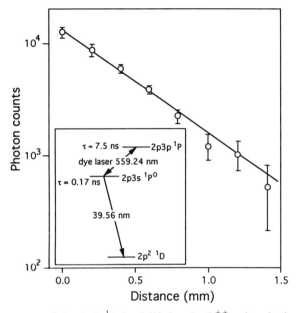

FIG. 6. Measurement of the $2p3s\,^1P$ level lifetime in $O^{++}$ using the beam-foil laser technique. The fluorescence at 39.56 nm from foil-excited $O^{++}$ ions was measured with and without the presence of cw laser radiation (559.24 nm) cycling ions between the $2p3s\,^1P$ and $2p3p\,^1P$ levels, as shown in the inset. The difference between the "laser on" and "laser off" fluorescence from the lower level ($2p3s\,^1P$) of the transition, as a function of downstream distance from the laser excitation region, is the sum of two exponentials corresponding to the upper and lower level lifetimes. The lower level lifetime was measured to be $0.17 \pm 0.01$ ns. Adapted with permission from Y. Baudinet-Robinet, P. D. Dumont, and H. P. Garnir, *Phys. Rev. A* **43**, 4022 (1991). © 1991 The American Physical Society.

## 12.6 Measurement of Long Lifetimes in Trapped Ions

### 12.6.1 Metastable Lifetimes

Direct measurement of lifetimes longer than about 10 μs through observation of the radiative decay is generally not possible in a beam or cell environment because of flight-out-of-view effects. Singly to triply ionized atoms have been confined in electromagnetic or electrostatic traps for long periods (relative to the radiative lifetime), eliminating motional error in lifetime measurements. In a typical trap, ions may be confined in a small volume for periods on the order of seconds or longer, permitting accurate measurement of radiative lifetimes in the ~0.1 ms to 1 s range. Lifetimes in this range typically result from magnetic dipole, electric quadrupole, or intercombination (spin-changing) radiative transitions out of metastable levels. Such transitions are important to the understanding of diffuse plasmas, such as those in outer stellar atmospheres [66], auroral phenomena [67], and parts of the interstellar medium [3]. Emission from metastable levels in such plasmas can be a sensitive probe of collisional processes, permitting inference of electron densities and temperatures, among other things [68].

## 12.6.2 General Techniques and Examples of Trapped Ion Lifetime Measurements

In the typical ion metastable lifetime measurement, a short (relative to the lifetime) excitation pulse excites a sample of trapped ground-level ions to the metastable level(s) of interest. The radiative lifetime is obtained from the subsequent time-resolved fluorescence decay. In practice, even at very low background pressures collisional quenching of the excited level is significant for the long lifetimes measured, and the radiative lifetime must be inferred from extrapolation of $\tau_{\text{eff}}^{-1}$ to zero pressure.

Thorough discussion of ion trapping techniques can be found in several excellent review articles [69] and in other chapters in this book. The three main types of traps—Paul (RF), Penning, and Kingdon (electrostatic)—have all been used for trapped ion lifetime measurements. Typical confinement volumes are on the order of $1–10 \text{ cm}^3$, containing $10^3–10^5$ ions in potential wells with effective depths of 10–50 eV. Confinement times may be several seconds or longer. The background pressure in the trap is typically held below $10^{-8}$ torr.

Usually the trap is filled with neutral atoms, which are then ionized and excited by pulsed electron bombardment. Gaseous or volatile substances (sub-millitorr vapor pressures are adequate) are introduced directly into the trap. Refractory elements can be introduced through atomic beams or sputtered vapors ionized by electron impact or at the surface of a hot filament [70]. Volatile molecules can be dissociated with moderate-energy electron beams (a few hundred electron volts) [71]. Electron bombardment has been used to produce ionization stages up to $+3$ [72]. Laser ablation of solid targets has-been used to fill traps with ionization stages up to $+4$ in tungsten and $+6$ in molybdenum [73], although the authors are not aware of any lifetime measurements beyond stage $+3$ at the time of this writing.

The electron pulse, typically about 1 ms long, broadly excites levels in the target ion. After a delay of several milliseconds to allow the excited ions to radiatively decay into the metastable levels and the ion cloud to stabilize, the time-resolved fluorescence decay curve is recorded with collection optics, a narrow-band interference filter, and a photon-counting PMT or channel plate detector. Signal levels are usually too low to permit use of a monochromator. After the spontaneous decay is observed for the equivalent of a few lifetimes, the trap is emptied of ions by applying an appropriate potential to the electrodes. The background signal is then collected, and the cycle repeated: ion production with an electron pulse, fluorescence decay observation, ion purging, background recording. Typically several hours are required to collect the $\sim 10^4$ counts sufficient for a decay curve with a good signal-to-noise ratio.

The nonselective excitation may produce blends or cascade repopulation of the target level. When two different decays with substantially different lifetimes are simultaneously recorded, it is usually possible to individually measure each lifetime with reasonable accuracy through a double exponential fit [74]. Cascade repopulation of one metastable level from another with a comparable lifetime is rarely a problem, but must be considered as a potential source of error.

Typical total uncertainty for trapped ion lifetime measurements are on the order of 5 to 10%. Often the major source of error is the absolute pressure measurement in the trap [75]. Even at typical trap background pressures of $\leq 10^{-7}$ torr, the lifetimes are long enough that collisional quenching of excited states is significant. The relatively large uncertainty in absolute pressure measurements in the $\leq 10^{-7}$ torr range limits the accuracy of the Stern–Volmer extrapolation. Other sources of error such as nonlinearity in the effective time base and statistical error, are typically small.

Prior [72] used an electrostatic trap to measure metastable level lifetimes in Ar III and Cu II in the 0.1–10 s range. Walch and Knight [76] used an RF trap to measure Kr II and Xe II lifetimes in the 5–10 ms range. Calamai and Johnson [74, 75] measured several lifetimes in the 1–100 ms range in doubly and triply ionized Ar, Kr, Xe, and Hg using electrostatic traps (see Figure 7).

Selective laser excitation schemes have been used in one set of experiments. A low-intensity atomic beam of Sr, Yb, or Pb [70, 77] passing through an RF trap was ionized at the surface of a hot filament. Ions confined in the trap were excited (through an $E1$ transition) with a pulsed tunable dye laser. The laser-excited upper level had a significant branch to a lower metastable level. Thus, after several laser shots (at 50 Hz), and after allowing time for the dipole-allowed transitions to depopulate the laser-excited level, essentially only the target metastable level was populated (beside the ground state). The LIF from the metastable was detected with time-resolved photon counting and the lifetime determined free from cascade or blends.

The ion-trapping group at NIST has conducted several interesting lifetime studies. In one experiment, a single Hg + ion was laser-cooled and trapped in a miniature RF trap [78]. Narrow-band 194 nm cw radiation was tuned slightly below the $5d^{10}6s\ ^2S_{1/2}$–$5d^{10}6p\ ^2P_{1/2}$ resonance to both excite the $^2P_{1/2}$ level and cool the ion. The lifetimes and branching fractions of the decay channels through the $5d^96s^2\ ^2D_{3/2}$ and $^2D_{5/2}$ levels to the ground state could be indirectly determined by observing the 194 nm fluorescence as the single ion cycled between the ground state and the $^2P_{1/2}$ or $^2D$ levels. The lifetime of the $^2P_{1/2}$ level was

a

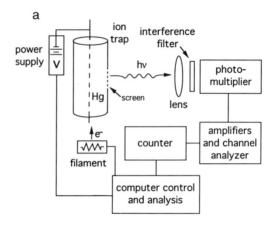

b

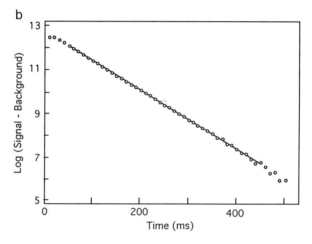

FIG. 7. (a) Schematic diagram of the electrostatic trap used by Calamai and Johnson to measure metastable level lifetimes in $Hg^+$ and $Hg^{++}$. Gaseous Hg is ionized and excited by a pulsed electron beam, and the ions are constrained to orbits around the central electrode of the trap for several seconds. The fluorescence from the desired level(s) is selected with a narrow band interference filter and recorded as a function of time since the excitation pulse. (b) Fluorescence decay curve of the $5d^9 6s^2 \, ^2D_{5/2}$–$5d^{10} 6s \, ^2S_{1/2}$ transition (282 nm) in $Hg^+$. The single-exponential fit (solid line) lifetime of this electric quadrupole transition was measured as $87 \pm 4$ ms. (c) Fluorescence decay curve of the $5d^9 6s^2 \, ^2D_{3/2}$–$5d^{10} 6s \, ^2S_{1/2}$ transition (198 nm) in $Hg^+$. This decay curve is contaminated by fluorescence (at 205 nm) from the second step of the $5d^9 6s 6p$ ($J = 9/2$)–$5d^9 6s 6p$ ($J = 7/2$)–$5d^9 6s^2 \, ^2D_{5/2}$ cascade. The lifetimes of the two detected components are sufficiently different to permit good fitting of the double exponential decay (solid line), giving lifetimes of $8.8 \pm 0.4$ ms and $60 \pm 6$ ms for the $^2D_{3/2}$ and $J = 9/2$ levels, respectively. Adapted with permission from A. G. Calamai and C. E. Johnson, *Phys. Rev. A.* **42**, 5425 (1990). © 1990 The American Physical Society.

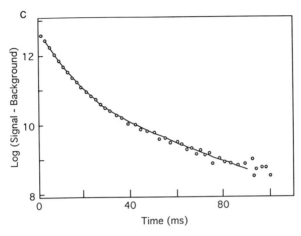

FIG. 7. *Continued*

determined as the inverse of the natural line width of the 194 nm transition as the laser was scanned through the transition in the cold (few millikelvin) ion.

## 12.7 Branching Fraction Measurements

Branching fractions must be combined with radiative lifetimes to determine absolute transition probabilities (unless the excited level has only one appreciable decay channel). Branching fractions are typically measured through emission spectroradiometry using a monochromator or Fourier transform spectrometer to measure relative emission intensities (photons/second) for transitions out of excited levels. Common spectral sources for neutral atoms and low ionization stages include HCD [79, 80] and inductively coupled plasmas (ICP) [80, 81] sources. Higher ionization-stage spectral sources include foil-excited ion beams and hot plasmas (such as laser-produced plasmas, fusion research plasmas, or occasionally astrophysical sources such as the solar atmosphere). Typical precision for branching fraction measurements ranges from ∼1% for strong transitions at nearby wavelengths in the near UV through near IR to ∼10% or worse for VUV or far IR transitions, weak transitions, or transitions at substantially different wavelengths. This section provides only the briefest introduction to techniques for branching fraction determination.

Branching fractions are determined from the relative emission intensities $I_{ul}$ (photons/s) measured with a spectrometer,

$$\phi_{ul} = A_{ul} \Big/ \sum_l A_{ul} = I_{ul} \Big/ \sum_l I_{ul}. \tag{12.11}$$

For levels with electric dipole-allowed ($E1$) transitions, it usually suffices to measure only the $E1$ relative intensities for the shorter-wavelength transitions, assuming that the higher-order multipole transitions are negligible and that the longest-wavelength (far IR) transitions are relatively unimportant [see Eq. (12.1)]. The estimation of completeness of the set of measured relative intensities—how much of the total transition intensity is accounted for by the measured branches—contributes to the uncertainty in the branching fraction measurements.

Branching fractions compare only the relative emission intensities of the transitions out of a particular excited level and are independent of any effects from source population distributions. The upper-level lifetime and the branching fractions accurately determine the absolute transition probabilities for transitions out of the upper level independent of the source's effective temperature, departure from thermodynamic equilibrium, distribution of ionization stages, etc.

## 12.7.1 Spectral Sources for Branching Fraction Measurements

HCD spectral sources tend to have relatively low atomic collision rates, giving narrow spectral lines, typically $<0.1 \, \text{cm}^{-1}$ wide (FWHM) [79, 80]. Doppler widths correspond to gas temperatures that are often only slightly greater than ambient temperatures. The sharp lines are advantageous for best resolution and identification of transitions in complex spectra, but can lead to self-absorption for the stronger transitions [8, 79]. To identify and compensate for self-absorption, the apparent branching fractions are measured as the discharge current and buffer gas pressure are varied. HCD sources typically have effective excitation temperatures in the 3000–5000 K range, although the excited-level population distribution generally varies substantially from a thermal (Boltzmann) distribution [82].

The ICP source is usually configured as an RF argon plasma torch burning in the ambient environment. ICP sources are relatively hot (typically ~7000–10,000 K) [8, 81]. The Doppler-broadened ICP spectral lines tend to be several times broader than the HCD lines, increasing the chance of blending or misidentification, but decreasing self-absorption. One advantage of the ICP source is that it can be operated in near-local the dynamic equilibrium (LTE), so that the excited-level population distribution closely approximates a Boltzmann distribution [8, 79]. The similarity between the gas temperature and the effective electron or excitation temperature is one indication that ICP discharges operate near LTE. Although LTE operation has no effect on the branching fraction measurement, it permits the ICP source to be used for absolute transition probability interpolation, as discussed later.

For emission spectroradiometry on higher ionization stages, the experimenter is forced to use less convenient sources such as laser-produced plasmas, foil-excited ion beams, tokamak plasmas, or observation of solar or stellar atmospheres. The laboratory sources usually suffer from short observation time, low signal-to-noise ratios reflecting low excited-level population densities, and great spectral complexity (multiple elements and/or ionization stages). Calibration of spectrometers in the very short wavelength range typical of high ionization-stage spectra is very difficult. Solar or stellar spectra can be recorded at high resolution, but these spectra are so rich that blends are very common, making it difficult to measure extensive sets of branching fractions. Much work remains to be done to improve the measurement of branching fractions in high ionization stages.

## 12.7.2 Spectrometers

The simplest spectroradiometric system is a scanning monochromator with a photoelectric detector such as a PMT or photodiode detector. Details of radiometric measurements with monochromators are widely available and will not be discussed here [83]. The biggest problems in using a monochromator are calibration of the instrumental spectral response and absolute wavelength, and the relatively long time required for recording a broad spectrum. Drifts in source temperature and/or excited-level populations during a long spectral scan alter the measured relative intensities of lines out of a given upper level.

Fourier transform spectrometers (FTS) have many advantages over monochromators for branching-fraction determinations [84]. The resolution of the FTS is determined by the mirror travel distance rather than diffraction through apertures, so high resolution with excellent throughput is possible. Excellent absolute wavelength accuracy is easily achieved over the entire spectrum without recourse to external standards. Because the entire spectrum is recorded simultaneously, source drifts may reduce the effective signal-to-noise ratios, but do not affect the accuracy of relative intensity measurements.

The one-meter McMath FTS of the National Solar Observatory at Kitt Peak, Arizona, has been heavily used for emission spectroradiometry [85]. This instrument provides spectral coverage from $\sim$230 to 18,000 nm (with different sets of optics) with high resolution, rapid data collection, and excellent absolute wavenumber calibration. A one-million–point spectrum can be recorded at a typical resolution of 0.01–0.05 cm$^{-1}$ in a few minutes with absolute wavenumber calibration better than 0.003 cm$^{-1}$. Figure 8 shows typical McMath FTS spectra from a HCD source, displaying the high resolution, excellent signal-to-noise ratio, and wide dynamic range that make the instrument so well-suited for branching fraction measurements. VUV FT spectrometers are being developed and used down to about 180 nm [86], and improved technology will undoubtedly result in good performance at shorter VUV wavelengths.

## 12.7.3 Calibration of Measurement Systems

Wavelength calibration for scanning monochromators is provided by accurately measured transfer standard wavelengths in several spectral regions [87].

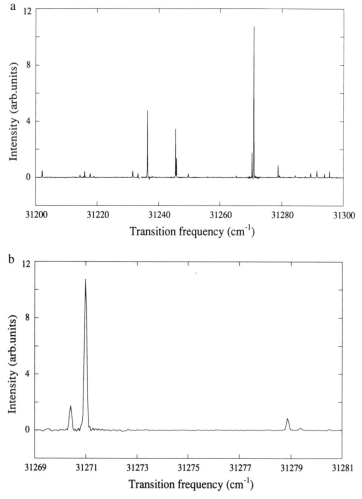

FIG. 8. Small portion of the hollow cathode emission spectrum of Fe (in Ar) recorded with the one-meter Fourier transform spectrometer at the National Solar Observatory in Kitt Peak, Arizona. Each figure (a–c) shows part of the original spectrum at increasing magnification, revealing the excellent resolution and large dynamic range provided by this instrument. See, for example, references [8, 79, 85].

Accurate calibration is necessary for line identification in complex spectra. More challenging is spectral response calibration for both FT and grating spectrometers. Various calibrated spectral radiance and irradiance lamps are provided by national standards laboratories. For example, NIST provides calibrated tungsten

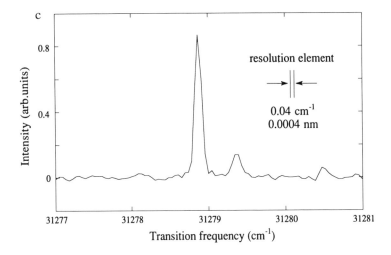

FIG. 8. *Continued*

filament lamps (225–2400 nm), deuterium lamps (200–350 nm), and argon and deuterium arc lamps (140–330 nm), with uncertainties quoted as ~0.5% at the longer wavelengths to ~5% or worse in the UV and VUV [88].

Ar I and II lines are prominent in Ar/HCD spectra and can serve as convenient internal calibration sources, recorded simultaneously with the target spectrum. Wiese *et al.* and Whaling *et al.* [89] measured and critically evaluated a unified set of transition probabilities in Ar I and Ar II covering the spectral range from about 210 to 5000 nm with uncertainties of ~5–10%. Helium transition probabilities have been calculated to high accuracy and used for internal spectral calibrations by mixing He with Ar in the discharge source. Useful He lines span the spectral range 400–3300 nm [90]. Continued expansion of the data base of accurately measured or calculated transition probabilities will provide additional calibration standards. High-accuracy VUV and far IR calibration standards are notably lacking at present.

Branching fraction measurements are accurate only when the complete set of (appreciable) decay channels has been recorded. Generally there will be transitions outside the recorded spectral range, particularly in the IR in complex spectra, that may contribute significantly to the total transition probability. Transition probability calculations, even if only order-of-magnitude estimates, may be sufficiently accurate to reveal whether the unobserved transitions contribute appreciably [8]. Contributions from unmeasured branches may be estimated in some cases through LIF experiments [91].

### 12.7.4 Transition Probability Measurements from Emission Spectra

The near-LTE operation of ICP sources has been exploited to infer large sets of absolute transition probabilities based on a smaller number of reference transition probabilities (with directly measured lifetimes and branching fractions). Whaling and Brault studied the Mo I spectrum using the McMath FTS with ICP and HCD sources [79]. From 67 published upper-level lifetimes and their measured branching fractions, they determined 659 absolute transition probabilities. From these transition probabilities, they found that the upper-level populations in the ICP closely approximated a Boltzmann distribution, and they interpolated many additional upper-level populations that did not have measured lifetimes. From the interpolated populations and the relative emission intensities, they inferred an additional ~2200 transition probabilities. Most of the ~2800 transition probabilities had reported uncertainties of about 10–15%. In a similar experiment on Fe I, O'Brian et al. measured 186 upper-level lifetimes ( ±5%), which were combined with branching fraction measurements made in ICP and HCD sources to generate ~1200 absolute transition probabilities densely spanning the entire Fe I excitation range [8]. From this large set of accurately measured (typically ~5–10% uncertainty) transition probabilities, it became clear that the ICP populations showed systematic departure from thermal distributions. However an additional 640 transition probabilities could be accurately interpolated from the ICP population distribution. The apparently thermal distribution of Mo I upper-level populations may have been caused by the relative sparsity of upper-level lifetimes available, particularly for the higher-lying levels. These methods allow accurate determination of nearly comprehensive sets of transition probabilities for neutral and singly ionized species. As further experience and insight is gained with ICP and other spectral sources, it may be possible to reduce the number of lifetimes needed for accurate analysis of transition probabilities from emission spectra.

## 12.8 Precision Measurement of Absorption Oscillator Strengths

Many other techniques are used to measure radiative lifetimes or oscillator strengths ($gf$ values) besides those discussed in this chapter. One that deserve special mention, because of its potential precision, accuracy, and broad applicability, is the measurement of relative $gf$ values through absorption spectroscopy. The absorptance integrated over the spectral line produced by a sample of $N_l$ atoms in lower state $l$ as they are excited to upper state $u$ can be described by the equivalent width $W_\lambda$, where

$$W_\lambda \propto N_l f_{lu} \Delta z. \tag{12.12}$$

In this expression $f_{lu}$ is the absorption oscillator strength, and $\Delta z$ is the effective path length. (See reference [92] for a more detailed explanation of equivalent widths.) This expression is valid for optically thin samples only. As the optical density of the sample increases, the expression for the equivalent width becomes more complicated, and the curve of growth (equivalent width as a function of absorbing atom density) must be measured or modeled [92]. Usually the path length is held constant and the relative lower-level population is inferred by assuming a Boltzmann distribution. The relative *gf* values are normalized with independently measured radiative lifetimes (or with absolute *gf* values determined by some other method) to yield absolute oscillator strengths.

In a typical absorption experiment, a well-stabilized furnace is used to produce a vapor in near-LTE with useful populations in the ground-level and lower-lying excited levels. A continuum source such as a high-pressure Xe arc lamp irradiates the sample, and the absorption signal is recorded with a monochromator and photodetector (usually a PMT or photodiode). The detection system is spectrally calibrated as discussed in Section 12.7.3. Accurate determination of relative lower-level populations requires both a close approximation to local thermodynamic equilibrium (LTE) and a precise temperature measurement.

The absorption method for determining relative oscillator strengths was first used by King and King in 1935 [93]. Although it has been frequently used since that time, one group's work deserves special discussion. Blackwell and collaborators at Oxford have refined the basic method and used it to make extensive measurements of unprecedented accuracy and precision (sometimes better than 1%) on the astrophysically important Fe group atoms [94].

The Oxford group used a modified King furnace designed for optimal temperature uniformity and stability. Matched high-resolution spectrometers were used to alternately measure absorption from a reference line and from a spectral line with unknown *gf* value. Great care was devoted to understanding and controlling all possible systematic errors. Temperature variations along the axis of the oven were mapped, and possible deviation from LTE investigated. For the highest-precision work (about 0.5% uncertainty), the absolute temperature was measured by optical pyrometry to a precision of about 1 K at temperatures in the 1500–2000 K range [95]. The continuum from the arc lamp was prefiltered to avoid any perturbation from thermal equilibrium in the metal vapor due to optical pumping. The power input and buffer-gas pressure were controlled to maintain stability of the oven temperature to within 0.5 K at 2000 K. Slight structure on the continuum from the arc lamp was measured and included in the analysis of the absorption data. A sophisticated data analysis and reduction technique was developed to insure the highest possible internal consistency of sets of oscillator strengths. Many other precautions were also taken.

High-precision measurements are very challenging, and there has been occasional controversy over whether the Oxford measurements have actually attained the claimed accuracy of as high as 0.5% (see, for example, reference [96]). More common are questions about the inferred absolute oscillator strengths due to errors in the lifetimes used for normalization [55]. However, no demonstrably superior measurements have been reported, and the Oxford relative *gf* data are widely considered as benchmark quality for Fe [94, 95, 97], Mn [94, 98], Cr [99], and other elements. Recent radiative lifetime measurements have further increased the value of the Oxford measurements by providing improved absolute normalization (see, for example, references [100] and [101]).

Relative absorption *gf* values are also inferred from stellar spectra. Absorption lines in the near-blackbody continuum of a star arise from atoms and ions in the cooler regions of stellar atmospheres and from the interstellar medium (see, for example, reference [102]). With appropriate assumptions about LTE (or modeled deviations from LTE) and appropriate optical densities, relative *gf* values are measured. The complexity of stellar spectra and uncertainty about the absorption conditions limits the accuracy of these measurements, but stellar absorption spectra may provide the only useful data for transitions from high-lying lower levels or in higher ionization stages.

## References

1. Shore, S. N. (1992). Chapter 2 in *Atomic and Molecular Data for Space Astronomy*, P. L. Smith and W. Wiese (eds.). Springer-Verlag, Berlin; Davidsen, A. F. (1993). *Science* **259**, 327.
2. Mathews, G. J., Meyer, B. S., Alcock, C. R., and Fuller, G. M. (1990). *Astrophys. J.* **358**, 36; Krause, L. M., and Romanelli, P. (1990). *Astrophys. J.* **358**, 47; Gilmore, G., Edvardsson, B., and Nissen, P. E. (1991). *Astrophys. J.* **378**, 17; Olive, K. A., Steigman, G., and Walker, T. P. (1992). *Astrophys. J.* **380**, L1.
3. Morton, D. C. (1991). *Astrophys. J. Suppl. Ser.* **77**, 119; Olive, K. A., and Schramm, D. N. (1992). *Nature* **360**, 439.
4. Lawler, J., Whaling, W., and Grevesse, N. (1990). *Nature* **346**, 635.
5. Denne, B. (1989). *Phys. Scripta* **T26**, 42.
6. Bogen, P., and Hintz, E. (1986). In *Physics of Plasma–Wall Interactions in Controlled Fusion*, D. E. Post and R. Bechrisch (eds.), *NATO ASI Series B* **131**, 211; Proud, J. M., and Luessen, L. H. eds. (1986). *Radiative Processes in Discharge Plasmas, NATO ASI Series B* **149**; Hutchinson, I. H. (1987). *Principles of Plasma Diagnostics*. Cambridge University Press, Cambridge.
7. Hunter, L. R., Krause, D., Jr., Berkeland, D. J., and Boshier, M. G. (1991). *Phys. Rev. A* **44**, 6140.
8. O'Brian, T. R., Wickliffe, M. E., Lawler, J. E., Whaling, W., and Brault, J. W. (1991). *J. Opt. Soc. Am. B* **8**, 1185.
9. Cowan, R. D. (1981). *The Theory of Atomic Structure and Spectra*. University of California Press, Berkeley.
10. Grevesse, N. (1984). *Phys. Scripta* **T8**, 49; Gustafsson, B. (1991). *Phys. Scripta* **T34**, 14.

11. Noecker, M. C., Masterson, B. P., and Weiman, C. E. (1988). *Phys. Rev. Lett.* **61**, 310; Blundell, S. A., Johnson, W. R., and Sapirstein, J. (1990). *Phys. Rev. Lett.* **65**, 1411.

12. Linsky, J. L. (Chapter 3) and Raymond, J. C. (Chapter 4) (1992). In *Atomic and Molecular Data for Space Astronomy*, P. L. Smith and W. Wiese (eds.). Springer-Verlag, Berlin.

13. Duquette, D. W., Salih, S., and Lawler, J. E. (1981). *Phys. Lett.* **83A**, 214; Salih, S., and Lawler, J. E. (1983). *Phys. Rev. A* **28**, 3653.

14. Schade, W., and Helbig, V. (1986). *Phys. Lett. A* **115**, 39; Duquette, D. W., and Lawler, J. E. (1982). *Phys. Rev. A* **26**, 330.

15. Duquette, D. W., Salih, S., and Lawler, J. E. (1981). *Phys. Rev. A* **24**, 2847.

16. Salih, S., and Lawler, J. E. (1990). *Astron. Astrophys.* **239**, 407.

17. O'Brian, T. R., and Lawler, J. E. (1991). *Phys. Rev. A* **44**, 7134.

18. O'Brian, T. R., and Lawler, J. E. (1992). *Astron. Astrophys.* **255**, 420.

19. Schade, W., Wolejko, L., and Helbig, V. (1993). *Phys. Rev. A* **47**, 2099.

20. Hannaford, P., and Lowe, R. M. (1981). *J. Phys. B* **14**, L5.

21. Hannaford, P., and Lowe, R. M. (1983). *Opt. Eng.* **22**, 532.

22. Duarte, F. J., and Hillman, L. W. (1990). *Dye Laser Principles, with Applications*. Academic Press, Boston.

23. Drexhage, K. H. (1990). Chapter 5 in *Dye Lasers*, F. P. Schafer, ed., 3rd Edition. Springer-Verlag, Berlin.

24. Shank, C. V. (1988). Chapter 2 in *Ultrashort Laser Pulses and Applications*, W. Kaiser, ed. Springer-Verlag, Berlin.

25. See, for example, Laubereau, A., and Seilmeier, A., eds. (1992). *Ultrafast Processes in Spectroscopy 1991*. Institute of Physics, Bristol.

26. Lawler, J. E. (1987). In *Lasers, Spectroscopy, and New Ideas*, W. M. Yen and M. D. Levenson, eds. Springer, New York.

27. O'Connor, D. V., and Phillips, D. (1984). *Time-Correlated Single Photon Counting*. Academic Press, London.

28. Engstrom, R. W. (1980). "Photomultiplier Handbook," RCA Tech. Ser. PMT-62, Lancaster, Pennsylvania.

29. Harris, J. M., Lytle, F. E., and McCain, T. C. (1976). *Anal. Chem.* **48**, 2095; Harris, J. M., and Lytle, F. E. (1977). *Rev. Sci. Instrumen.* **48**, 1469; Naqvi, K. R., Haggquist, G. W., Burkhart, R. D., and Sharma, D. K. (1992). *Rev. Sci. Instrum.* **63**, 5806.

30. Carlsson, J. (1988). *Z. Phys. D* **9**, 147.

31. Lopez-Delgado, F., Tramer, A., and Munro, I. H. (1974). *Chem. Phys.* **5**, 72.

32. O'Connor, D. V., Ware, W. R., and Andre, J. C. (1979). *J. Phys. Chem.* **83**, 1333.

33. Demas, J. N. (1983). *Excited State Lifetime Measurements*. Academic Press, New York.

34. Carlsson, J., Sturesson, L., and Svanberg, S. (1989). *Z. Phys. D* **11**, 287; Carlsson, J., Jonsson, P., and Sturesson, L. (1990). *Z. Phys. D* **16**, 87.

35. van der Veer, W. E., van Diest, R. J. J., and Donszelmann, A. (1993). *Z. Phys. D* **25**, 201.

36. Meister, E. C., Wild, U. P., Klein-Bolting, P., and Holzwarth, A. R. (1988). *Rev. Sci. Instrumen.* **59**, 499.

37. Marsden, G. C., Den Hartog, E. A., Lawler, J. E., Dakin, J. T., and Roberts, V. D. (1988). *J. Opt. Soc. Am. B* **5**, 606.

38. Stern, O., and Volmer, M. (1919). *Phys. Z.* **20**, 183.

39. Carrington, C. G., and Corney, A. (1971). *J. Phys. B* **4**, 849.

40. Haroche, S., Paisner, J. A., and Schawlow, A. L. (1973). *Phys. Rev. Lett.* **30**, 948.

41. Hannaford, P., Lowe, R. M., Grevesse, N., and Noels, A. (1992). *Astron. Astrophys.* **259**, 301.
42. Bard, A., Kock, A., and Kock, M. (1991). *Astron. Astrophys.* **248**, 315; Holweger, H., Bard, A., Kock, A., and Kock M. (1992). *Astron. Astrophys.* **249**, 545.
43. Bergeson, S. D., and Lawler, J. E. (1993). *Astrophys. J.* **408**, 382.
44. Bashkin, S., Meinel, A. B., Malmberg, P. R., and Tilford, S. G. (1964). *Phys. Lett.* **10**, 63.
45. Morgenstern, R., Niehaus, A., der Heer, F. J., Drentje, A. G., eds. (1987). "Proceedings of the Conference on Physics of Multiply-Charged Ions," *Nucl. Instrum. Methods* **B23**, 1.
46. Sorenson, G. (1976). In *Beam–Foil Spectroscopy*, I. A. Sellin and D. J. Pegg, eds. Plenum, New York; Poulsen, O., Andersen, T., Bentzen, S. M., and Koleva, I. (1982). *Nucl. Instrum. Methods* **202**, 139; Zeigler, J. F., Biersack, J. P., and Littmark, U. (1985). *The Stopping and Range of Ions in Solids*. Pergamon, Oxford.
47. Smith, P. L., and Whaling, W. (1969). *Phys. Rev.* **188**, 36.
48. Engstrom, L., and Bengtsson, P. (1991). *Phys. Scripta* **43**, 480.
49. Curtis, L. J., Berry, H. G., and Bromander, J. (1971). *Phys. Lett. A* **34**, 169; Curtis, L. (1976). Chapter 3 in *Beam–Foil Spectroscopy*, S. Bashkin, ed. Springer-Verlag, Berlin; Engstrom, L. (1989). *Phys. Scripta* **40**, 17.
50. Engstrom, L. (1982). *Nucl. Instrum. Methods* **202**, 369.
51. Macek, J., and Burns, D. (1976). Chapter 9 in *Beam–Foil Spectroscopy*, S. Bashkin, ed. Springer-Verlag, Berlin.
52. Sellin, I. A., and Pegg, D. J., eds. (1975). *Beam–Foil Spectroscopy* (2 volumes). Plenum, New York.
53. Cocke, C. L. (1976). Chapter 5.1 in *Methods of Experimental Physics*, Volume 13, Part B, D. Williams (ed.). Academic Press, New York.
54. Bashkin, S., ed. (1976). *Beam–Foil Spectroscopy*. Springer-Verlag, Berlin.
55. Beideck, D. J. *et al.* (1993). *Phys. Rev. A* **47**, 884.
56. Leckrone, D. S., Wahlgren, G. M., and Johansson, S. G. (1991). *Astrophys. J. Lett.* **377**, L37.
57. Maniak, S. T., Trabert, E., and Curtis, L. J. (1993). *Phys. Lett. A* **173**, 407.
58. Andra, H., Gaupp, A., and Wittman, W. (1973). *Phys. Rev. Lett.* **31**, 501.
59. Guo, B., Ansbacher, W., Pinnington, E. H., Ji, Q., and Berends, R. W. (1992). *Phys. Rev. A* **46**, 641.
60. Tanner, C. E. *et al.* (1992). *Phys. Rev. Lett.* **69**, 2765.
61. Weiman, C. E., and Hollberg, L. (1991). *Rev. Sci. Instrum.* **62**, 1.
62. Ansbacher, W., Li, Y., and Pinnington, E. H. (1989). *Phys. Lett. A* **139**, 165.
63. Gosselin, R. N., Pinnington, E. H., and Ansbacher, W. (1988). *Phys. Rev. A* **38**, 4887.
64. Andra H. J. (1982). *Nucl. Instrum. Methods* **202**, 123.
65. Baudinet-Robinet, Y., Garnir, H. P., Dumont, P. D., and El Himdy, A. (1989). *Phys. Scripta* **39**, 221; Dumont, P. D., Garnir, H. P., Baudinet-Robinet, Y., and El Himdy, A. (1988). *Nucl. Instrum. Methods* **B35**, 191; Baudinet-Robinet, Y., Dumont, P. D., Garnir, H. P., and El Himdy, A. (1989). *Phys. Rev. A* **40**, 6321; Baudinet-Robinet, Y., Dumont, P. D., and Garnir, H. P. (1991). *Phys. Rev. A* **43**, 4022; Bastin, T., Baudinet-Robinet, Y., Garnir, H. P., and Dumont, P. D. (1992). *Z. Phys. D* **24**, 343.
66. Doschek, G. A., Feldman, U., Van Hoosier, M. E., and Bartoe, J. D. (1976). *Astrophys. J. Suppl. Ser.* **31**, 417; Moos, H. W. *et al.* (1983). *Astrophys. J.* **275**, L19; Judge, P. G. (1986). *Mon. Not. Roy. Astron. Soc.* **221**, 119; Kastner, S. O., Bhatia, A. K., and Feibelman, W. A. (1989). *Mon. Not. Roy. Astron. Soc.* **237**, 487; Keenan, F.

P. *et al.* (1991). *Astrophys. J.* **379**, 406; Dufton, P. L. *et al.* (1991). *Mon. Not. Roy. Astron. Soc.* **253**, 474.

67. Meng, C. I., Rycroft, M. J., and Frank, L. A. eds. (1991). *Auroral Physics.* Cambridge University Press, New York; Ono, T., and Hirasawa, T. (1992). *J. Geomag. Geoelec.* **44**, 91; Gladstone, G. R. (1992). *J. Geophys. Res. Space Phys.* **97**, 1377; Hubert, D., and Kinzelin, E. (1992). *J. Geophys. Res. Space Phys.* **97**, 4053.

68. Doschek, G. A. (1985). In *Autoionization*, A. Temkin (ed.), p. 171. Plenum, New York.

69. Dehmelt, H. G. (1967). *Adv. At. Mol. Phys.* **3**, 53; Dehmelt, H. G. (1969). *Adv. At. Mol. Phys.* **5**, 109; Wineland, D. J., Itano, W. M., and Van Dyck, R. S., Jr. (1983). *Adv. At. Mol. Phys.* **19**, 136.

70. Roth, A., Gerz, C., Wilsdorf, D., and Werth, G. (1989). *Z. Phys. D* **11**, 283.

71. Calamai, A., Han, X., and Parkinson, W. H. (1992). *Phys. Rev. A* **45**, 2716.

72. Prior, M. H. (1984). *Phys. Rev. A* **30**, 3051.

73. Kwong, V. H. S. (1989). *Phys. Rev. A* **39**, 4451.

74. Calamai, A., and Johnson, C. E. (1990). *Phys. Rev. A* **42**, 5425.

75. Calamai, A., and Johnson, C. E. (1992). *Phys. Rev. A* **45**, 7792.

76. Walch, R. A., and Knight, R. D. (1988). *Phys. Rev. A* **38**, 2375.

77. Gerz, C., Hilberath, T., and Werth, G. (1987). *Z. Phys. D* **5**, 97; Gerz, C., Roths, J., Vedel, F., and Werth, G. (1988). *Z. Phys. D* **8**, 235.

78. Itano, W. M., Bergquist, J. C., Hulet, R. G., and Wineland, D. J. (1987). *Phys. Rev. Lett.* **59**, 2732.

79. Whaling, W., and Brault, J. W. (1988). *Phys. Scripta* **38**, 707.

80. Whaling, W., Hannaford, P., Lowe, R. M., Biemont, E., and Grevesse, N. (1984). *J. Quant. Spectros. Radiat. Transfer* **32**, 69.

81. Thompson, M., and Walsh, J. N. (1989). *Handbook of Inductively Coupled Plasma Spectrometry.* Blackie, Glasgow.

82. Humphrey, J. N., Adams, D. L., and Whaling, W. (1984). *J. Quant. Spectrosc. Radiat. Transfer* **31**, 1; Hudson, R. S., Skrumeda, L. L., and Whaling, W. (1987). *J. Quant. Spectrosc. Radiat. Transfer* **38**, 1.

83. See, for example, Wyatt, C. (1978). *Radiometic Calibration: Theory and Methods.* Academic Press, New York; Thorne, A. P. (1988). *Spectrophysics*, 2nd Edition. Chapman and Hall, London.

84. See for example, Bell, R. J. (1972). *Introductory Fourier Transform Spectroscopy.* Academic Press, New York; Griffiths, P. R. (1986). *Fourier Transform Infrared Spectrometry.* Wiley, New York.

85. Brault, J. W. &(1976). *J. Opt. Soc. Am.* **66**, 1081; Brault, J. W. (1979). *Osserve. e Mem. dell Osserv. Astr. di Arcetri* **106**, 33.

86. Thorne, A. P., Harris, C. J., Wynne-Jones, I., Learner, R. C. M., and Cox, G. (1987). *J. Phys. E* **20**, 54; Thorne, A. P. (1991). *Anal. Chem.* **63**, 57A.

87. Rao, K. N. (1966). *Wavelength Standards in the Infrared.* Academic Press, New York; Reader, J., and Corliss, C. H. (1980). "Wavelength and Transition Probabilities for Atoms and Atomic Ions, Part 1," *NSDRS-NBS* **68**, U.S. Government Printing Office, Washington; Kelly, R. L. (1987). *J. Phys. Chem. Ref. Data* **16**, Suppl. 1.

88. Simmons, J. D. (1991). "NIST Calibration Services Users Guide 1991," NIST Special Publication 250, U.S. Government Pringint Office, Washington, D.C.

89. Wiese, W., Brault, J. W., Danzmann, K., Helbig, V., and Kock, M. (1989). *Phys. Rev. A* **39**, 2461; Whaling, W., Carle, M. T., and Pitt, M. L. (1993). *J. Quant. Spectrosc. Radiat. Transfer* **50**, 7.

90. Fernley, J. A., Taylor, K. T., and Seaton, M. J. (1987). *J. Phys. B* **20**, 6457.

91. Duquette, D. W., Den Hartog, E. A., and Lawler, J. E. (1986). *J. Quant. Spectrosc. Radiat. Transfer* **35**, 281.
92. Thorne, A. P. (1988). *Spectrophysics*. Chapman and Hall, London.
93. King, R. B., and King, A. S. (1935). *Astrophys. J.* **82**, 377.
94. Blackwell, D. E., and Collins, B. S. (1972). *Mon. Not. Roy. Astron. Soc.* **157**, 255; Blackwell, D. E., Ibbetson, P. A., and Petford, A. D. (1975). *Mon. Not. Roy. Astron. Soc.* **171**, 195.
95. Blackwell, D. E., Ibbetson, P. A., and Petford, A. D. (1979). *Mon. Not. Roy. Astron. Soc.* **186**, 633; Andrews, J. W., Coates, P. B., Blackwell, D. E., Petford, A. D., and Shallis, M. J. (1979). *Mon. Not. Roy. Astron. Soc.* **186**, 651; Blackwell, D. E., Petford, A. D., and Shallis, M. J. (1979). *Mon. Not. Roy. Astron. Soc.* **186**, 657.
96. Kock, M., Kroll, S., and Schnehage, S. (1984). *Phys. Scripta* **T8**, 84.
97. Blackwell, D. E., Petford, A. D., and Simmons, G. J. (1982). *Mon. Not. Roy. Astron. Soc.* **201**, 595; Blackwell, D. E., Ibbetson, P. A., Petford, A. D., and Willis, R. B. (1976). *Mon. Not. Roy. Astron. Soc.* **177**, 219; Blackwell, D. E., and Shallis, M. J. (1979). *Mon. Not. Roy. Astron. Soc.* **186**, 669; Blackwell, D. E., Petford, A. D., Shallis, M. J., and Simmons, G. J. (1980). *Mon. Not. Roy. Astron. Soc.* **191**, 445; Blackwell, D. E., Petford, A. D., Shallis, M. J., and Simmons, G. J. (1982). *Mon. Not. Roy. Astron. Soc.* **199**, 43.
98. Booth, A. J., Blackwell, D. E., Petford, A. D., and Shallis, M. J. (1984). *Mon. Not. Roy. Astron. Soc.* **208**, 147.
99. Blackwell, D. E., Menon, S. L. R., and Petford, A. D. (1984). *Mon. Not. Roy. Astron. Soc.* **207**, 533.
100. Lowe, R. M., and Hannaford, P. (1991). *Z. Phys. D* **21**, 205.
101. Bergeson, S. D., and Lawler, J. E. (1993). *J. Opt. Soc. Am. B* **10**, 794.
102. Federman, S. R., Beideck, D. J., Schectman, R. M., and York, D. G. (1992). *Astrophys. J.* **401**, 367; Thevenin, F. (1989). *Astron. Astrophys. Suppl. Ser.* **77**, 137.

# 13. DOPPLER-FREE SPECTROSCOPY

## James C. Bergquist

National Institute of Standards and Technology
Boulder, Colorado

## 13.1 Introduction

Spectroscopy is an important tool for investigating the structure of physical systems such as atoms or molecules. Thermal motion of free atoms and molecules gives rise to Doppler broadening of the characteristic spectral transitions, which often blurs important details of the spectra and prevents a deeper understanding of the underlying physics. Long before the appearance of lasers, well before their use in nonlinear, high-resolution spectroscopy, long before laser-cooling, atomic and molecular beams had been used for high-resolution spectroscopy. And, perhaps not surprisingly, beam techniques remain prominent in spectral studies. In the course of this chapter, we will seek a rudimentary appreciation of the methods used to reveal Doppler-free spectra. But first we will begin with a brief discussion of motional effects that produce line broadening and shift. before we address their countermeasures.

## 13.2 Spectral Line-Broadening Mechanisms

Atoms or molecules in gases can be relatively free and undisturbed, but their spectral lines are spread out over a range of frequencies by the Doppler effect because they are moving in all directions with high thermal velocities. Those particles moving toward an observer absorb light at lower frequencies than those at rest; those receding absorb at higher frequencies. The resulting Doppler broadening can often mask spectral fine and hyperfine structure, even though each individual absorber still retains its typically much narrower natural line width. The spectral line of a moving particle is shifted away from its rest-frame frequency $\omega_0$ to $\omega_0 + \mathbf{k} \cdot \mathbf{v}$, where $\mathbf{k}$ is the wave vector of the interrogating radiation and $\mathbf{v}$ is the velocity of the particle. Since the particles move isotropically in all directions, the Doppler shift is different for each particle; this is an example of inhomogeneous broadening. At thermal equilibrium the velocity distribution is Maxwellian:

EXPERIMENTAL METHODS IN THE PHYSICAL SCIENCES
Vol. 29B

$$w(v) = \left(\frac{1}{\sqrt{\pi}u}\right) \exp(-v^2/u^2), \qquad u = (2k_B T/m)^{1/2}, \qquad (13.1)$$

where $m$ is the particle's mass, $T$ is the temperature (in kelvins), and $k_B$ is Boltzmann's constant. The spectral lineshape of the ensemble is obtained by the convolution of the lineshape of an individual particle with a Maxwellian distribution of velocity-projected frequency shifts. The result is the Voigt profile with a characteristic spectral width at half maximum (FWHM) given by

$$\Delta\omega_D = (2\omega_0/c)((2kT/m)\ln 2)^{1/2}. \qquad (13.2)$$

For quantum absorbers with an atomic mass near 100 uu and near room temperature (300 K), the Doppler width $\Delta\omega_D \sim 10^{-6}\omega_0$. For optical frequencies $\omega_0 \sim 2\pi(5 \times 10^{14}\,\text{Hz})$, Doppler broadening ranges from $10^8$ to $10^{10}$ Hz. Laser-cooling routinely produces atomic temperatures below $10^{-3}$ K, but the Doppler width even at 1 μK remains substantial, ranging from $10^4$ to $10^6$ Hz. For high-resolution spectroscopy of ultracold free particles, Doppler-free interrogation is still important.

The motion of atoms or molecules broadens and shifts spectral features through other mechanisms as well. The finite interaction time of the particles with the radiation field produces line broadening due to the energy–time uncertainty principle. We may solve for the lineshape as the Fourier time transform of an undamped harmonic oscillator viewed for a finite time interval $\Delta t$. The magnitude of the FWHM is

$$\Delta\omega_t \simeq 2\pi(0.89)/\Delta t. \qquad (13.3)$$

This broadening effect very often results from the travel time $\Delta t = L/v$ of a freely moving quantum system through a radiation region of finite extent $L$. If the radiation field is not uniform, but rather is a nearly planar Gaussian beam that probes the atoms perpendicular to their velocity, then the spectral lineshape has a linewidth

$$\Delta\omega_t = (2\sqrt{2}\ln 2)(v/w_0), \qquad (13.4)$$

where $2w_0$ is the Gaussian beam diameter. This spectral width is nearly the same as that obtained for the finite uniform light field with $L = 2w_0$. The wave surface of a Gaussian beam is strictly not planar, but curved away from its beam waist. The curvature of the wave surface through which the atom passes restricts the effective transit time and broadens the line. The line width is [1]

$$\Delta\omega_R = (2v\sqrt{2}\ln 2)(w_0^{-2} + w_R^{-2})^{1/2}, \qquad (13.5)$$

where $w_R = R\lambda/\pi w_0$, and $R$ is the radius of curvature of the Gaussian light beam. In order that the curvature of the probe field not significantly broaden the line above the transit time limit, $R \gg \pi w_0^2/\lambda$. For thermal velocities near $T = 300$ K, transit time linewidths typically vary from $10^3$ to $10^5$ Hz.

above the transit time limit, $R \gg \pi w_0^2/\lambda$. For thermal velocities near $T = 300$ K, transit time linewidths typically vary from $10^3$ to $10^5$ Hz.

Another pernicious effect of atomic motion is the spectral line broadening caused by interparticle collisions or collisions with the walls of the vessel that confines the gas. Such collisions intermittently alter the phase of the resonance frequency of the particle and sometimes cause the cessation of the atomic precession (state decay). State decay can occur in a collision if the energy difference between atomic levels is not large compared to the kinetic energy of the collision partners. Hence, collisions causing state decay are relatively rare in the case of optical transitions in atoms, but are not uncommon for mid-infrared transitions in molecules. Usually, collisions that cause state decay are treated phenomenologically as a modification to the natural decay rate, so the spectral line is broadened, but not shifted. The amplitude of the atomic phase change caused by a random collision depends on the relative impact parameter and velocity of the colliding partners. If the phase change $\Delta\theta \ll \pi$, then the spectral feature is not broadened, but it is shifted by the average phase shift $\langle\theta\rangle$ to $\omega_0 + \langle\theta\rangle/\Delta t_c$ where $t_c$ is the mean duration time between collisions. For stronger collisions, $\Delta\theta \gg \pi$, and the phase of the atomic oscillation is abruptly changed during the short collision time. The atomic resonance is not shifted in this process, but it is broadened to [1]

$$\Delta\omega_c \simeq 2/\Delta t_c. \tag{13.6}$$

The interparticle broadening due to collisions is pressure dependent with typical molecular values that range from 8 to 240 kHz/Pa (1–30 MHz/torr). If the pressure is reduced until the mean free path exceeds the dimensions of the gas cell, then wall collisions dominate. Particles that collide with the cell wall can also state-decay, but most certainly suffer velocity changes and phase shifts that generally terminate their contribution to the signal. The line broadening is inversely proportional to the transit time of the particle across the cell. However, in the next section we will show that, in some cases, collisions can lead to Doppler narrowing!

## 13.3 Spectral Line-Narrowing Techniques

Limitations in optical spectroscopy caused by the broadening and shift due to thermal motion have been reduced and even eliminated by numerous techniques developed over the past half-century. The first generally productive counter-measure to Doppler broadening was the method of atomic beams developed in the 1930s and still widely used [2]. The idea is straightforward: $\mathbf{k} \cdot \mathbf{v}$ is minimized by probing *all* atoms in a direction that is nearly perpendicular to the atomic motion. The direction of the velocity vector is restricted to a small cone by a collimator (or a sequence of collimators) placed in front of an effusive source.

The particles travel out from the source through the collimator into a high-vacuum region, where they are intercepted by the probe radiation. The practical degree of collimation ranges from 10 to $10^3$, which reduces the Doppler broadening proportionally. For oven temperatures near 300 K or higher, Doppler widths of atomic beams remain stubbornly large ($\geq 10^7$ Hz). Higher collimation is possible but lowers the particle flux and thereby the signal. The loss in signal-to-noise ratio is tolerable if substructure can be revealed. Laser-cooling of the atomic source makes possible intense beams of cold atoms (even near monovelocity beams of atoms). Cold atomic beams are discussed elsewhere in this volume, but as stated earlier, temperatures near $10^6$ K still give Doppler broadening near $10^3$–$10^5$ Hz for atomic transitions in the visible.

The high intensity per unit bandwidth characteristic of lasers makes it possible to virtually eliminate the first-order Doppler broadening in gas samples. This can be done with either of two nonlinear techniques, saturated absorption [1, 3–6] or two-photon spectroscopy [1, 3, 4, 8, 9]. In saturation spectroscopy, a group of absorbers in a narrow interval of axial velocities is labeled by their nonlinear interaction with a monochromatic traveling laser wave, and a Doppler-free spectrum of these velocity-selected absorbers can then be observed with a second laser beam used as a probe. In order to be sensitive to the natural resonant frequency in the laboratory frame, parallel, counter-running waves of the same frequency (which may or may not be coincident spatially) are used. Since the Doppler shifts have the opposite sign for the two beams, the two traveling-wave frequencies are viewed as equivalent only by absorbers essentially free of velocity along the light axis. Thus, if the laser is tuned to the absorber's resonant frequency, the two waves work together to saturate the same "zero velocity" class of absorbers, which then produces a narrow spectral resonant feature [5–7].

In two-photon spectroscopy, the transition energy of the atom or molecule is provided by the simultaneous absorption of two photons. If the photons each are absorbed from one of two counter-traveling waves that have the same optical frequency, then the Doppler shift $\mathbf{k} \cdot \mathbf{v}$ of one photon is precisely cancelled by the Doppler shift of the other photon. The condition for resonance is that twice the field frequency equals the unshifted frequency of the two-quantum transition. Independent of their velocities (ignoring second-order Doppler effects), all atoms contribute to the two-photon Doppler-free signal at $\omega = \omega_0$. This is in contrast to saturation spectroscopy, where only a small slice of the velocity distribution that is essentially normal to the laser beam can interact with the radiative driving field. Two-photon spectroscopy has other advantages compared to saturated absorption spectroscopy [1, 3, 4].

The resonance is not broadened or shifted by wavefront curvature of the light beams (*provided* that the wavefronts of the counter-running beams are matched), because the two photons are absorbed simultaneously. In saturated absorption studies, the wavefronts needed to be flat in order to well define the narrow velocity distribution needed for high resolution. Since the direction of atomic motion is not important in two-photon spectroscopy, long interaction times are made possible by using an atomic beam directed along (or at a small angle to) the optical axis. In saturation spectroscopy, the atomic motion must be orthogonal to the light beam; for long interaction times, large-aperture optics of high quality (expensive) are needed to give a resolving power limited by transit-time broadening. Two-photon spectroscopy is not without its problems. Residual first-order Doppler effects will remain unless the wavefronts of the counter-running beams are perfectly matched, that is, the wavevectors must be oppositely directed everywhere. Wavevectors can be arbitrarily well opposed by carefully mode-matching the laser beam into an optical cavity that also serves to enhance the intensity in the light beams. A relatively large broadening and shift are produced by the optical Stark effect. In the general case, the coefficients for the optical Stark shift and for the probability for the two-photon transition are nearly the same. Therefore, a compromise must be struck between signal strength and Stark shift.

In Section 13.2, we remarked that particle collisions change the velocity of the quantum radiators interacting with the light field, which can *reduce* Doppler broadening. This narrowing can result if the particle's motion is altered, but otherwise there is little or no change in the phase of the oscillator during the collision. In other words, it is important that the precession frequencies of the particles remain phase-coherent (or nearly so). If there are many collisions during the effective decay time of the particle's oscillation, then the Doppler-shifted frequency $\omega_0 + \mathbf{k} \cdot \mathbf{v}$ is frequently altered and essentially averaged to zero. This is termed Dicke narrowing after the author who first described it [10]. Significant narrowing occurs if the collisions confine the particles to volumes with dimensions smaller than the wavelength of the interrogating light and the decay rate of the oscillating dipole remains small ($\gamma \ll 1/\Delta t_c$). Until recently, this has been difficult at optical frequencies, where collisional broadening simply overwhelms any Dicke narrowing. However, laser-cooling and trapping of atoms [11] and ions [12] have revived interest in Dicke's method applied to Doppler-free spectroscopy; we will give an example and a more complete discussion of Dicke narrowing in the last section of this chapter.

Armed with these and other Doppler-free techniques, spectroscopists have pushed fractional resolution in the optical domain to nearly $1 \times 10^{-15}$. Pressure broadening and shift and geometrical broadening and shift offer a practical challenge, but in many experiments, these have been virtually eliminated. The spectral width of naturally narrow optical transitions is most often dominated by transit-time broadening and second-order Doppler effects (*and* by the spectral purity of the laser, but that is a topic to be treated in the third volume of this series, *Electromagnetic Radiation*). Second-order Doppler effects can be partially reduced by labeling and using one velocity class of absorbers, but only with a loss in signal. Clearly, most broadening and shift problems would be reduced if the particle temperature were lowered. Particles at rest have no motional effects. It is understandable, then, that laser-cooling of free and trapped particles has become a prominent tool in the research and spectral studies of many atomic physicists.

For room-temperature gas samples, one quickly encounters practical limits (and in the case of saturated absorption, loss of signal) when the interaction time is increased. A partial circumvention of the practical limitations imposed by transit-time broadening uses the interference that arises from the sequential interactions of the induced dipole of a quantum absorber with radiation fields separated in time [13–15]. Transit-time broadening is then determined by the temporal separation of the light fields. In our first example, we will look more closely at the Ramsey method of separated atom–field interaction regions and why it is particularly useful in saturated absorption.

## 13.4 Saturated Absorpion with Ramsey Interference Fringes

With the elimination of first-order Doppler effects, attention turned to the resolution limitation imposed by interaction-time broadening. To be transit-time limited in saturated absorption spectroscopy, the projection of atomic motion onto the radiation-field axis must be such that $\mathbf{v} \cdot \mathbf{k}$ is less than the inverse of the transit time through the beam. Hence, the greater the resolution, the stricter the velocity selection, and so fewer atoms contribute to the signal. A particularly simple yet powerful scheme to greatly reduce interaction-time broadening and attain spectrally narrow lines at microwave frequencies was proposed by Ramsey [13]. He suggested intercepting a beam of atoms with spatially separated radiation fields (derived from the same source) so that independent phase evolution of atoms and radiation field could occur during the free flight of the atoms between fields. Temporal modulation of the field could also be achieved by pulsing the field on and off, provided that the atoms were slow enough to remain in the interaction region or were otherwise confined spatially. The interaction of absorber and field in the first radiation beam (pulse) produces a coherent superposition of upper and lower states. The induced atomic polarization precesses at the natural resonance frequency in the field-free region between the radiation zones (between pulses). The effect of the second light field (pulse) on the atomic transition probability depends on the relative phase of the radiation field to the absorber's oscillation, so the absorbers passing through the second field (pulse) will either be further excited or returned to the ground state. Thus, an interference results, because the quantum absorption transition probability no longer depends only on the frequency of the driving field, but also on the phase-evolution difference of the atom and the field during the free flight of the atom between zones (between pulses). The phase difference is proportional to the product of the frequency detuning of the radiation field from resonance ($\omega_0 - \omega$) and the interzone (interpulse) transit time $T$. For spatially separated beams, $T = L/v_x$, where $L$ is the interzone spacing and $v_x$ is the atomic speed orthogonal to the axis $\mathbf{z}$ of the light beams and in the plane containing them. In the pulsed case, the free precession period is independent of the atomic velocity. Note that on resonance, all atoms constructively contribute to the signal independent of their velocities (ignoring, for the moment, second-order Doppler effects). The central fringe of the Ramsey absorption profile is centered at $\omega = \omega_0$, with a spectral width determined by the time between the radiation regions (pulses), rather than by the transit time through each region (pulse time). However, synthesis of Ramsey's idea and saturated absorption in the optical domain is difficult because the wavelength is shorter than the dimensions of the interaction region. Even in a well-collimated atomic beam, there will be a spread of the residual Doppler velocity projections $v_z$ on the direction of light wave propagation. Likewise, in saturated absorption spectroscopy, the criterion that $\mathbf{v} \cdot \mathbf{k}$ be less than the inverse of the interaction time in the first light beam (pulse) does not prevent a spread of the atoms over several wavelengths in the second light beam.

By way of example, consider a beam of quantum absorbers crossing a collimated laser beam of waist size $w_0$. Near perpendicular incidence, there is a narrow acceptance angle, $\delta\theta \equiv v_z/v_x \approx \lambda/\pi w_0$, in which the accumulated phase difference between the field frequency and the Doppler-shifted frequency (seen by the moving absorber) never exceeds $\pi$ radians. That is, absorbers within this slice enter and leave the radiation field before any mismatch of their frequency to the field frequency is recognized (transit-time broadening exceeds the residual Doppler broadening). Adding a second optical interaction zone a distance L downstream does not lead to strong Ramsey fringes because the angular slice defined by the first interaction maps into a large extension, $\Delta z = L \cdot \delta\theta = L\lambda/3w_0$, in the second field interaction region. The condition for increased resolution by the Ramsey interference, $L/w_0 \gg 1$, is just the condition that the dipoles originating at one spatial position in the first light field will be spread out over several wavelengths in the second field region. The Ramsey signal is spatially averaged to zero, since particles with the *same* field-free precession time experience *different* phases of the second driving field dependent on their spatial entry position into the second zone. The same arguments apply to the pulse interrogation method.

Baklanov and his colleagues introduced the idea of a third, equally spaced interaction zone as a method to recover the Ramsey fringes [14]. However, as a way to better see the phase relationships leading to the interference signal, it is more convenient to assume four interaction zones (pulses) [15–17]. In saturation spectroscopy, the narrow resonances arise from four conceptually separate time-ordered interaction processes: lower state population to dipole, dipole to upper state population, upper state population to dipole, and finally, dipole dotted with the field to project out the excitation probability. The saturated absorption Ramsey interference signal develops when the interactions occur sequentially, one per zone (pulse). As in the RF and microwave case, the dipole prepared by the first interaction will precess at its own natural frequency $\omega_0$ and decay with a dipole decay rate $\gamma_{ab}$ in the darkness between the first two light beams (pulses). In the second interzone (interpulse) space, the excited-state population will only decay, with the population decay rate $\gamma_b$. In the third interzone (interpulse) space, the system again carries a dipole moment and precesses and decays as before. Assuming the interzone (interpulse) times to be $T$, $aT$, $bT$, respectively and ignoring the field phase (assumed fixed) leads to the expression for the total phase of the Ramsey signal [15]:

$$\exp[i\phi_{\text{total}}] = \exp[-ikv_z(b-1)T]\exp[-i(\omega-\omega_0)(b+1)T]$$
$$\times \exp[-\gamma_{ab}(b+1)T]\exp[-\gamma_b aT]. \tag{13.7}$$

Because of the distribution of $v_z$ values, the Ramsey fringes appear only for the case $b = 1$, that is, when the dipole free precession times are equal. The second field-free region does not improve spectral resolution, but it can enhance the Ramsey signal [16, 17]. We can also show that the fringes arise only for the case of two separated interactions with parallel, co-running beams followed by two separated interactions with parallel, oppositely running beams. The spatial modulation in the atomic response produced by the first two field interactions can be exactly unfolded in the probe process to reveal the narrow Ramsey signal, but only if the second two interactions are separated by the same time interval as the first two. Unfolding the spatial modulation to reveal the narrow Ramsey structure has two important consequences:

- Essentially all atoms for which the broadening due to the field interaction time exceeds the residual Doppler broadening contribute to the Ramsey interference signal. If the radiation field filled the entire effective aperture $2L$ (or if the pulse duration extended for the effective free precession time $2T$), then only those absorbers with $v_z \leq \lambda v_x/2\pi L$ ($v_z \leq \lambda/2\pi T$) would contribute to the signal. Thus, the Ramsey method permits significant line narrowing with essentially no loss in signal. In fact, it is best to shorten the field interaction time to increase the number of contributing atoms.
- Large aperture optics are not necessary.

Saturated absorption optical Ramsey fringes were first observed in the experiment shown in Figure 1 [15], where a fast ($v/c \approx 10^{-3}$), monovelocity ($\Delta v/v \approx 10^{-4}$) beam of metastable $^{20}$Ne atoms sequentially interacted with spatially separated light beams from a frequency-stabilized dye laser. The

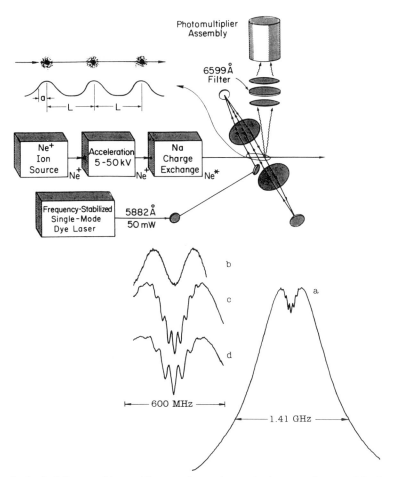

FIG. 1. Optical Ramsey fringes. The experimental setup is shown at the top of the figure. A fast ($v/c \sim 10^{-3}$) monovelocity ($\Delta v/v \sim 10^{-4}$) beam of metastable neon atoms sequentially interacts with three spatially separated, standing-wave light beams formed by the opposition of two cat's-eye retroreflectors. A dye laser at 588 nm drives the $1s_5$–$2p_2$ transition and the excitation probability is monitored by fluorescence emission on the $2p_2$–$1s_2$ transition at 660 nm. The fluorescence signals are shown at the bottom of the figure. Curve a shows most of the beam Doppler profile, the saturation dip and the narrow Ramsey fringes. Curves b–d show the fluorescence signals obtained near resonance for two separated laser beams (b), for three beams (c), and for four beams (d), respectively (the distance between first and last zones is the same for c and d). (From reference[15].)

opposition of two cat's-eye retroreflectors to produce the parallel radiation zones intrinsically provides a stable spatial phase relationship from beam to beam [15]. Several fluorescent profiles are included in Figure 1. Curve (a) shows most of the beam Doppler profile, the saturated absorption dip, and the fringes due to the atom's interaction with three equally spaced standing-wave radiation zones. Curve (b) is the fluorescence profile produced when the Ne beam interacts with only two standing waves. Consistent with our discussion earlier, the optical Ramsey fringes are averaged to zero. The spectral profiles of curves (c) and (d) show the interference fringes obtained when the atomic beam interacts with three and four equally spaced standing waves, respectively. Spectral line narrowing with little loss in signal (compared to the Lamb dip) is apparent. To a good approximation [13], the fringe pattern for equal relaxation constants $\gamma$ is an exponentially damped cosine proportional to

$$(1/v^2)\exp[-(\omega-\omega_0')^2(w_0/v_r)^2]\cos[(\omega-\omega_0')(2v_xL/w_0^2-\gamma)(w_0/v_r)^2)], \quad (13.8)$$

where $v^2 = v_r^2 + v_z^2$, $v_r^2 = v_x^2 + v_y^2$, $\omega_0' = \omega_0(1 - v^2/2c^2 \pm \hbar k/2mc)$ and where $c$ is the speed of light and $2\pi\hbar$ is Planck's constant. The shift in the natural resonance frequency due to photon recoil $\pm \hbar k/2mc$ and due to the second-order Doppler shift $v^2/2c^2$ has been explicitly included in the expression for $\omega_0'$. Usually, these terms are ignored but with long-lived transitions that afford higher resolution, subtle shifts and distortions of the saturated absorption resonance caused by these terms can be seen. Conservation of 4-momentum (momentum and energy) requires that, in absorption, the atom absorb the energy and momentum of the photon, thereby changing not only the atom's internal energy, but also its motional energy. Similarly, in the emission process the atom must change its kinetic and internal energy to provide for the emitted photon's energy and momentum. Thus, the frequency of the absorbed photon is blue-shifted relative to the Doppler-shifted natural resonance frequency of the atom and, correspondingly, the emitted photon is red-shifted. In linear spectroscopy, the frequency shift due to the photon recoil cannot be directly detected, but in saturation spectroscopy, photon recoil manifests itself as a spectral doublet. There are two frequencies of anomalously high transmission for the probe wave. One occurs at the frequency where the atom's ground state density is reduced by the power wave interacting with the same set of absorbers. The second transmission peak comes at the frequency where the power wave interacts with the same velocity class of excited-state absorbers. The two peaks are symmetrically displaced about the Bohr frequency, with a frequency splitting $\delta\omega = \hbar k^2/m$. Early high-resolution experiments in saturation spectroscopy [18–21] fully resolved the recoil doublet and demonstrated the possibility of a direct frequency measurement of $\hbar/m$ for atomic systems. Laser-cooling has renewed interest in the precise measurement of atomic recoil (better described now in terms of optical atomic interferometers with single-photon or two-photon beamsplitters [22]) both for an alternative determination of the fine structure constant through $\alpha^2 = (2R_\infty/c)(h/m)(m/m_e)$ [23] and for a new mass standard [24] ($R_\infty$ is the Rydberg and $m_e$ is the mass of the electron).

The line center and spectral shape become highly sensitive to resolving power when the fractional frequency of resolution approaches that of the second-order Doppler shift [19, 25, 26]. This is particularly true for Ramsey spectroscopy with thermal beams. The resonance frequency for a particular velocity $v$ is shifted by $-(v/c)^2/2$, but the interfringe spacing is proportional to $v/2L$ (assuming $v_x \simeq v$). Velocity averaging washes out the fringe structure away from the spectral line center, because each cosine, with period dependent on $v$, contributes coherently only at line center, but adds with random phase elsewhere. When the interference fringe spacing begins to approach the same order of magnitude as the second-order Doppler shift, then there is no frequency, or "line center," where all velocity contributions add coherently. Rather, there is an arbitrary frequency position which receives the greatest portion of constructive interference that depends on the interzone spacing and on the atomic beam velocity distribution. Experiments in which a thermal beam of calcium ions was probed by the saturated absorption Ramsey method using widely separated beams revealed lineshapes that were distorted by second-order Doppler effects [19, 25]. But recent, beautiful work with cold atoms and Ramsey interrogation with pulsed beams has virtually eliminated problems due to second-order Doppler shifts [27].

## 13.5 Two-Photon Spectroscopy

Doppler-free two-photon spectroscopy is not as hampered by transit-time effects as linear- and saturated-absorption spectroscopy. If two photons of the same energy are absorbed from counter-running beams whose wavevectors are strictly opposed, then there is no momentum transfer between the electromagnetic field and the atom, and consequently no Doppler broadening, independent of the particles' motion. Long interaction regions are possible with simple optics and reasonably collimated atomic beams. Power broadening and shift and second-order Doppler effects tend to be the obstructions to higher resolution in two-photon experiments. The most familiar source of power broadening is saturation, which occurs whenever the incident laser intensities are sufficiently strong to equalize the populations of the two states connected by the nonlinear transition. This effect only broadens the lineshapes. There will also be a second-order shift between any pair of energy levels connected by an electric dipole interaction with an square of the oscillating electric field. In general, the light shift (AC Stark shift) is proportional to the square of the on-resonance Rabi rate divided by the detuning, $(\boldsymbol{\mu} \cdot \mathbf{E}/\hbar)^2/(\omega - \omega_0)$ [3, 4], where $\boldsymbol{\mu}$ is the atomic dipole moment and $\mathbf{E}$ is the amplitude of the field. In linear or saturation spectroscopy, the light shift goes unnoticed because the atomic transition frequency is moved symmetrically about the line center. For red detuning, the coupled energy states are pushed apart; for blue detuning, the levels are pushed together. The magnitude of the light shift is the same but oppositely directed for equal detunings away from resonance. Hence, the observed spectral feature is unshifted. Unfortunately, light shifts can never be entirely eliminated in two-photon spectroscopy, because the transition between levels is made by virtual, dipole-allowed transitions through intermediate states that are usually far from resonance. To find the quadratic frequency shift of either atomic level connected by the two-photon transition, it is necessary to sum over all intermediate states [3, 4]:

$$\Delta w_i = (1/4\hbar^2) \sum \{|\mathbf{\mu}_{ni} \cdot \mathbf{E}|^2/(\omega_{ni} - \omega) + |\mathbf{\mu}_{ni} \cdot \mathbf{E}|^2/(\omega_{ni} + \omega)\}. \qquad (13.9)$$

The difference of the individual shifts to the initial and final states makes up the optical Stark shift for the two-photon transition. Equation (13.9) shows that the effect is proportional to intensity, while the two-photon absorption signal is proportional to the square of the intensity. The persistence of a light shift in two-photon spectroscopy for arbitrary light intensities makes difficult the precise determination of the unperturbed line center. Nonuniform field intensities and different atom–field interaction histories further complicate the spectral shape. Experiments done at several intensity levels help extrapolate the Stark shift to zero intensity, but often brute-force computer analysis is needed to decipher the unshifted position of the line center [28].

One proposal to reduce the quadratic shift and at the same time preserve the signal strength was to use the Ramsey method of separate interrogation regions, whereby the atom spends most of its time freely precessing between sequential interactions with the high-intensity field(s). It was expected that the Stark shift would be diluted by the ratio of the dark-precession time to the field-interaction time. But, while this effect was seen in the shift of the line-center position of the Ramsey signal relative to that of the two-photon signal from a single interaction region [26], properly the comparison should be made between a Ramsey two-photon signal and a single-zone two-photon signal that are of equal amplitude *and* resolution. Because the probability of two-photon excitation depends on the product of the interaction time and the square of the field intensity, the single-zone intensity can be proportionately lowered if the interaction time is increased. Hence, no difference is expected in the magnitude of the Stark shift if the resolution and probability for excitation remain the same. Unfortunately, Doppler-free two-photon spectroscopy is never free of AC Stark effects.

Second-order Doppler effects also persist for moving atoms in two-photon spectroscopy. The spectral line shape is largely unaffected if the effective line width is much larger than the second-order shift $\omega_0(v/c)^2$. The shift remains, but it is probably not important in this case. When the spectral resolution is increased until it is comparable to or exceeds the second-order shift for the mean thermal velocity, then the line shape is also distorted. For a Maxwell–Boltzmann velocity distribution, Minogin [29] finds a sharply asymmetrical line profile whose peak height is red-shifted by $(\omega_0/2)(u/c)^2$ (the line center is difficult to define).

The first demonstrations of Doppler-free two-photon absorption were done by several groups using atomic sodium [1, 3, 4]. Since then, two-photon spectroscopy has been used to study a great many transitions between states of the same parity and to determine term values, pressure shifts, broadening coefficients, and other parameters. Without question, however, the example for two-photon spectroscopy that is rich in physics and storybook in character is the continuously increasing precision spectroscopy of hydrogen that has been orchestrated by Professor Hänsch over the past two decades. Spectral studies of hydrogen have already played an important role in the development of atomic physics and quantum mechanics, from the Balmer spectrum and the Bohr model to the Lamb shift and QED. Largely because of its simplicity, it remains a dominant tool in Hänsch's research. Much of his work centers around the narrow two-photon transition from the $1S$ ground state to the metastable $2S$ state ($\gamma_{2s} \simeq 7$ Hz). A direct frequency measurement of this transition has produced the most accurate measurement of the Rydberg constant [30]. The frequency difference between the $1S$–$2S$ two-photon transition in hydrogen and in deuterium has provided a precise measure of the RMS structure radius of the deuteron [31]. In addition, the direct frequency comparison of the $1S$–$2S$ transition to the $2S$–$4S$ two-photon transition has furnished a measurement of the ground state Lamb shift to an accuracy of 6 ppm [32], which exceeds the best RF measurements of the $2S$ Lamb shift. Better precision is still expected. Figure 2 [33] shows spectra of the $F = 1$ hyperfine component of the hydrogen $1S$–$2S$ two-photon transition at different nozzle temperatures. The asymmetry and shift due to the second-order Doppler effect are readily apparent. Cooling the atoms to a temperature near 8.6 K brings a fractional resolution of $1 \times 10^{-11}$ (presently limited by laser frequency fluctuations). An enjoyable summation of these results and future possibilities can be found in reference [33].

## 13.6 Trapped Particle Spectroscopy at the Dicke Limit

Doppler-free spectroscopy is possible if the spatial excursions of an atom are constrained to less than the wavelength $\lambda$ of the transition probed [10]. In the RF and microwave region, this condition, referred to as the Lamb–Dicke criterion, is easily satisfied because the radiation wavelengths are several centimeters and longer. At shorter wavelengths, tightly confining particles generally introduces other perturbations which are detrimental to line narrowing. Ion traps, in which ions undergo approximately simple harmonic motion, constrain particles virtually without perturbation [34]. In the optical region, the Dicke criterion was finally satisfied for a single, laser-cooled ion tightly, but benignly, confined in a miniature RF Paul trap [35, 36]. (Coulomb repulsion between ions makes it difficult, but not impossible, to meet the Lamb–Dicke criterion when more than one ion is in the trap [37].) Spectral narrowing by this method can be understood by transforming the electric field vector into the rest frame of the atom

$$\mathbf{E}(\mathbf{r}(t), t) = \mathbf{E}_0 \cos(\omega t + \mathbf{k} \cdot \mathbf{r}(t) + \phi). \qquad (13.10)$$

Recall that $|\mathbf{k}| = 2\pi/\lambda$. If $|\mathbf{r}(t)| \ll \lambda$, then $\mathbf{k} \cdot \mathbf{r}(t) \to 0$ independent of the relative direction of motion. The atom begins to accumulate (lose) phase during the first half of its cycle, but before gaining (losing) $\pi$ radians and losing coherence with the field, it reverses its direction and returns to the original phase setting. Hence, the only significant response of the atomic system occurs at $\omega = \omega_0$.

The second-order Doppler shift of an ion that is laser-cooled to about 1 mK (which is near the Doppler-cooling limit for most ions using a strongly allowed transition) is around one part in $10^{18}$. It also appears that, for a single ion shifts of resonance frequencies owing to electric and magnetic fields could be as small as one part in $10^{18}$, since the ion is trapped in a region where the electric field approaches zero. Signal-to-noise ratio suffers when the sample under observation consists of only a single ion, or at most a few. However, once an ion has been trapped and cooled, its presence can be detected easily by laser-induced fluorescence, since a strongly allowed transition can scatter as many as $10^8$ photons per

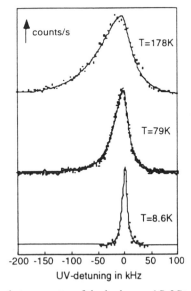

FIG. 2. Doppler-free two-photon spectra of the hydrogen 1$S$–2$S$ transition observed in an atomic beam. The $F = 1$ hyperfine component has been recorded at different nozzle temperatures. The asymmetry and shift caused by second-order Doppler effects are greatly reduced at $T = 8.6$ K. (From reference [33].)

second. More importantly, a transition with a narrow natural linewidth can be detected with nearly unit efficiency by the presence or absence of the strong fluorescence [39]. Consider an atom that has both a strongly allowed transition and a weakly allowed transition that share the ground state. Then the narrow transition is detected as follows: The atom is assumed to be initially in the ground state. Radiation at a frequency near the weak resonance is pulsed on, possibly causing a transition to the long-lived upper level. Next, light with a frequency near that of the strong transition is pulsed on. If the atom had made a transition in the previous step, no fluorescence will be observed; otherwise, fluorescence will be observed at an easily detectable level. The detection of the presence or absence of fluorescence from the strongly allowed transition is much easier than attempting to detect the one photon that eventually is emitted when the meta-stable state decays. Detecting each transition to the metastable state makes it possible to achieve a signal-to-noise ratio that is limited only by the quantum statistical fluctuations in making the weak transition [36, 38].

Figure 3 [36] shows the fully resolved recoilless optical resonance and motional sidebands of the narrow $S-D$ transition at $\lambda = 282$ nm on a single, laser-cooled $^{198}Hg^+$ ion confined in a miniature RF Paul trap. Each *single-photon* transition to the electric-quadrupole–allowed metastable $D$ state ($\tau > 0.1$ s) was

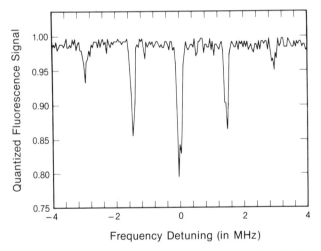

Frequency Detuning (in MHz)

FIG. 3. Quantized signal showing the electric-quadrupole–allowed $S-D$ transition ($\lambda = 282$ nm) in a single, laser-cooled $^{198}Hg^+$ ion. On the vertical axis is plotted the probability measure of observing fluorescence from the $S-P$ first resonance transition at 194 nm as a function of the relative detuning of the 282-nm laser from line center. The $S-D$ transition and the $S-P$ transition are probed sequentially to avoid light shifts and broadening on the narrow $S-D$ transition. The integration time per point is about 16 s. Clearly resolved are the Doppler-free absorption resonance and the Doppler sidebands due to residual secular motion of the laser-cooled ion. (From reference [36].)

detected with nearly unit efficiency by monitoring for the presence (no transition made) or absence (weak transition made) of fluorescence from the strongly allowed $S$–$P$ transition at 194 nm. Note the absence of background or any instrumental noise. The fractional resolution of this spectrum already exceeded $3 \times 10^{-11}$.

More recently [40], the spectral linewidth observed for this transition, limited by the frequency fluctuations in the probe laser, was less than 80 Hz, corresponding to a fractional resolution of less than $1 \times 10^{-13}$. Laser cooling to the zero-point energy of the trap's harmonic well has reduced the spatial excursions of a trapped ion to less than 2.5 nm [41]. Very recently, laser-cooled neutral atoms have also been confined to the Lamb–Dicke regime in optical wells generated by crossed laser beams [42, 43]. However, neutral trapping can only be done by exchange of internal and kinetic energy; therefore, by its nature, it must perturb the internal energy level structure of the atom.

Trapped ions that are laser-cooled to the Doppler-cooling limit (which may not produce a spatial confinement less than an optical wavelength) offer perhaps the highest potential for spectral resolution and accuracy. Atom–field interaction times can be arbitrarily long (days), which not only eliminates interaction-time broadening, but also makes it possible to saturate weakly allowed transitions. For example, the electric-quadrupole–allowed, $S$–$D$ transition in mercury can be saturated with as few as 10 photons per second focused to about $\lambda^2$, if the laser line width is less than the natural linewidth. (This transition is also two-photon allowed, but remember that two-photon spectroscopy is burdened by light shifts.) Doppler effects are eliminated, or greatly reduced, to all orders. Perturbations due to collisions, already extremely small in room-temperature traps at pressures below $10^{-8}$ Pa ($10^{-10}$ torr), can be virtually eliminated in a cryogenic environment. There are no first-order electric field shifts, and higher-order electric field shifts, as well as magnetic field shifts, can be reduced to below one part in $10^{18}$. Two-photon spectroscopy can complement linear spectroscopy of trapped ions to reach states of like parity, which may be difficult, if not impossible, to reach otherwise. If the ions are cooled nearly into the Lamb–Dicke regime, the two-photon resonance will be Doppler-free independent of the light-beam directions (note that only a single beam is now necessary!). Prospects for all the methods seem exciting.

# References

1. Shimoda, K., ed. (1976). *High-Resolution Laser Spectroscopy*. Springer-Verlag, Berlin.
2. Ramsey, N. F. (1956). *Molecular Beams*. Oxford University Press, London.
3. Letokhov, V. S., and Chebotayev, V. P. (1977). *Nonlinear Laser Spectroscopy*. Springer-Verlag, Berlin.

4. Levenson, M. D. (1982). *Introduction to Nonlinear Laser Spectroscopy*. Academic Press, New York.
5. Bennet, W. R., Jr. (1962). *Phys. Rev.* **126**, 580.
6. Lee, P. H., and Skolnick, M. L. (1967). *Appl. Phys. Lett.* **10**, 303.
7. Hänsch, T. W., and Toschek, P. (1968). *IEEE J. Quant. Electron.* **QE-4**, 467.
8. Vasilenko, L. S., Chebotayev, V. P., and Shishaev, A. V. (1970). *JETP Lett.* **12**, 113.
9. Cagnac, B., Grynberg, G., and Braben, F. (1974). *Phys. Rev. Lett.* **32**, 643; Levenson, M. D., and Bloembergen, N. (1974). *Phys. Rev. Lett.* **32**, 645.
10. Dicke, R. H. (1953). *Phys. Rev.* **89**, 472.
11. Hänsch, T. W., and Schawlow, A. L. (1975). *Opt. Commun.* **13**, 68.
12. Wineland, D. J., and Dehmelt, H. (1975). *Bull. Am. Phys. Soc.* **20**, 637.
13. Ramsey, N. F. (1950). *Phys. Rev.* **78**, 695.
14. Baklanov, Ye. V., Dubetsky, and Chebotayev, V. P. (1976). *Appl. Phys.* **9**, 171.
15. Bergquist, J. C., Lee, S. A., and Hall, J. L. (1977). *Phys. Rev. Lett.* **38**, 159.
16. Bordé, Ch. J., Salomon, Ch., Avrillier, S., van Lerberghe, A., Bréant, Ch., Bassi, D., and Scoles, G. (1984). *Phys. Rev. A* **30**, 1836.
17. Helmcke, J., Zevgolis, D., and Yen, B. U. (1982). *Appl. Phys. B* **28**, 83.
18. Hall, J. L., Bordé, Ch. J., and Uehara, K. (1976). *Phys. Rev. Lett.* **37**, 1339.
19. Bergquist, J. C., Barger, R. L., and Glaze, D. J. (1979). In *Laser Spectroscopy IV*, H. Walther and K. V. Rothe (eds.), p. 120. Springer-Verlag, Berlin.
20. Barger, R. L., Bergquist, J. C., English, J. C., and Glaze, D. J. (1979). *Appl. Phys. Lett.* **34**, 850.
21. Helmcke, J., Ishikawa, J., and Riehle, F. (1989). In *Frequency Standards and Metrology*, A. De Marchi (ed.), p. 270. Springer-Verlag, Berlin.
22. Borde, Ch. J. (1989). *Physics Letters A* **140**, 10.
23. Weiss, D., Young, B., and Chu, S. (1993). *Phys. Rev. Lett.* **70**, 2706.
24. Wignall, J. W. S. (1992). *Phys. Rev. Lett.* **68**, 5.
25. Barger, R. L. (1981). *Opt. Lett.* **6**, 145.
26. Lee, S. A., Helmcke, J., and Hall, J. L. (1979). In *Laser Spectroscopy IV*, H. Walther and K. V. Rothe (eds.), p. 130. Springer-Verlag, Berlin.
27. Sengstock, K., Sterr, U., Hennig, G., Bettermann, D., Müller, H. and Ertmer, W. (1993). *Opt. Commun.* **103**, 73 (1993).
28. Biraben, F., Garreau, J. G., Julien, L. and Allegrini, M. (1989). *Phys. Rev. Lett.* **62**, 621.
29. Minogin, V. G. (1976). *Kvantovaya Elektronika* **3**, 2061.
30. Andreae, T., König, W., Wynands, R., Leibfried, D., Schmidt-Kaler, F., Zimmermann, C., Meschede, D., and Hänsch, T. W. (1992). *Phys. Rev. Lett.* **69**, 1923.
31. Schmidt-Kaler, F., Liebfried, D., Weitz, M., and Hänsch, T. W. (1993). *Phys. Rev. Lett.* **70**, 2261.
32. Weitz, M., Huber, A., Schmidt-Kaler, F., Liebfried, D., and Hänsch, T. W. (1994). *Phys. Rev. Lett.* **72**, 328.
33. Hänsch, T. W. (1994). In *Atomic Physics 14*, D. J. Wineland, C. E. Weiman, and S. J. Smith (eds.), p. 63. American Institute of Physics, New York.
34. Dehmelt, H. S. (1967). *Advan. Atomic and Mol. Physics* **3**, 53; (1969). *Advan. Atomic and Mol. Physics* **5**, 109.
35. Janik, G., Nagourney, W. and Dehmelt, H. (1985). *J. Opt. Soc. Am. B* **2**, 1251.
36. Bergquist, J. C., Itano, W. M., and Wineland, D. J. (1987). *Phys. Rev. A* **36**, 428.
37. Raizen, M. G., Gilligan, J. M., Bergquist, J. C., Itano, W. M., and Wineland, D. J. (1992). *Phys. Rev. A* **45**, 6493.

38. Wineland, D. J. (1984). *Science* **226**, 395.
39. Dehmelt, H. G. (1975). *Bull. Am. Phys. Soc.* **20**, 60.
40. Bergquist, J. C., Itano, W. M., and Wineland, D. J. (1994). In *Frontiers in Laser Spectroscopy*, T. W. Hänsch and M. Inguscio (eds.), p. 359. North Holland, Amsterdam.
41. Diedrich, F., Bergquist, J. C., Itano, W. M., and Wineland, D. J. (1989). *Phys. Rev. Lett.* **62**, 403.
42. Jessen, P., Serz, C., Lett, P., Phillips, W., Rolston, S., Spreeuw, R., and Westbrook, C. (1992). *Phys. Rev. Lett.* **68**, 3861.
43. Phillips, W. D. (1995). In *Atomic Physics 14*, D. J. Wineland, C. E. Weiman, and S. J. Smith (eds.), p. 211 and references therein. American Institute of Physics, New York.

# 14. MICROWAVE SPECTROSCOPY

## R. D. Suenram and Anne M. Andrews

National Institute of Standards and Technology
Molecular Physics Division
Gaithersburg, Maryland

## 14.1 Introduction

It has been 17 years since the last review article on microwave spectroscopy appeared in this series [1]. That article was written by Johnson and Pearson of the National Bureau of Standards. It covered the state-of-the-art techniques in use at that time, which included a number of innovations in the field that were pioneered and developed at NBS. In the introduction of that article it was pointed out that the field of microwave spectroscopy, which resulted from advanced military research initiated during the latter part of the Second World War, "was rapidly reaching maturity and the challenge to future research would be to find new areas of science where the experience and technology previously developed could have a truly useful impact." The last decade has seen rapid spectroscopic advances in this field which have been led by the development of molecular beam techniques. During this time, several fundamentally different molecular beam microwave spectrometers have been developed which allow the study of hydrogen-bonded and van der Waals clusters, as well as larger molecular species whose spectra are too complex to decipher using conventional spectroscopic techniques.

In this chapter we will look at the state-of-the-art techniques being used in conventional microwave spectroscopy today, as well as explore three new molecular beam microwave techniques that are the current hotbed of activity in the field. It is these latter techniques which are revolutionizing the field once again by allowing the study of species that would have been considered much too difficult a decade ago. The cooling effect attained by the rapid expansion of the molecular beams in these spectrometers greatly simplifies the rotational spectra of large species, resulting in spectra that are readily assignable because only the lowest energy levels of the species are populated. In addition, the expansion process permits species in the molecular beam to cluster with the carrier gas or other species in the beam; the spectra of these clusters provide structural information about van der Waals complexes and hydrogen-bonded dimers, trimers, and tetramers.

EXPERIMENTAL METHODS IN THE PHYSICAL SCIENCES
Vol. 29B

There are a number of good textbooks on the subject of microwave spectroscopy which give background information on the technique. These texts provide useful information for the different types of spectrometers and techniques discussed throughout this chapter [2–10].

## 14.2 Microwave Spectroscopy with Conventional Spectrometers

### 14.2.1 Description of the Spectrometer

Figure 1 shows a block diagram for the microwave spectrometer currently in use at NIST. The essential elements consist of a frequency source, an absorption cell, a detector, and some sort of modulation scheme to enhance the sensitivity of the spectrometer. The operational range of a microwave spectrometer has classically been in the 8 to 40 GHz region. A number of different types of frequency sources, absorption cells, detectors, and modulation schemes that have been developed over the years were adapted for a certain area of research or type of molecular species being studied. A number of these variations are discussed in the following sections. In the system depicted in Figure 1, the frequency source used most often is a reflex klystron. We currently have reflex klystrons or solid-state sources that allow spectral coverage to 120 GHz at NIST. Data acquisition with this system is straightforward once the frequency source has been phase-locked. The phase-sensitive detection system is referenced to the square-wave generator and the preamplifier. Once the frequency source has been stepped, the output of the phase-sensitive detection system is sampled through an analog-to-digital input board in the computer. This provides a simple data array of frequency vs. voltage which is simultaneously recorded in computer memory and displayed on the computer screen as the scan progresses.

Early work in the field relied on spectrometers that were essentially home-made from some commercially available components and some components that were built in the laboratory. In the mid-1960s, Hewlett-Packard[1] introduced a commercial microwave spectrometer which employed synthesizer-driven, phase-locked backward wave oscillator (BWO) sources [11]. An improved version was introduced in 1970 which covered the 8 to 40 GHz spectral window using four different waveguide bands. A number of these instruments are still in use today, although emerging technologies, primarily in the development of

---

[1] From time to time throughout this chapter, certain companies and products will be referred to by name. This is necessary to adequately describe the products or services and is not to be construed as an endorsement by the NIST.

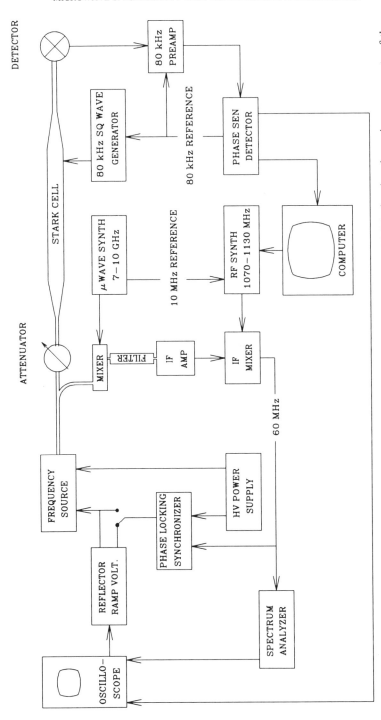

Fig. 1. Block diagram of the conventional microwave spectrometer currently in use at NIST, showing the major components of the spectrometer.

solid-state frequency sources (microwave synthesizers), are supplanting these older units.

## 14.2.2 Frequency Sources

For spectroscopic purposes, the most frequently used sources are reflex klystrons and backward wave oscillators (BWOs). These sources have characteristically low noise and generally good spectral purity, typically delivering stable signals that are less than 1 kHz wide when phase-locked to a reference oscillator. Reflex klystrons are more narrow-banded devices than BWOs and typically have a tuning range that is 10% of the absolute frequency of the tube. Klystron tuning is accomplished in two ways: The klystrom can be electrically tuned over tens or hundreds of megahertz by applying a ramp voltage to the reflector of the tube, or it can be mechanically tuned through the full operating range by changing the cavity length of the tube. For quick, broad-banded spectral searches, the tube is swept electrically by varying the reflector voltage with a ramp voltage derived from an oscilloscope. On a typical sweep, the frequency of the klystron can be varied from a few megahertz to over 100 MHz. This is convenient for observing microwave transitions and studying the Stark effects of the transitions in question. Once a spectral region has been scanned, the cavity of the klystron can be mechanically changed to move the frequency of the klystron to the new spectral window. This can be done in a repetitive fashion until the entire frequency range of the klystron has been covered.

In this free-running mode of klystron operation, there is usually some short-term jitter associated with the tube, so precise spectral line measurement requires disengaging the ramp voltage and phase-locking the klystron to some harmonic of a reference oscillator (microwave synthesizer). With this mode of operation, a frequency range of up to 60 MHz can be scanned under computer control. This is convenient when weak signals are being sought (signal averaging) or when numerous 60 MHz segments are needed to piece together a linear scan of a broader spectral region for assignment purposes.

Frequency measurement is achieved by mixing some of the microwave radiation from the frequency source with a harmonic of the microwave synthesizer. If the frequency range being scanned is near 100 GHz or higher, the harmonics may be as high as 10–15. An intermediate frequency (IF) near 1 GHz is taken from the harmonic mixer, filtered and amplified, and then mixed in a second, lower-frequency mixer with the output of a radio-frequency synthesizer. The 60 MHz IF from the second mixer is split and sent to the input of the modified 60 MHz phase-locking synchronizer [12] and to a spectrum analyzer tuned to 60 MHz. The output of the spectrum analyzer is then sent to the dual-trace oscilloscope for use as a frequency marker during free-running scans of the reflector voltage.

BWOs are more broad-banded devices than klystrons, and they can typically cover a full waveguide band. They can be swept in a free-running mode by varying the voltage applied to the device. For accurate frequency measurements, they can also be phase-locked to a reference oscillator in a fashion that is very similar to klystrons. Recent work in Russia has led to the development of synthesizer-driven BWO systems that allow computer-controlled frequency scans to be made over full waveguide bands. These systems are commercially available in several models that operate up to 118 GHz [13] with 10–20 milliwatts of output power and experimental models exist that permit phase-locked operation to nearly 400 GHz [14]. Analogous systems have been developed using solid-state sources but the output power of these devices is marginal for most spectroscopic applications [15].

Solid-state oscillators are used as the fundamental microwave sources in microwave synthesizers, where they are referenced to harmonics of extremely stable quartz oscillators (which typically operate at 10 MHz). At higher frequencies, solid-state Gunn diodes are used for frequency generation [16]. These devices are easily incorporated into phase locking schemes using the same methods shown in Figure 1. The one drawback is that this type of source tends to be somewhat noisier than the more conventional klystrons and BWOs.

## 14.2.3 Modulation Techniques

In practically all conventional microwave spectrometers, Stark-effect modulation is used to allow synchronous detection of the molecular signals. This type of modulation was first described by Hughes and Wilson in 1947 [17] and remains the most sensitive technique in use today. In this scheme, a high-voltage square wave is applied between two plates in the microwave absorption cell. Because microwave experiments require only a low sample pressure (0.13–13 Pa = 1–100 μm Hg), it is generally possible to apply up to several kilovolts to the plates before electrical arcing occurs in the absorption cell. The square-wave voltage applied has a 50% duty cycle, with the high voltage being on for half the cycle and off for the other half. On the *off* part of the cycle, the voltage must be zero-based to ensure that the zero-field frequency of the molecular transition is not shifted by any residual field.

The energy levels of most asymmetric top polar molecules have a second-order dependence on the applied electric field which obeys the equation $\Delta v = C \cdot \mu^2 E^2$. In this equation, $\Delta v$ is the frequency shift in megahertz with applied electric field $E$ in volts/centimeter. The constant $C$ is dependent on the structural parameters of the molecule which can be calculated in a straightforward manner from the moments of inertia [18]. With typical values for electric dipole moment on the order of $3.3 \times 10^{-30}$ C · m [1 D] and $C$ in the range of $10^{-3}$ with the appropriate conversion factors to give units of megahertz on $\Delta v$ for $\mu$ in D, values of the electric field must be on the order of 1 kV/cm to affect a frequency shift of several megahertz. This shift is adequate because the rotational transitions usually have full width at half the peak intensity (FWHM) of 0.1 to 1 MHz.

Using Stark modulation with synchronous detection at the modulation frequency permits each rotational transition to be observed both at zero field on one phase of the detection cycle and with the energy levels split by the Stark field on the opposite phase of the cycle. The advantage of Stark modulation is twofold: First, it permits one to use an extremely sensitive AC detection scheme to observe rotational transitions, and second, it serves as an assignment aid in that the number and relative intensities of Stark-field split transitions are indicative of the rotational transition being investigated.

Of all the components used in the construction of a microwave spectrometer, the most difficult to come by is a high-voltage square-wave generator. This is one piece of instrumentation whose circuitry is still better suited to electron tube technology than to solid-state technology. Several recent papers in the literature by Bernius and Chutjian describe the construction of a solid-state square-wave generator that will operate to more than 1000 volts [19, 20]. After doing extensive searching, the authors have been able to locate only two companies that will construct a high-voltage square-wave generator. One is Directed Energy, Inc., in Fort Collins, Colorado and the other is Montech, a small company associated with Monash University in Australia.

Other modulation techniques are sometimes used when Stark modulation is not suitable. Experiments involving electric discharges for producing radicals or ions in particular are not amenable to the use of Stark modulation due to electrical breakdown of the gases in the absorption cell. In some cases the Stark effect of the rotational transitions is not sufficiently fast to allow the transitions to be modulated. This generally occurs for transitions that have high-$J$ and low-$K$ quantum numbers. For applications such as these, some type of source modulation is commonly used. The main drawback of source modulation is that the baseline is difficult to keep flat, and thus baseline subtraction methods are also needed to attain high sensitivity. One good source modulation technique that has been used in a number of experiments at NIST is the tone-burst technique developed by Pickett [21]. When the tone is on, the source frequency is shifted by the frequency of the tone, which places it outside the line width of the transition being measured. When the tone is off, the transition appears at the unshifted frequency. With this technique the uneven baseline problem is minimized, and the source can still be phase-locked and driven under computer control. In the implementation of this technique at NIST, a high-frequency tone (1–5 MHz) is modulated with a low-voltage square wave (up to several volts) at the standard 80 kHz frequency at which the Stark modulation normally operates. This enables the use of all the same components in the detection system.

## 14.2.4 Waveguide Cells and Parallel Plates

In a conventional microwave spectrometer, the absorption cell most often used is a standard X-band (8–12 GHz) waveguide cell. Centered in the narrow waveguide dimension is a collinear Stark septum which is electically isolated from the waveguide by two pieces of grooved Teflon. Electrical connection to the septum is accomplished by drilling a hole in the center of the narrow face of the waveguide. A hermetically sealed, high-voltage connector is attached to the outside of the waveguide and the center pin is connected to the septum either by soldering or preferably by a metal ball-and-spring arrangement which allows for movement of the septum during heating or cooling cycles. The length of the cell varies considerably, but is usually one to four meters for spectroscopic applications up to 40 GHz. For specialized applications, the inner surface of the waveguide can be gold-plated to retard corrosion.

The versatility of a microwave spectrometer can be greatly increased by using absorption cells of different designs that are tailored to the chemical species or frequency range of interest. One design that has been utilized extensively at NIST for a wide variety of projects is a one-meter Stark cell in P-band waveguide (12–18 GHz) that is constructed entirely from stainless steel and Teflon parts. This cell has been used for the study of corrosive species such as chlorine nitrate [22] and $(HF)_2$ [23]. The absorption cell has been fitted with an outer stainless steel cooling cylinder which allows it to be cooled with liquid nitrogen to 77 K. A drawing of this absorption cell is shown in Figure 2. Experiments involving the unstable dioxirane [24] and ethylene primary ozonide (1, 2, 3-trioxolane) [25]

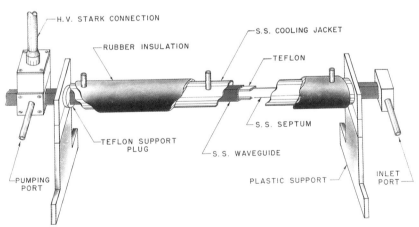

FIG. 2. Drawing of the one-meter stainless steel Stark absorption cell with integral outer stainless steel cooling cylinder. (Reprinted with permission from: *R. D. Suenram, F. J. Lovas, J. Am. Chem. Soc.,* **100,** 5117–5122 (1978). Copyright 1978 American Chemical Society.)

species were successfully carried out using this absorption cell. These experiments involve condensing ozone and ethylene[2] on the cell walls and then observing the reaction products as the cell warms to 193 K. A further advantage of this particular absorption cell design is that it propagates a broad range of microwave frequencies, working well at least to 130 GHz.

A second type of absorption cell that is often used in microwave spectrometers is a parallel-plate absorptiion cell. This type of cell design is extremely versatile in that it allows experiments to be carried out on transient species or high-temperature molecules. We have designed a number of different parallel-plate absorption cells at NIST. These range in length from 10 cm to over a meter. A particularly useful design is given by Johnson and Pearson [1] (Figure 6, page 117) in their previous article on this subject. This cell design is well suited for the study of transient species produced by RF or microwave discharge tubes or ovens mounted externally to the cell. Fast pumping through the fairly open parallel plate structure enables one to flow unstable species rapidly through the cell and detect their spectra before they decompose or condense on collisions

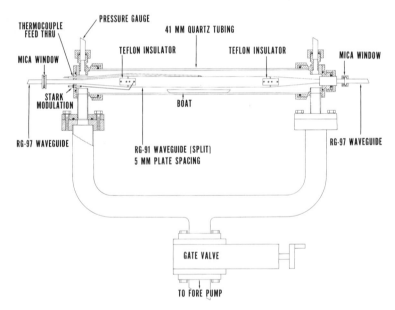

Fig. 3. Small parallel-plate cell capable of operation at 1000°C. (Reprinted with permission from: *R. D. Suenram and F. J. Lovas, J. Mol. Spectrosc.*, **72**, 372–382 (1978). Copyright 1978, Academic Press.)

[2] Caution is to be observed when condensing ozone with liquid nitrogen, as it is unstable and explosions can result if too much is condensed at one time.

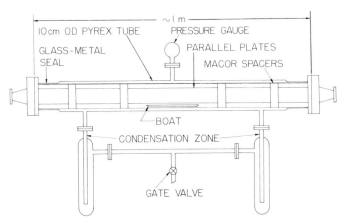

FIG. 4. One-meter parallel-plate cell used for organic compounds of low volatility. (Reprinted with permission from: *R. D. Suenram and F. J. Lovas, J. Am. Chem. Soc.*, **72,** 7180–7184 (1980). Copyright 1980, American Chemical Society.)

with the cell walls. Numerous radicals [26–28] and unstable species [29, 30] have been studied over the years in this type of cell at NIST.

Several other parallel-plate cells have been constructed at NIST to allow *in situ* studies of high-temperature species. One of these cells employs small K-band (18–26.5 GHz) size plates that are approximately 50 cm long (see Figure 3). They are housed in a quartz tube 4 cm in diameter which permits the entire cell to be inserted into an oven and heated to 1000°C. This cell has been used for the spectroscopic investigations of small, high-temperature molecules.

A larger parallel plate cell was developed for the study of low-volatility molecular species. This cell (shown in Figure 4) is approximately one meter in length, and it is designed so that even the ends of the cell can be heated with heating tapes to keep materials from condensing near the ends of the parallel plates. The cell operates up to approximately 200°C, so it is convenient for studying organic compounds of low volatility. In this cell, the compound of interest is inserted into the vacuum envelope in a boat placed directly below the parallel plates. The entire cell can be preheated before the actual heating of the sample is initiated. With this arrangement, a single sample of 5 to 10 g (or 5 to 10 mL) can be used for at least a week before it is entirely used up. Furthermore, the sample can be warmed and cooled daily without condensation on the parallel plates. We have used this cell for the microwave investigations of glycine [31] indole [32] sulfuric acid [33] and $BH_3NH_3$ [34].

## 14.2.5 Detectors

Early experiments relied on crystal detectors to monitor the microwave radiation emanating from microwave absorption cells. Commercial or homemade devices similar to Figure 9 on page 124 of reference [1] were commonly used until the mid-1970s. At this point Schottky diodes became commercially available. These new devices were extremely efficient at converting the microwave radiation to a DC signal with typical responsivities of nearly 1.5 volt/milliwatt of microwave radiation. Schottky diodes consist of an array of diodes on a chip. A fine tungsten whisker is contacted to one of the diodes on the wafer [35]. The entire assembly is mounted on a metal wafer which mounts into a waveguide tuning stub. The modern Schottky diodes are quite robust and require only minimal care to avoid accidental static electricity–type shocks, which can cause the whisker to lose contact with the diode. Although the metal wafers on which the diodes are mounted are fairly expensive, they can be replaced in the field.

For routine use in the laboratory, the Schottky diodes are the detector of choice because they are low-noise, sensitive, and require no liquid-nitrogen or liquid-helium cooling to make them operational. In most cases the amount of microwave power coming through the cell is sufficient to bias the diodes so no additional circuits are needed to provide a DC bias to the diode. In cases where extremely low microwave power ($<1$ milliwatt) is available, external bias circuits can be used. This generally occurs at higher frequencies ($>200$ GHz) where fundamental power from microwave oscillators is usually low or where harmonic generation is being used.

The most sensitive microwave detectors are liquid-helium–cooled bolometers. There are a number of different types of detectors in this category, but fast, indium–antimonide detectors are generally used. A fast response-time detector is necessary in order to pick up the Stark-modulated signals, which typically are in the 5 to 100 kHz frequency region. This type of detector is commercially available and can be mounted in a commercial liquid helium dewar. Integral preamplifiers can be used with this detector package and, depending on the type of filters used on the input, the detector can be used throughout the far-infrared spectral region. State-of-the-art devices have responsivities of 4–5 V/milliwatt of microwave power and noise equivalent power in the vicinity of $10^{-13}$–$10^{-14}$ watts/$\sqrt{\text{Hz}}$. In practice, these devices afford an increase in signal-to-noise of 2–3 compared to a room-temperature Schottky diode detector.

## 14.3 Molecular Beam Electric Resonance Opto-thermal Spectrometers

### 14.3.1 Description of the Spectrometer

A new type of microwave molecular beam spectrometer has recently been developed which incorporates a helium-cooled bolometer detector into a modified molecular beam electric-resonance type spectrometer (MBER spectrometer) [36]. The new spectrometer is an electric-resonance optothermal spectrometer (EROS). A schematic diagram for a spectrometer of this type which was constructed at NIST is shown in Figure 5. This spectrometer employs a 60 μm diameter continuous-flow nozzle with a backing pressure of 0.1–0.2 MPa (1–2 atm). Helium is generally used as the inert backing gas. The compound or compounds of interest are blended with helium at the 1–10% level and expanded through the nozzle. Approximately 2.5 cm downstream from the nozzle, the molecular beam is skimmed by a 1 mm diameter conical skimmer. In the zone between the nozzle exit and the skimmer entrance, the molecular beam is traversed by microwave radiation. After the skimmer, the molecular beam passes through a state-selecting field of quadrupolar symmetry. This field is 56 cm long, with a beam stop located at the center of the field. The field assembly consists of four parallel 6.4 mm diameter stainless-steel rods separated so that the centers of the rods are equally spaced on a 12.7 mm diameter circle coincident with the molecular beam axis. This gives an entrance aperture of 6.4 mm for the quadrupole field. Two of the rods on the diagonal are energized at a positive voltage, and two are energized at a negative voltage. With this arrangement, approximately 30 kV can be applied before electrical breakdown occurs across the plastic spacers which act as a support frame to hold the rods.

After the molecular beam exits the quadrupole field, it impinges on a liquid-He–cooled bolometer detector held at 1.6 K by pumping on the liquid

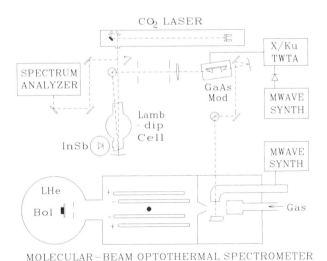

MOLECULAR – BEAM OPTOTHERMAL SPECTROMETER

FIG. 5. Drawing of the electric-resonance opto-thermal spectrometer used at NIST. (Reprinted with permission from: *G. T. Fraser, A. S. Pine, W. A. Kriner, and R. D. Suenram, Chemical Physics,* **156,** 523–531 (1991). Copyright 1991, Elsevier Science Publishers BV, Academic Publishing Division.)

helium. The detector element is constructed of silicon which is attached to a $2 \times 6$ mm diamond target that is partially shielded from the molecular beam by a 1.6 K, 3 mm diameter aperture. In addition, the detector is nearly completely shadowed from the nozzle by the 1 mm diameter glass-rod beam stop, which is located at the middle of the quadrupole field.

The nonpolar components of the beam and the majority of the carrier gas miss the detector because it is blocked by the beam stop. Only molecules with positive Stark effects are focused into the aperture of the detector. Molecules with negative Stark effects are defocused from the detector by the quadrupole field. Thus, the detector is bombarded by a constant steady-state flux of the polar molecules in the beam. When the microwave frequency is tuned across a rotational transition, a change in the steady-state flux is recorded by a temperature change of the sensitive bolometer. The steady-state flux can be changed in either direction because it will depend on whether the rotational transition in question has greater focusing or defocusing properties.

The microwave source indicated in Figure 5 can be any of the sources discussed in Section 14.2.2. As in the case of the conventional Stark spectrometer, some type of modulation is needed in order to facilitate the use of lock-in detector techniques. When a microwave synthesizer or synthesizer-driven BWO is being used, the microwaves are 100% amplitude-modulated using a microwave switch with an 80 dB on/off ratio. If klystrons or Gunn diodes are used as frequency sources, they are usually swept via computer control using the same techniques as described for conventional spectrometers. In this case, however, tone burst modulation is used [21]. All modulation frequencies are low (200–400 Hz) because the response time of the bolometer is slow.

### 14.3.2 Applications of the EROS Spectrometer

This instrument has several desirable properties in that the frequency range of the instrument is limited only by the availability of frequency sources. It works equally well in the microwave, far-infrared, infrared, and visible spectral regions. This allows one to easily perform double- or even triple-resonance experiments. In double-resonance experiments, one can even use radio frequencies. The other convenient aspect is that since the detection system is detecting molecules and not radiation, the noise of the source does not enter into the picture; thus, there are no background problems that sometimes hinder more conventional techniques. On the other hand, with this type of spectrometer, one loses the transition identification information which Stark modulation provides.

Shortly after the EROS instrument was constructed at NIST, it was used to locate transitions of the elusive $A_1^{\pm}$ state of the $(H_2O)_2$ molecule [37]. In order to accomplish this, use was made of the extensive frequency coverage possible with the instrument. Figure 6 shows the $7_{07}$–$6_{06}$ $A_1^{-}$–$A_1^{+}$ transition, which was recorded

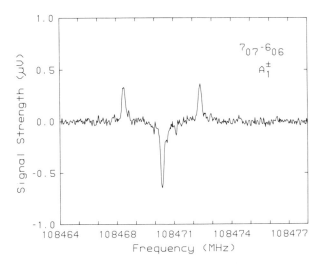

FIG. 6. Spectral trace of the $7_{07}-6_{06}$ $A_1^- -A_1^+$ transition of $(H_2O)_2$. (Reprinted with permission from: *G. T. Fraser, R. D. Suenram, and L. H. Coudert, J. Chem. Phys.,* **90,** 6077–6085 (1989). Copyright 1989, American Institute of Physics.)

using a computer-controlled, phase-locked klystron modulated with Pickett's tone-burst modulation scheme [21]. The $A_1^\pm$ states were difficult to observe in the $(H_2O)_2$ spectrum because of their low statistical weight of 1 (compared with statistical weights of 3 and 6 for the other states) and the absence of the $B_1^\pm$ partner states, which have zero statistical weight for $(H_2O)_2$. After the assignment was made with transitions measured from 26.8 GHz to 108.5 GHz, it was clear why this state had not been previously observed using the more sensitive Fourier-transform microwave spectrometer described in the next section: No transitions of the $A_1^\pm$ state occur within the limited frequency range ($\sim$4 GHz to 26.5 GHz) of the FTMW type of spectrometer.

Microwave-infrared double resonance was used extensively in work on the water–ammonia hydrogen-bonded complex [38]. In this case, a tunable microwave–side-band $CO_2$ laser was used to observe the infrared spectrum of the $NH_3$ umbrella fundamental vibration of $HOH–NH_3$ at a resolution of $\sim$3 MHz. Although a number of transitions were observed in this work, it was not possible to obtain an assignment because of missing spectral windows caused by the incomplete frequency coverage of the $CO_2$ side-band system. Following the observations of the infrared spectrum, each infrared transition was checked for double-resonance possibilities using known pump frequencies of previously assigned microwave transitions [39, 40]. With this technique, it was possible to assign most of the transitions observed in the infrared spectrum. Furthermore, once the infrared assignment was obtained, it was possible to lock the $CO_2$

side-band system at the frequency of a known infrared transition and, by scanning the microwave oscillator, observe and assign rotational transitions in the vibrational excited state in the cold molecular beam environment.

In some recent work involving the gauche–gauche conformer of 2-fluoroethanol [41], an interesting background state or "dark state" was discovered and the spectroscopy worked out using several double-resonance and triple-resonance schemes with the NIST EROS instrument. During the infrared investigation utilizing the tunable microwave–side-band $CO_2$ laser system, it was observed that the $4_{13}$ upper state was "split" into a doublet through a 50–50 mixing with a $J = 4$ level of a background vibration. Thus, a pathway was provided to the ladder of dark states, which were probed via double- and triple-resonance techniques. The observed energy-level diagram is shown in Figure 7. The unperturbed levels are shown in the left-hand side of the drawing, while the perturbed energy level pattern is shown in the right half of the drawing after the perturbation is "turned on." As shown in the drawing, various combinations of double and triple resonances were used to probe the spectroscopy of the dark state and determine the rotational constants of this state, which is located approximately 1000 cm$^{-1}$ above the ground state. The dark state is a $4_{31}$ level of

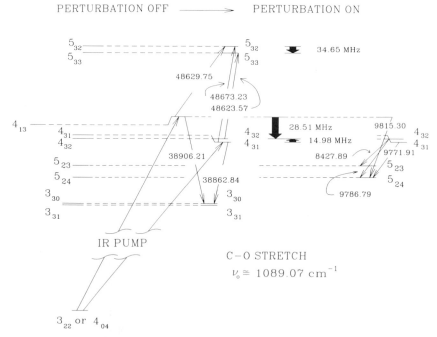

FIG. 7. Energy-level diagram for 2-fluoroethanol, showing how the "dark" state spectroscopy can be unraveled.

an excited torsional state of the gauche–gauche conformer, which is nearly coincidental in energy with the $4_{13}$ level.

## 14.4 Fourier Transform Microwave Spectroscopy

### 14.4.1 Fabry–Perot Cavity Spectrometers

The pulsed molecular beam Fabry–Perot cavity Fourier transform microwave (FTMW) spectrometer was developed by Professor W. H. Flygare and co-workers at the University of Illinois in the late 1970s [41, 42]. Since that time, similar machines have been built in about 20 laboratories around the world, where various improvements and innovations have been made.

The central feature of the FTMW spectrometer is a high-Q Fabry–Perot cavity made of two concave spherical mirrors. When the cavity is stimulated by microwave radiation within its bandwidth (~1 MHz), a standing wave is formed. This standing wave can then interact with a molecular sample placed in the microwave cavity. If there is a molecular transition within the bandwidth of the cavity, a population change occurs which induces a bulk polarization of the gas sample that can be described by Bloch-type equations [43]. When the stimulating pulse is turned off, the bulk polarization decays and the sample emits radiation at the frequency of the molecular transition.

A schematic of the pulsed molecular beam FTMW spectrometer at NIST is shown in Figure 8 [44]. The evacuated chamber contains a narrow-banded (~1 MHz), high-Q (10,000–30,000) Fabry–Perot resonant cavity. By adjusting the position of the mirrors, the cavity is pretuned to the desired frequency. A gas pulse containing about 1–2% of the species of interest seeded in 0.1–0.2 MPa (1–2 atm) of Ar or Ne carrier gas is injected into the chamber. Following a delay of about 1 ms to allow the gas pulse to travel to the center of the Fabry–Perot cavity, a burst of approximately 1 milliwatt of microwave radiation is coupled into the cavity via an L-shaped antenna passing through one of the mirrors. After a delay of 1–10 μs to allow the initial microwave pulse to die out, the free induction decay of the bulk polarization is monitored with a heterodyne detection system. Because microwave frequencies are too high to be monitored in the time domain by even the fastest digitizers, the frequency of the free induction decay (6–26 GHz) is down-converted through a series of mixers to the difference between the free induction decay and the initial microwave pulse frequency (a few hundred kilohertz). This time domain signal is digitized and Fourier-transformed to the frequency domain. Figure 9 shows a typical free induction decay and its Fourier transform.

The spectral range of most FTMW spectrometers currently in use ranges from approximately 3 GHz to 26.5 GHz. The spectral window covered by these instruments is convenient because the low-$J$ transitions which are populated at

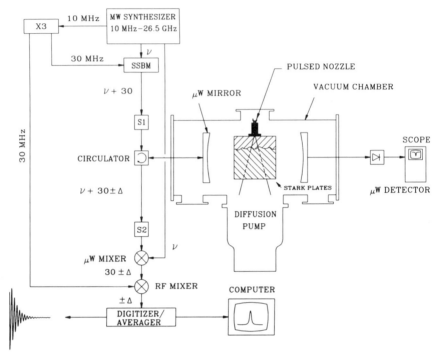

FIG. 8. Schematic of the essential components of a pulsed Fabry–Perot cavity Fourier transform microwave spectrometer.

the low molecular beam temperatures of 1 K occur in this region for most molecular species. In some instances, however, it would be advantageous to develop an instrument that operates at higher frequency. This is particularly true for the laser vaporization experiments described in Section 14.4.5, because current interests are centered on diatomic and triatomic species and species that have non-singlet $\Sigma$ ground electronic states. Many of these species have no transitions in the operational range of the spectrometers currently in use. The range is presently being expanded on the high-frequency side to enable operation to 40 GHz [45]. Kolbe and Leskovar have described a pulsed FTMW spectrometer that could easily be adapted for use with a pulsed nozzle that would operate at 140 GHz [46].

## 14.4.2 Recent Instrumental Improvements

Recent advances in microwave techniques have allowed many improvements to be made in the operational features of the FTMW spectrometer. These advances have been described in various publications scattered throughout the

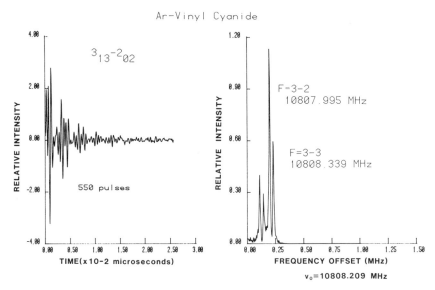

FIG. 9. Free induction decay and the Fourier transform spectrum that is typical of the output of a FTMW spectrometer. The spectrum shows two $^{14}N$ hyperfine components of the $3_{13}-2_{02}$ rotational transition of the Ar · vinyl cyanide van der Waals complex. Each $^{14}N$ hyperfine component is doubled because of Doppler effects in the beam. Note that one hyperfine component is below the microwave pump frequency and one is above the pump frequency. (Reprinted with permission from: R. D. Suenram and F. J. Lovas, *J. Chem. Phys.*, **87**, 4447–4455 (1987). Copyright 1987, American Institute of Physics.)

literature, but a number of them are important enough that they should be documented here for future workers in the field.

The original configuration of the FTMW spectrometer employed standard waveguide components to couple radiation in and out of the Fabry–Perot cavity and to connect all the various microwave components together. At present, coaxial components are used for the most part, since they are much smaller and easier to configure in the frequency range in which the spectrometer operates. Coaxial components continue to become available at higher frequency, with many components now being marketed that operate to 60 GHz and beyond.

The double heterodyne mixing scheme in the FTMW spectrometer originally employed two separate microwave oscillators, a master oscillator and a phase-locked oscillator which was slaved to the master, typically at a 30 MHz offset in frequency. This made operation of the instrument somewhat tedious, especially if the master oscillator was not a microwave synthesizer, since then two microwave oscillators needed to be phase-stabilized in order to make a frequency measurement. At NIST a scheme was developed that allowed the use of only one microwave oscillator (a microwave synthesizer) to be used in conjunction with a single–side-band modulator. This scheme eliminates one of the microwave oscillators and the associated phase-locking equipment. The new scheme is more amenable to automation, since the microwave synthesizer is easily computer-programmable.

Even though the sensitivity of the FTMW spectrometer is one of its major strong points, further improvements in signal-to-noise have been obtained through the use of state-of-the-art microwave components. One of the major sources of noise in the FTMW spectrometer is the noise temperature associated with the microwave mixer in the system (see Figure 8). One way to lower the system noise is to place a low-noise microwave amplifier between the microwave switch labeled S2 and the microwave mixer. In addition to lowering the system noise, the amplifier serves to buffer switching transients which often saturated the microwave mixer. At NIST we have used amplifiers in this position for several years, and recently we have tested a cryogenic microwave amplifier which offers some additional improvement in signal-to-noise.

System noise can be lowered by a factor of two by using an image rejection microwave mixer instead of the conventional double–side-band mixer. With this type of mixer, the noise from the second side band is greatly reduced through the use of appropriate electronic components in the mixer. Further improvements in signal-to-noise could be obtained by using an integrated receiver system, all of which was cryogenically cooled. Ultimate sensitivity would be obtained by cooling the entire Fabry–Perot cavity to cryogenic temperatures in addition to using a cryogenically cooled integrated receiving system.

In the original design of the instrument, the microwave radiation was fed in through one of the cavity mirrors and the emission signal observed from a detection system through the second mirror. Some increase in signal-to-noise is obtained using the arrangement shown in Figure 8, where the detection system has been moved to the same mirror that is used to stimulate the cavity. This is accomplished using a microwave circulator, as shown in the figure. With this arrangement, it is convenient to monitor the cavity modes via a Schottky detector from an antenna on the second mirror.

Another recent improvement, introduced by Grabow and Stahl [47] involves mounting the pulsed nozzle near the center of one of the mirrors of the Fabry–Perot cavity. A 1 mm diameter hole is drilled through the mirror so that the molecular beam from the nozzle is directed down the length of the cavity axis. As a result, the molecules in the beam remain in the Fabry–Perot cavity longer, and the signal is enhanced by up to a factor of five over the conventional orientation where the pulsed beam is directed across the cavity (see Figure 9). In addition, the molecular signal emanating from the cavity occurs for a longer time, which Fourier-transforms to extremely narrow transitions with full-width at half intensity (FWHM) of 2 kHz. An example of the spectra obtainable with the collinear orientation is shown in Figure 10.

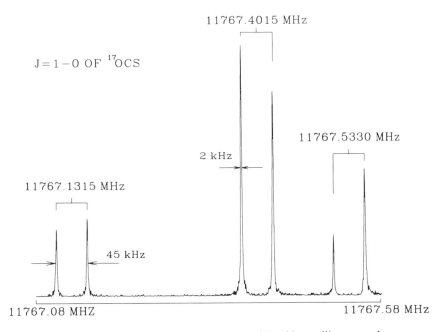

FIG. 10. An example of the 2 kHz linewidths attainable with a collinear nozzle arrangement in the FTMW spectrometer. The three hyperfine components arise from the $^{17}O$ nuclear quadrupole moment and the 45 kHz splitting is due to Doppler effects of the beam. (Reprinted with permission from: J.-U. Grabow and W. Stahl, *Z. Naturforsch.*, **45a**, 1043–1044 (1990). Copyright 1990, Zeitschrift für Naturforsch.)

The Stark plates shown in Figure 8 were also an addition to the original design of the spectrometer, which did not provide Stark-effect data. At NIST, Stark plates have been installed by attaching them to the rods on which the mirrors ride. The plate separation is large (26 cm), but with this spacing the plates do not interfere with the microwave propagation of the Fabry–Perot cavity and thus can be left in the vacuum chamber all the time. Since the spacing is so large, there are paths to ground that are shorter than the plate spacing. In order to make the electric field symmetric with respect to ground at the center of the cavity, one plate is energized to a ( + ) DC voltage and the other to a ( − ) DC voltage. Several similar designs using a Faraday cage or steel mesh plates are in use in other laboratories. Although this experimental arrangement delivers electric fields which are considerably lower than those discussed in Section 14.2.3 regarding Stark modulation techniques with conventional spectrometers, they are sufficient to extract Stark-effect data from FTMW spectrometers. Because of the narrower transitions ($\sim$10–15 kHz), the FTMW spectrometers do not require shifts as large as in waveguide spectrometers, with shifts of 1 MHz or less usually being adequate for determination of electric dipole moments. The addition of Stark-effect data has been essential in the determination of molecular dipole moments and in the assignments of complex spectra, particularly in asymmetric rotors and other spectra where the transitions are split into complex multiplets from internal motions.

### 14.4.3 Utility of the Spectrometer

The FTMW spectrometer offers a combination of high resolution and high sensitivity that makes it a powerful spectroscopic tool. This is achieved through a combination of molecular beam methods and time domain Fourier transform spectroscopy.

The molecular beam is generally produced using a pulsed pinhole nozzle of about 1 mm diameter with a stagnation pressure of 0.1–0.2 MPa [1–2 atm]. When the nozzle is fired, vibrational, rotational, and translation degrees of freedom are efficiently cooled in the adiabatic expansion by collisions with the rare gas carrier, resulting in rotational and translational temperatures on the order of 1 K. The vibrational cooling is not as efficient, but generally the vibrational degrees of freedom are cooled to the extent that excited vibrational states are difficult to observe in the beam. The favorable rotational partition function greatly simplifies spectra by reducing congestion, and it significantly enhances sensitivity. The narrow velocity distribution reduces pressure and Doppler contributions to inhomogeneous line widths, permitting very high resolution.

There are advantages inherent in time domain spectroscopy which are exploited by the FTMW technique. The first is the ease of signal averaging, which permits the detection of very weak molecular signals—for example, $^{13}$C or $^{34}$S in natural abundance. The pulsed nozzles in the FTMW spectrometers typically operate at 10–50 Hz, allowing thousands of cycles to be averaged in a few minutes. The same feature also allows a simple background subtraction by monitoring the microwave signals with the gas pulse on and off. Further, time domain signals are power-independent, resulting in no contribution to line widths from power broadening and better resolution.

The primary limitation of the FTMW spectrometer arises from the narrow-bandedness of the instrument. This requires the operator to set the microwave frequency and tune the Fabry–Perot cavity at small steps (typically 250–500 kHz), making searching for transitions time–consuming and tedious. This has been recently addressed at the University of Kiel, the University of Illinois [48] and NIST by introducing computer-controlled scanning, where a computer is used to step the synthesizer, step the mirrors and initiate the pulse sequence. A log is kept of signal vs. frequency. This can in principle be used to scan a cavity mode up to the length of travel of the mirrors: For 5 cm travel, this corresponds to several hundred megahertz scanning range.

### 14.4.4 Dimers, Trimers, and Beyond

The molecular beam aspect of the FTMW spectrometer makes it well-suited to the observation of molecular complexes, which are formed in relatively large numbers by three-body collisions in the expansion. A large number of dimers, a handful of trimers, and fewer still higher complexes have been characterized, with FTMW studies contributing structural, binding, and dynamical information.

The goal of the study of molecular complexes has been to provide accurate and detailed information on bimolecular interactions, and then to relate that information to trimolecular and higher clusters and finally to condensed phases. In the last 15 years, significant progress has been made along these lines, largely with the use of FTMW spectrometers. Initial research focused primarily on complexes of rare gases with diatomic and triatomic molecules, many of which are now well understood. In recent years these studies have expanded to look at interactions of larger and more chemically interesting molecules. For example, a number of mixing nozzles have been developed to study van der Waals complexes of reactive molecules. Systems such as ethylene $\cdot$ ozone [49] and ketene $\cdot$ ethylene [50] have been studied at NIST, while Gutowsky *et al.*, have used a mixing nozzle to study complexes of $NH_3$ with HCN and HF [51]. Efforts have also been made to study complexes whose strength is intermediate between van der Waals molecules and covalently bonded molecules. These have included studies of $BF_3$ with various binding partners [52] and $SO_2$ with a number of amines [53].

Progress has also been made, led by Gutowsky *et al.*, at Illinois, in expanding the studies of larger clusters. Beginning with the observation of $Ar_2 \cdot HF$ [54], they have studied a number of related trimers and higher complexes, with $Ar_4 \cdot HF$ [55] being the largest reported to date. In addition, they have recorded the spectra of the mixed trimer $NH_3 \cdot HCN \cdot HF$ [51]. The NIST group has also observed several interesting trimers, namely, $H_2O-H_2O-CO_2$ [56], $CO_2 \cdot CO_2 \cdot H_2O$ [57], and $CO \cdot CO \cdot H_2O$ [58]. It can be safely said, however, that the study of large clusters is in its infancy and will present one of the major challenges to the FTMW technique in the near future.

### 14.4.5 Laser Vaporization and High-Temperature Species

There has been considerable interest in expanding the use of the FTMW technique to substances which do not have appreciable vapor pressures at ambient temperatures. Efforts have been concentrated in the use of high-temperature nozzles and lasers to vaporize high-boiling and refractory materials. These endeavors stem from the fact that the expansion process cools the molecules to near absolute zero, providing a spectrum of a high-temperature species that does not suffer from the complexity usually encountered in the study of spectra of a nonvolatile substance at elevated temperature where numerous

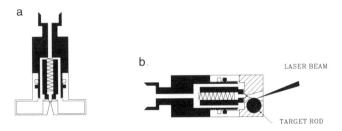

FIG. 11. (a) Heated reservoir nozzle used at NIST to study numerous species of low volatility, (b) Laser vaporization nozzle used for refractory species. (Figure 11a reprinted with permission from: R. D. Suenram, G. T. Fraser, C. W. Gillies, and J. Zozom, *J. Chem. Phys.*, **88**, 722–729 (1988). Copyright 1988, American Institute of Physics.)

energy levels are populated. While both techniques have demonstrated a number of successes, each technique has some shortcomings associated with it.

Various heated nozzles similar to the design shown in Figure 11a have been constructed in different laboratories. For the most part, these have involved the use of a heated reservoir near the nozzle tip to vaporize the material, which is then swept along in a rare gas stream. Several designs include an extended plunger and water cooling. These nozzles have been successfully used to observe spectra of such substances as indole [59], carbazole [60], Ar–Hg [61], formamide–water, formamide–methanol [62] and formaldehyde dimer [63]. However, the operation of these nozzles has been confined to temperatures below about 250°C. This is primarily due to the requirement of a pulsed gas source and the damage caused by overheating a solenoid valve. More recently, Legon *et al.*, have introduced a new design which separates the heat from the valve by introducing the nonvolatile material through a constantly flowing heated capillary tube into a pulsed rare gas expansion. This design has been used up to 400°C in the study of $NH_4I$ [64].

Laser vaporization has been used to produce beams of refractory materials in a number of types of experiments and has been employed with some success in the FTMW spectrometer. A rod of the material is placed in a machined-out section of an extended nozzle tip, as shown in Figure 11b. A pulse from a Nd:YAG laser is focused on the rod, and the vaporized material is swept along by a pulse of carrier gas. This technique has been used at NIST to observe the spectra of a number of refractory materials. The first experiments with this technique were carried out on the $SiC_2$ species [65]. In this case, a target rod of SiC was used to observe the $J = 1_{11}-0_{00}$ transition of $SiC_2$. Subsequent work involving rare earth oxides has been carried out at NIST. YO, LaO, ZrO, and HfO were observed by ablating rods of $Y_2O_3$, $La_2O_3$, $ZrO_2$ and $HfO_2$, respectively [66]. For other molecules, target rods were painted with thick coatings of the materials of interest, which were then ablated. Spectra observed by this method include SrO and VO, which were ablated from rods painted with $SrCO_3$ and $V_2O_5$, respectively [67, 68]. Some species have also been observed by reacting hot atoms ablated from a rod of pure metal with a molecule seeded in the gas stream. For example, NbO has been observed by ablating a rod of pure Nb into a stream of 1% $CO_2$ in Ar; however, the signals were a factor of four weaker than those observed by ablating a rod of $Nb_2O_5$ [73]. Recently, Gerry *et al.*, at the University of British Columbia (UBC) have observed several metal chlorides by ablating Ag and sweeping the surface with a $Cl_2$ in Ne mixture. We have recently observed MgNC and CaNC at NIST using metal rods and adding cyanogen to the carrier gas stream [69].

### 14.4.6 Radicals and Unstable Species in Discharge Experiments

Electric discharges have been used for many years in the study of radicals and ions using conventional microwave and other types of spectroscopy, and recently several groups have reported successes with discharge experiments in the FTMW technique. Generally, the experiment is set up with a hot wire either inside the nozzle orifice, which discharges when the nozzle fires, or directly downstream from the nozzle. Endo *et al.* have observed radicals including $HC_9N$ [70] and $C_3S$ [71] using a pulsed discharge just in front of the valve. Additionally, with this system they have recorded the first FTMW spectrum of a complex containing a radical in the Ar–OH van der Waals complex [72]. Using a DC discharge, a number of radicals and unstable species have been observed at NIST and the University of Tokyo including SO, $C_2S$, $C_3S$, $c$-$HC_3$, $CH_2CC$, $c$-$C_3H_2$, $CH_2CN$, and HCCN [73]. At the UBC, similar experiments have produced spectra of $SiF_2$, FCCCN, FNO, and $FNO_2$. Stahl *et al.*, at Kiel have used discharges to produce rotationally cold but vibrationally hot beams. OCS has been observed with up to 6000 cm$^{-1}$ of vibrational excitation.

## 14.5 Slit-Jet Microwave Spectrometers

Another recent addition to the growing line of microwave molecular beam spectrometers was developed by Bumgarner and Blake [74] and first described in 1989. This type of spectrometer incorporates a slit-jet nozzle approximately 25 μm wide × 10 cm long. This type of spectrometer has been used extensively in the infrared [75] and far-infrared spectral regions [76] but this is the first attempt at use in what is normally considered the microwave region. The essential components of the spectrometer are nearly identical with the conventional spectrometer (Figure 1), except that the absorption cell has been replaced with the slit-jet source. The microwave radiation is focused across the slit-jet with a lens and collected after the slit with a second lens. The initial experiments of Bumgarner and Blake employed a frequency-modulated microwave synthesizer with frequency-doubled and tripled output. The output radiation was detected using a Schottky diode detector (Section 14.2.5) and lock-in detection. Recently, Stark plates have been added to enable Stark modulation to be employed. The plate spacing is large, so limited electric fields can be obtained, but for pseudo–first-order Stark behavior and some faster second-order transitions, modulation can easily be obtained. This has improved the signal-to-noise of the instrument by a factor of five and has removed some of the baseline problems inherent with source modulation [77]. In principle this type of spectrometer can employ any of the frequency sources described in Section 14.2.2, which makes it an excellent instrument for broad-banded scans throughout the microwave frequency region. The one drawback of this instrument is the copious amount of sample required to form the slit jet. Instruments of this type typically require pumping systems consisting of a large roughing pump and Roots blower combination that are capable of handling 70 $m^3$/min of gas on a continuous basis. This is not a serious problem for most chemical species, and the user-friendly, broad-banded scanning capabilities more than offset this inconvenience. With this instrument it is possible to obtain a rough rotational assignment on a molecular complex in a few days. Once the assignment has been made, follow-up work can be accomplished using a FTMW instrument which requires only small amounts of sample. The power of this combination of instruments has recently been demonstrated in the study of the benzene–water [78, 79], benzene–ammonia [80] and methanol–water [81] weakly bound complexes.

# References

1. Johnson, D. R., and Pearson, R. (1976). In *Spectroscopy*, D. Williams (ed.), *Methods of Experimental Physics*, Vol. 13, Part B, Chapter 4.3. Academic Press, New York.
2. Gordy, W., Smith, W. V., and Trambarulo, R. F. (1953). *Microwave Spectroscopy*. Wiley, New York. (1966). Dover, New York.
3. Townes, C. H., and Schawlow, A. W. (1955). *Microwave Spectroscopy*. McGraw-Hill, New York. (1975). Dover, New York.
4. Gordy, W., and Cook, R. L. (1970). *Microwave Molecular Spectra*. Wiley (Interscience), New York.
5. Kroto, H. W. (1975). *Molecular Rotation Spectra*. John Wiley and Sons, New York.
6. Delucia, F. C., Krupnov, A. F., and Burenin, A. V. (1976). In *Molecular Spectroscopy: Modern Research*, K. N. Rao (ed.), Vol. 2, Chapter 2. Academic Press, New York.
7. Carrington, A. (1974). *Microwave Spectroscopy of Free Radicals*. Academic Press, New York.
8. Sugden, T. M., and Kenney, C. N. (1965). *Microwave Spectroscopy of Gases*. D. van Nostrand, New York.
9. Wollrab, J. E. (1967). *Rotational Spectra and Molecular Structure*. Academic Press, New York.
10. Harmony, M. D. (1972). *Introduction to Molecular Energies and Spectra*, Chapter 7. Holt, Rinehart and Winston, New York.

11. Karasek, F. W. (1972). *Research and Development* **23**, 38–40.
12. Pickett, H. M. (1977). *Rev. Sci. Instrum.,* **48**, 706–707.
13. Alekshin, Yu. I., Altshuller, G. M., Pavlovsky, O. P., Karyakin, E. N., Krupnov, A. F., Pavliev, D. G., and Shkaev, A. P. (1990). *Int. J. of Infrared and Millimeter Waves* **11**, 961–971.
14. Karaykin, E. N. (1992). Private communication.
15. Fortunato, M. P., and Ishikawa, K. Y. *Microwaves Magazine*, May 1982.
16. Button, K. J., ed. (1979). *Infrared and Millimeter Waves; Sources of Radiation*, Vol. 1, Chapter 2. Academic Press, New York.
17. Hughes, R. H., and Wilson, E. B., Jr. (1947). *Phys. Rev.* **71**, 562–563; other related papers include Low, W., and Townes, C. H. (1949). *Phys. Rev.* **75**, 529–530; Hedrick, L. C. (1949). *Rev. Sci. Instrum.* **20**, 781–783; McAfee, K. B., Jr., Hughes, R. H., and Wilson, E. B., Jr. (1949). *Rev. Sci. Instrum.* **20**, 821–826.
18. A good description of this is given in Chapter 10 of Townes, C. H., and Schawlow, A. W. (1955). *Microwave Spectroscopy*. McGraw-Hill, New York.
19. Bernius, M. T., and Chutjian, A. (1989). *Rev. Sci. Instrum.* **60**, 779–782.
20. Bernius, M. T., and Chutjian, A. (1990). *Rev. Sci. Instrum.* **61**, 925–927.
21. Pickett, H. M. (1980). *Appl. Opt.* **19**, 2745–2749.
22. Suenram, R. D., and Johnson, D. R. (1977). *J. Molec. Spectrosc.* **65**, 239–248.
23. Lafferty, W. J., Suenram, R. D., and Lovas, F. J. (1987). *J. Molec. Spectrosc.* **123**, 434–452.
24. Suenram, R. D., and Lovas, F. J. (1978). *J. Am. Chem. Soc.* **100**, 5117–5122.
25. Gillies, J. Z., Gillies, C. W., Suenram, R. D., and Lovas, F. J. (1988). *J. Am. Chem. Soc.* **110**, 7991–7999.
26. Lovas, F. J., Suenram, R. D., and Evenson, K. M. (1983). *Ap. J.* **267**, L131–L133.
27. Powell, F. X., and Johnson, D. R. (1969). *J. Chem. Phys.* **50**, 4596.
28. Amano, T., Saito, S., Hirota, E., Morino, Y., Johnson, D. R., and Powell, F. X. (1969). *J. Molec. Spectrosc.* **30**, 275–289.
29. Lovas, F. J., Suenram, R. D., and Stevens, W. J. (1983). *J. Molec. Spectrosc.* **100**, 316–331.
30. Suenram, R. D., Lovas, F. J., and Stevens, W. J. (1983). *J. Molec. Spectrosc.* **112**, 482–493.
31. Suenram, R. D., and Lovas, F. J. (1980). *J. Am. Chem. Soc.* **102**, 7180–7184.
32. Suenram, R. D., Lovas, F. J., and Fraser, G. T. (1988). *J. Molec. Spectrosc.* **127**, 472–480.
33. Kuczkowski, R. L., Suenram, R. D., and Lovas, F. J. (1981). *J. Am. Chem. Soc.* **103**, 2561–2566.
34. Throne, L. R., Suenram, R. D., and Lovas, F. J. (1983). *J. Chem. Phys.* **78**, 167–171.
35. A number of companies have excellent application notes that describe in detail the solid-state physics employed in the manufacture of these devices; cf. Farran Technology Application Note 12. Farran Technology Limited, Ballincollig, Cork, Ireland.
36. English, T. C., and Zorn, J. C. (1972). In *Methods of Experimental Physics*, 2nd Edition, Vol. 3, D. Williams (ed.). Academic Press, New York.
37. Fraser, G. T., Suenram, R. D., and Coudert, L. H. (1989). *J. Chem. Phys.* **90**, 6077–6085.
38. Fraser, G. T., and Suenram, R. D. (1992). *J. Chem. Phys.* **96**, 7287–7297.
39. Herbine, P., and Dyke, T. R. (1985). *J. Chem. Phys.* **83**, 3768–3774.
40. Stockman, P. A., Bumgarner, R. E., Suzuki, S., and Blake, G. A. (1992). *J. Chem. Phys.* **96**, 2496–2510.

41. Miller, C. C., Philips, L. A., Andrews, A. M., Fraser, G. T., Pate, B. H., and Suenram, R. D. (1994). *J. Chem. Phys.* **100**, 831–839.
42. Balle, T. J., and Flygare, W. H. (1981). *Rev. Sci. Instrum.* **52**, 33–43.
43. Balle, T. J., Campbell, E. J., Keenan, M. R., and Flygare, W. H. (1980). *J. Chem. Phys.* **72**, 922–932.
44. Campbell, E. J., Buxton, L. W., Balle, T. J., and Flygare, W. H. (1981). *J. Chem. Phys.* **74**, 813–828.
45. Lovas, F. J., and Suenram, R. D. (1987). *J. Chem. Phys.* **87**, 2010–2020.
46. Stahl, Wolfgang. Private communication.
47. Kolbe, W. F., and Leskovar, B. (1985). *Rev. Sci. Instrum.* **56**, 97–102.
48. Grabow, J.-U., and Stahl, W. (1990). *Z. Naturforsch.* **45a**, 1043/1044.
49. Chung, C., Hawley, C. J., Emilsson, T., and Gutowsky, H. S. (1990). *Rev. Scient. Instrum.* **61**, 1629–1635.
50. Gillies, C. W., Gillies, J. Z., Suenram, R. D., Lovas, F. J., Kraka, E., and Cremer D. (1991). *J. Am. Chem. Soc.* **13**, 2412–2421.
51. Lovas, F. J., Suenram, R. D., Gillies, C. W., Gillies, J. Z., Fowler, P. W., and Kisiel, Z. (1994). *J. Am. Chem. Soc.* **116**, 5285–5294.
52. Emilsson, T., Klots, T. D., Ruoff, R. S., and Gutowsky, H. S. (1992). *J. Chem. Phys.* **93**, 6971–6976.
53. Dvorak, M. A., Ford, R. S., Suenram, R. D., Lovas, F. J., and Leopold, K. R. (1992). *J. Am. Chem. Soc.* **114**, 108–115.
54. Oh, J.-J., LaBarge, M. S., Matos, J., Kampf, J. W., Hillig II, K. W., and Kuczkowski, R. L. (1991). *J. Amer. Chem. Soc.* **113**, 4732.
55. Gutowsky, H. S., Klots, T. D., Schuttenmaer, C. A., and Emilsson, T. (1987). *J. Chem. Phys.* **86**, 569–576.
56. Gutowsky, H. S., Chuang, C., Klots, T. D., Emilsson, T., Ruoff, R. S., and Krause, K. R. (1988). *J. Chem. Phys.* **88**, 2919–2924.
57. Peterson, K., Suenram, R. D., and Lovas, F. J. (1991). *J. Chem. Phys.* **94**, 106–117.
58. Peterson, K. I., Suenram, R. D., and Lovas, F. J. (1989). *J. Chem. Phys.* **90**, 5964–5970, Erratum (1992) *S. Chem. Phys.* **92**, 5166..
59. Peterson, K. I., Suenram, R. D., and Lovas, F. J. (1995). *J. Chem. Phys.* **102**, 7807–7816.
60. Suenram, R. D., Lovas, F. J., and Fraser, G. T. (1988). *J. Molec. Spectrosc.* **127**, 472–480.
61. Lovas, F. J., Fraser, G. T., and Marfey, P. S. (1988). *J. Mol. Struct.* **19**, 135–141.
62. Ohshima, Y., Lida, M., and Endo, Y. (1990). *J. Chem. Phys.* **92**, 3990–3991.
63. Lovas, F. J., Suenram, R. D., Fraser, G. T., Gillies, C. W., and Zozom, J. (1988). *J. Chem. Phys.* **88**, 722–729.
64. Lovas, F. J., Suenram, R. D., Coudert, L. H., Blake, T. A., Grant, K. J., and Novick, S. E. (1990). *J. Chem. Phys.* **92**, 891–898.
65. Legon, A. C., and Stephenson, D. (1992). *J. Chem. Soc. Farad. Trans.* **88**, 761–762.
66. Suenram, R. D., Lovas, F. J., and Matsumura, K. (1989). *Ap. J. Lett.* **342**, L103–L105.
67. Suenram, R. D., Lovas, F. J., Fraser, G. T., and Matsumura, K. (1990). *J. Chem. Phys.* **92**, 4724–4733.
68. Blom, C. E., Hedderich, H. G., Lovas, F. J., Suenram, R. D., and Maki, A. G. (1992). *J. Mol. Spec.* **152**, 109–118.
69. Suenram, R. D., Fraser, G. T., Lovas, F. J., and Gillies, C. W. (1991). *J. Mol. Spectrosc.* **148**, 114–122.

70. Scurlock, C., Steimle, T., Suenram, R. D., and Lovas, F. J. (1994). *J. Chem. Phys.* **100,** 3497–3502.
71. Iida, M., Ohshima, Y., and Endo, Y. (1991). *Ap. J.* **371,** L45.
72. Ohshima, Y., and Endo, Y. (1992). *J. Molec. Spectrosc.* **153,** 627–634.
73. Ohshima, Y., Iida, M., and Endo, Y. (1991). *J. Chem. Phys.* **95,** 7001–7003.
74. Lovas, F. J., Suenram, R. D., Ogata, T., and Yamamoto, S. (1992). *Ap. J.* **399,** 325–329.
75. Bumgarner, R. E., and Blake, G. A. (1989). *Chem. Phys. Lett.* **161,** 308–314.
76. Lovejoy, C. M., and Nesbitt, D. J. (1987). *Rev. Sci. Instr.* **58,** 807–811.
77. Busarow, K. L., Blake, G. A., Laughlin, K. B., Cohen, R. C., Lee, Y. T., and Saykally, R. J. (1987). *Chem. Phys. Lett.* **141,** 289–291.
78. Blake, G. A. Private communication.
79. Suzuki, S., Green, P. G., Bumgarner, R. E., Dasgupta, S., Goddard, W. A. III, and Blake, G. A. (1992). *Science* **257,** 942–945.
80. Gutowsky, H. S., Arunan, E., and Emilsson, T. (1993). *J. Chem. Phys.* **99,** 4883–4893.
81. Rodham, D. A., Suzuki, S., Suenram, R. D., Lovas, F. J., Dasgupta, S., Goddard, W. A. III, and Blake, G. A. (1993). *Nature* **362,** 735–737.
82. Stockman, P. A., Blake, G. A., Suenram, R. D., and Lovas, F. J. Submitted for publication.

# 15. FAST BEAM SPECTROSCOPY

## Linda Young

Argonne National Laboratory
Argonne, Illinois
and
Joint Institute for Laboratory Astrophysics
University of Colorado
Boulder, Colorado

## 15.1 Introduction

Fast atomic and molecular beams are extremely versatile tools for high-resolution spectroscopy. In this chapter "fast" is taken to mean super-thermal, or with acceleration energies in excess of 200 eV. This versatility stems from a number of properties of fast beams. First, beams can be readily isolated and identified by their mass-to-charge ratio, $m/q$. Second, the beam constituents are typically in a collision-free environment (i.e., beam densities of $<10^7/cm^3$), thus permitting perturbation-free measurements to be made. Third, precise control of the velocity of the beam, and hence Doppler shift enables precision spectroscopy to be done with a fixed-frequency laser. Fourth, the mass-selected beams can be generated in a variety of charge states (positive, neutral, and negative) either through direct extraction from an appropriate source, or through charge exchange. Fifth, the flight time to the detection region can be made small ($\approx 1\ \mu s$) so that short-lived and metastable species can be readily detected. Sixth, the beams are of sufficiently high energy that even neutral photodissociation products of molecular species can be directly detected without recourse to a secondary ionization or laser.

In comparison to other techniques for the spectroscopic study of ions, e.g., traps and spectroscopy in discharges, the fast beam technique offers an easily obtained field-free, very low-density (hence, perturbation-free) environment. While elegant techniques of cooling ions in traps produce the ultimate in resolution, the convenient access to highly excited metastable states and rare, unstable isotopes makes the fast beam a more generally applicable spectroscopic tool. In the study of dissociation processes, the aforementioned advantages which permit very selective excitation with a laser are combined with kinematic benefits which allow extremely high resolution and efficiency in the direct detection of fragmentation processes.

EXPERIMENTAL METHODS IN THE PHYSICAL SCIENCES
Vol. 29B

Since the area of fast-beam spectroscopy is so diverse, it is impossible to cover all aspects within this chapter. In particular, only fast-beam spectroscopic studies in which a laser is involved will be reviewed. Within this restriction, an attempt has been made to pick examples of the various techniques which span a wide variety of species (from multiply ionized atoms to complex molecules) in order to illustrate the many applications of fast-beam spectroscopy. Some areas of fast-beam laser spectroscopy have been covered in detail in earlier reviews [1, 2] and accordingly are treated more sparingly here. In addition, beam-foil spectroscopy has been omitted because of the large number of excellent review articles and books in this field [3–5].

The chapter is organized as follows. A summary of fast-beam kinematics will first be given in order to point out some advantages of the fast-beam method for spectroscopy. It will be followed by a short discussion of the methods of generation of specific beams. Section 15.3 is devoted to measurements on atomic systems: high-resolution laser spectroscopy, radio-frequency and microwave spectroscopy, lifetime measurements, and electron spectroscopy. Section 15.4 concentrates on measurement techniques for molecular systems: photofragmentation and photodetachment spectroscopy. Section 15.5 discusses the application of fast beams to metrology.

## 15.2 Background

### 15.2.1 Fast-Beam Kinematics

The optical absorption lines of an accelerated species can be made extremely narrow because of a simple kinematic effect [6]. Consider an ensemble characterized by a thermal velocity of $v_0$, and thereby an energy spread before acceleration, $E_0$, of

$$E_0 = \tfrac{1}{2}mv_0^2. \tag{15.1}$$

After acceleration in the $z$ direction through a potential $E$ to velocity $v = \sqrt{2E/m}$, the energy spread remains the same; however, the velocity spread in the $z$-direction, $dv$, is now

$$dv = dE/mv = E_0/mv = v_0^2/v = v_0\sqrt{E_0/4E}. \tag{15.2}$$

That is, the velocity spread has been compressed by a factor of $2\sqrt{E/E_0}$. For a typical energy spread in the ion source of $dE = E_0 = 2$ eV and an acceleration energy of $E = 50$ keV, this corresponds to a $\approx 300$-fold reduction in the velocity spread!

The Doppler spread for optical excitation in the $z$ direction is much reduced. The rest frame resonance frequency for the accelerated atom, $\nu_0$, is related to the laser frequency, $v_L$, by the following:

$$v_0 = v_L \gamma(\beta)[1 - \beta \cos(\theta)], \tag{15.3}$$

where $\beta$ is $v/c$, $\gamma(\beta) = (1 - \beta^2)^{-1/2}$ and $\theta$ is the angle of intersection between the laser and ion beams ($\theta = 0°$ for copropagating, $\theta = 180°$ for counterpropagating). Neglecting all other sources of broadening, the Doppler width in the collinear direction ($z$) is

$$\Delta v_z = v_0(v_0/c)\sqrt{E_0/4E}. \tag{15.4}$$

Acceleration in the $z$ direction does not alter the thermal velocity in the $x$ and $y$ directions. Thus, there is a minor line-width contribution of $\Delta v \approx v_0(v_0/c)\Delta\theta$ due to the wavefront curvature of the ion beam, where $\Delta\theta$ is the FWHM angular divergence of the beam, as well as a contribution of $\Delta v \approx v_0(v/c)(\Delta\theta)^2/8$ due to the small angle intersection of the laser with the divergent beam. In practice, Doppler widths for absorption in the collinear direction ($z$) of tens of megahertz in the visible region can be obtained, allowing direct resolution of hyperfine structure. The optical resolution obtainable for perpendicular incidence of the laser is usually much lower and is dominated by the angular divergence of the beam:

$$\Delta v_\perp = v_0(v/c) \sin \Delta\theta. \tag{15.5}$$

Both collinear and perpendicular fast-beam/laser geometries can yield appreciable probabilities for excitation which can be estimated if the cross-section, $\sigma$, for the process is known. For small excitation probabilities, the probability of excitation is $P_e \approx \sigma It$, where $I$ = laser power density (photons/cm$^2$) and $t$ is the dwell time of the particle in the laser field. In using this method, $\sigma$ must properly account for the overlap of the laser and transition bandwidths [7]. The excitation probability can be considerably higher in the collinear geometry, because the dwell time can be made longer if good laser/ion beam overlap is maintained.

Collinear excitation is preferred for high-resolution spectroscopy (because of the kinematic compression of the Doppler spread) and for the observation of low–cross-section processes (because of the increased interaction time). Another feature of the collinear geometry is the possibility of Doppler-tuning the beam into resonance with the laser. However, for processes where a precise $t = 0$ initiation point is desired, perpendicular or angle tuned excitation offer advantages.

## 15.2.2 Fragmentation Kinematics

The use of fast beams to study dissociative processes (photodissociation of molecules or photodetachment of negative ions) offers a number of advantages

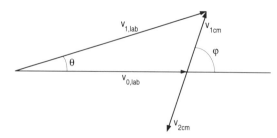

FIG. 1. Velocity diagram for a dissociation event. A parent ion of mass $m_1 + m_2$ dissociates into two fragments $m_1$ and $m_2$ at an angle of $\varphi$ in the c.m. frame. The c.m. energy release is $W$, and the c.m. velocity for $m_1$ ($m_2$) is $v_{1cm}$ ($v_{2cm}$). The laboratory velocities for the parent and $m_1$ fragment are $v_{0,lab}$ and $v_{1,lab}$, respectively.

which stem from the kinematics. Consider the dissociation event pictured in Figure 1. A molecule having initial kinetic energy $E$ in the laboratory frame dissociates into two fragments of mass $m_1$ and $m_2$ with a total center of mass (c.m.) kinetic energy release $W$. The laboratory energy of $m_1$, ejected at a c.m. angle $\varphi$ (or laboratory angle $\theta$) with respect to the beam direction, is given by the following expression [8]:

$$E_1 = \frac{m_1 m_2}{m_1 + m_2} \left[ \frac{E}{m_2} + 2\sqrt{\frac{EW}{m_1 m_2}} \cos \varphi + \frac{W}{m_1} \right]. \tag{15.6}$$

This general expression is also applicable to electron detachment with $m_1 = m_e$.

It is important to note that these kinematics lead to a large magnification of the c.m. energy release in the lab frame when $E \gg W$. For dissociation of a homonuclear diatomic molecule, from Eq. (15.6), the energy of a fragment in the lab frame is $E_1 = E/2 \pm (EW)^{1/2} + W/2$. For a $W$ of 300 meV and parent ion beam energy of 3 keV, the fragment energy in the lab frame appears at $\approx 1500 + 30$ eV. Thus, a resolution of 1 eV in the lab frame corresponds to an uncertainty of $\pm 10$ meV in the dissociation energy, showing that very high-resolution studies are possible.

From Fig. 1, the laboratory angle $\theta$ is related to the c.m. angle $\varphi$ by the following:

$$\theta = \tan^{-1}[v_{1cm} \sin \varphi / (v_{0,lab} + v_{1cm} \cos \varphi)]. \tag{15.7}$$

The laboratory angle is compressed, allowing the design and fabrication of detectors which can monitor essentially all c.m. fragmentation angles simultaneously.

## 15.2.3 Generation of Beams

A vast array of fast beams have been generated and subsequently utilized for laser-based spectroscopy. Many of the studies have used ions, including multiply charged ions [9] and metallic cluster anions [10], directly extracted from an ion source. Typically the ions extracted from a source have substantial internal energy content, so that many metastable states are populated and thus available for study. The interested reader is referred to Chapter 3 on ion sources.

In addition to direct extraction, an invaluable method for producing negative ion beams is charge exchange in a vapor. In particular, successive charge transfer to a positive ion from a suitably chosen vapor is a reliable method of production for *metastable* atomic negative ions [11]. The efficiency of the double charge-exchange process depends greatly on the projectile–target combination, and the highest efficiency occurs when the energy defect is minimized in both electron-capture processes [12]. Efficiencies for the production for a variety of negative ions by exchange in Na and Mg vapor range from 0.4 to 90%. The efficiencies increase almost monotonically with projectile electron affinity [12]. As an example of metastable negative ion production. the optimal target thickness of Li vapor for the production of $Ca^-$ from 80 keV $Ca^+$ was found to be $\approx 10^{16}$ Li atoms/cm$^2$ when $\approx 0.5\%$ of the incident beam was converted to $Ca^-$. At higher thicknesses, multiple scattering reduces both the negative ion fraction and beam quality [11].

Fast neutral beams also are formed by charge exchange, and this has been extensively studied for atomic species, both experimentally [13–15] and theoretically [16, 17]. Total charge-transfer cross-sections for the process

$$A^+ + X \rightarrow A + X^+ + \Delta E,$$

where $A^+$ is the incident ion and X the target alkali atom, are typically $10^{-16}$–$10^{-15}$ cm$^2$ for intermediate ion velocities ($10^5/\sqrt{\mu} < v < 10^8$ cm/s, where $\mu$ is the reduced mass of the collision pair in amu). The neutral beam, A, is formed in a variety of final states [15, 18, 19], with those having small energy defect $\Delta E$ being formed preferentially. Fortunately, the energy transfer to the target is of the order or $\Delta E^2/E$, so that the velocity distribution of the neutralized beam remains virtually unaltered and suitable for high-resolution spectroscopy.

Because the cross-sections for electron transfer are enhanced at small $\Delta E$, one can prepare the target X such that $\Delta E$ is optimized for capture into a specific excited state of A. The "preparation" can be done either by a judicious choice of target and projectile final state [15, 20] or by laser excitation of the target [21, 22]. Very recent studies [22] demonstrate a *substantial* ($10^4$) enhancement of the population in selected Rydberg states by charge exchanging with a laser-excited target as compared to a gas target.

## 15.3  Studies of Atoms

Much of fast beam spectroscopy of atoms is motivated by the possibility of studying exotic species in a perturbation-free environment, e.g., rare and unstable isotopes, metastable states of atoms, multiply ionized atoms, and negative ions. In the following sections the adaptation of techniques, often initially demonstrated on thermal samples to fast beam studies is discussed.

### 15.3.1 High Resolution Laser Spectroscopy

Because of the kinematic compression of the velocity spread, the resolution in a collinear fast-beam laser interaction is relatively high. This enables direct optical measurements of fine and hyperfine structures using linear spectroscopic techniques. A standard setup for a collinear fast-beam laser spectroscopy experiment is shown in Figure 2.

Ions created in a suitable source are accelerated through an electrostatic potential, mass-separated with a magnetic field, and sent down a straight section containing a post-acceleration tube where the interaction with the laser field occurs. The basic parameters of the beam are a divergence of $\Delta\theta \approx 0.001$ and an energy spread of $dE/E \approx 10^{-4}$. The post-acceleration tube allows the beam energy to be fine-tuned and also confines the resonant beam/laser interaction to a well-defined region. A resonant beam/laser interaction is detected by the fluorescence emitted from the beam using photomultipliers in conjunction with interference filters or a grating spectrometer. The wavelength discrimination of the fluorescence photons is important to reduce noise from beam collisions with residual background gas and scattered light from the laser. A wavelength scan is achieved either by tuning the laser or by adjusting the post-acceleration voltage and thus the Doppler shift. The usual diagnostics for the laser, i.e., absolute and relative wavelength determination, are typically employed.

In addition to a standard laser-induced fluorescence (LIF) scan, where a spectrum of fluorescence vs. wavelength is recorded, transition energies can be

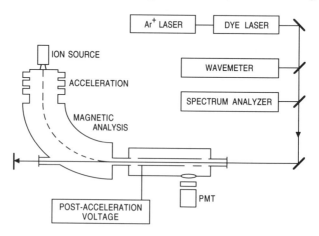

FIG. 2. A generic fast ion laser spectroscopy apparatus.

measured with great absolute accuracy in the collinear geometry without precise measurement of the beam velocity because the first- and second-order Doppler shifts exactly cancel when the geometric mean of the resonance frequencies for co- and counterpropagating excitation is taken. If $v_0$ is the transition frequency in the rest frame of the atom, and $v_b$ and $v_r$ the transition frequencies for parallel and anti-parallel laser excitation, Eq. (15.3) yields

$$v_0 = v_r \gamma(\beta)[1 + \beta] \quad \text{and} \quad v_0 = v_b \gamma(\beta)[1 - \beta] \quad (15.8)$$

and thus

$$v_0 = \sqrt{v_b v_r}. \quad (15.9)$$

To illustrate this technique, we examine the absolute wavelength measurements made of the $2s \rightarrow 2p$ transitions in two-electron systems. These measurements provide rigorous tests of relativistic quantum theory, quantum electrodynamics (QED), and electron correlations in a multiparticle system [23, 24]. Relativistic and QED effects scale as $Z^4$ and higher. Thus, studies over a wide range of $Z$ are desirable. Measurements have been made on $Li^+$ [25–27], $Be^{2+}$ [28], and $B^{3+}$ [29], as well as on many higher-$Z$ systems using either beam-foil or discharge sources combined with spectrometer techniques (see references in [23]). Beam-foil measurements have absolute wavelength accuracy at the $\approx 10$ ppm level, whereas the laser-based measurements on $Li^+$ approach one part in $10^8$. Unfortunately, laser-based experiments are limited at high $Z$ by the availability of cw sources in the ultraviolet region.

Absolute wavelength measurements of the $1s2s\,^3S \rightarrow 1s2p\,^3P$ transition in $Li^+$ have been made using laser-induced fluorescence (LIF) spectroscopy with an $11°$ intersection angle on an intermediate velocity ion beam (4.1–6.3 keV), and with colinear geometry and higher velocity beam (50 keV). The Doppler compressed FWHM widths were 950 MHz and 100 MHz, respectively. The collinear fast-beam geometry allowed the determination of the absolute wavelength of the $2^3S_1$–$2^3P_0$ transition to $\approx 1$ part in $10^8$, where over 70% of the total systematic error was due to accelerator drift during the time interval necessary to change the laser wavelength from $v_r$ to $v_b$. This can be eliminated by measuring the red- and blue-shifted transition frequencies simultaneously with saturation spectroscopy on the collinear fast-beam [30]. Saturated absorption and two-photon absorption in fast beams is discussed elsewhere [2].

A vastly increased sensitivity to the detection of isotopes of small abundance is made possible by the application of the photon burst method [31–33] to fast beams [34]. The basis of the method, shown in Figure 3, is that an $n$-photon burst of resonantly scattered light occurs as a single atom or ion crosses the laser beam. Because of the nonlinearity of $n$-event Poisson statistics, high discrimination against random background and the wings of nearby lines is obtained as $n$ is increased [35]. In combination with the mass-selectivity available in fast beams,

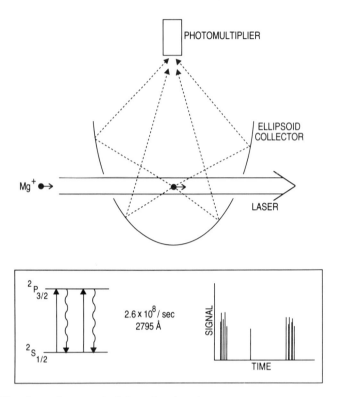

FIG. 3. The photon burst method for enhancing the detection of single atoms [33]. The bursts are due to multiple excitation–de-excitation as a single Be$^+$ ion passes through the laser beam. The isolated count, due to random background such as scattered light, can be easily distinguished.

measurement of atomic isotope ratios in the $10^{-11}$ to $10^{-15}$ range may be possible. The isobaric discrimination, typically poor in a conventional mass spectrometer, can be extremely high using a photon burst detector, since isobars of other elements are not usually resonant with the laser light. Recently, a selectivity of $10^8$ has been demonstrated in Mg$^+$ [36].

On initial examination, the technique appears useful only for a few atoms which have cycling transitions that are capable of being excited with current laser sources. The laser source limitation can be partially overcome by the use of transitions from metastable atomic states produced by selective charge transfer. The optical pumping problem results because radiative decay may occur to a level other than the original and thus terminate the photon burst prematurely. The problem may be alleviated somewhat by judicious choice of transitions, and photon burst transitions have been found for roughly half the elements in the periodic table [33].

In addition to the somewhat specialized measurements just outlined, LIF spectroscopy on fast beams has been widely used to measure hyperfine structure of atoms [37]. Much of this work, reviewed in references [1] and [38], has been devoted to the study of rare and unstable isotopes, where measurements of hyperfine structure and isotope shifts are used to determine systematics in nuclear properties, e.g., nuclear spin, magnetic dipole moment, electric quadrupole moments, and change in the nuclear mean square charge radius. More accurate hyperfine structure information can, however, be obtained using radio-frequency spectroscopy.

## 15.3.2 Radio-Frequency and Microwave Spectroscopy

Radio-frequency (microwave) spectroscopy in a fast beam is a direct descendant of the atomic beam magnetic resonance method pioneered by Rabi [39]. Its application typically leads to a $10^3$-fold improvement in the precision ($\approx 1$ kHz) with which small energy intervals can be determined over that obtained by laser spectroscopy alone ($\approx 1$ MHz). This improvement is due to the fact that small energy intervals are measured directly, rather than by the differencing of two optical frequencies. Two recent reviews of laser-RF spectroscopy include many details and samples of the application to thermal beams [40] and a comparison to fast beams [41].

The principle of the method is illustrated in Figure 4a. The fast beam passes through three regions sequentially: the pump, RF, and probe regions. The laser is tuned to be on resonance with the same transition in both the pump and probe regions. In the pump region, the population of a particular hyperfine level is depleted through optical pumping, such that in the probe region the resonance LIF is substantially reduced. If, in the intervening RF region, a transition from a different hyperfine level is induced, the population of the initially depleted state, and thus the LIF signal, will be restored. The probability of a transition in the RF region is given by the Rabi two-level formula [42]:

$$P(\omega) = \frac{\omega_R^2}{(\omega - \omega_0)^2 + \omega_R^2} \sin^2\{\tfrac{1}{2}[(\omega - \omega_0)^2 + \omega_R^2]^{1/2}T\}, \qquad (15.7)$$

where $\omega$ is the applied RF frequency, $\omega_0$ is the resonance frequency, $\omega_R$ is a parameter for the strength of the transition, and $T$ is the transit time through the RF region.

Figure 4b shows an example of an RF resonance obtained by sweeping the RF frequency while fixing the laser on resonance and observing the LIF signal in the probe region [43]. The line drawn through the data is a fit to the Rabi two-level formula, with the addition of normalization and a constant background. The width of the resonance is transit-time limited to be $0.8T$; thus, slower beams yield higher precision. The fitted resonance frequency must be corrected for the

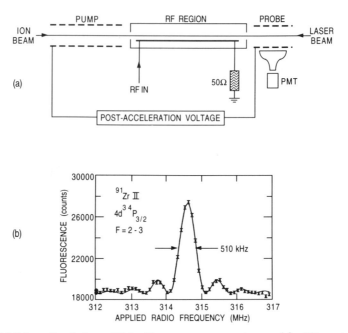

FIG. 4. (a) Schematic of a laser-RF double-resonance apparatus, used for RF spectroscopy of ions. (b) A typical RF resonance with a fit to the Rabi two-level formula shown [43].

first-order Doppler shift, typically by making measurements with the RF radiation both co- and counterpropagating, as discussed for optical measurements. In the collinear geometry shown earlier, the systematic error introduced by light (AC Stark) shifts caused by the presence of the laser field in the RF interaction region must be considered [44]. The addition of a static magnetic field in the interaction region permits the measurement of $g$-factors with a precision of $\approx 1\%$ [45].

A much more specialized RF measurement apparatus, shown in Figure 5, has been used to measure fine structure intervals in high-$L$, $n = 10$ Rydberg states of

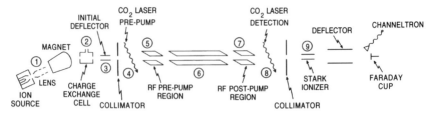

FIG. 5. Schematic of the apparatus for radio-frequency spectroscopy of neutral He [46].

helium at subkilohertz levels [46]. Such measurements permit the observation of retardation, QED, and relativistic effects on these physically large ($r \geq 100a_0$) atoms. High-Rydberg He atoms are formed by charge exchange in Ar gas. A 10 kV/cm electrostatic field is used to deflect parent ions and field ionize/deflect high-$n$ ($n \geq 20$) atoms. The laser for the pump and probe regions, a cw single frequency $CO_2$ laser, is tuned by varying the angle of intersection between the laser and neutral beam and is used to excite atoms in the $n = 10$ state to the $n = 30$ state. The resulting $n = 30$ atoms are Stark-ionized and detected by a channeltron. The higher-$L$ laser transitions are unresolved by the laser. Thus, to detect a transition between high-$L$ states, two auxiliary RF regions are used to transfer the population of the high-$L$ state to a lower-$L$ state which is resolved by the laser. Extreme care is taken in the construction of the RF region, which is long to reduce transit time broadening and is, as usual [47], designed as 50 ohm transmission line, to minimize reflections and stray fields. Amplitude modulation of the RF, coupled with phase-sensitive detection of the Stark-ionized current, is used to increase sensitivity.

Stimulated Raman transitions can also be used to measure intervals with RF precision [48]. This approach relies on population trapping in the two lower levels of a three-level "lambda" system (two lower levels and common upper level) induced by two coherent optical fields [49]. Two optical frequencies offset by a known RF frequency are generated by appropriate modulation of a cw laser beam. One of the optical frequencies is placed on exact resonance with a transition (only necessary for fluorescence detection), while the other is scanned under precise RF control. When the frequency offset matches the interval between two lower levels, population can be trapped in the lower levels and a decrease in the resonance fluorescence observed. The method yields excellent precision as the width is transit-time limited, $0.9T$. However, the accuracy is limited by AC Stark shifts.

For ultimate precision in RF spectroscopy, one can apply the Ramsey separated oscillatory field (SOF) method to fast beams [42]. This has been used to measure the Lamb shift in hydrogen [50]. The principal experimental obstacle is the 100 MHz natural line width of the $2\,^2S_{1/2}$–$2\,^2P_{1/2}$ transition, due to the 1.6 ns lifetime of the $^2P_{1/2}$ state. A variant of the standard SOF technique was used to obtain resonances significantly *narrower* than the natural line width of the transition. This was accomplished by changing the relative phase of the RF fields in the two separated regions in order to isolate the interference term. The interference term depends on the amplitude for the atom to make a transition to the $2^2P_{1/2}$ state in the first region, and back to the $^2S_{1/2}$ state in the second region. Thus, by increasing the separation or the RF fields, one essentially selects those atoms in the $2^2P_{1/2}$ state that live for the flight time between regions. Using this method, a FWHM of $\approx 28$ MHz, less than one-third of the natural width, was achieved.

### 15.3.3 Precision Lifetime Measurements

An early application of lasers to fast beam spectroscopy was to make cascade-free lifetime measurements [51]. This method is a direct descendant of the beam-foil method [3–5] where passage through a thin foil excites atoms in the fast beam at a precise point, defined as $t = 0$. The exponential decay of the fluorescence is monitored as a function of distance downstream from the excitation point, using a spectrometer to isolate the transition of interest. Since foil excitation is not state-selective, cascades from higher-lying levels can severely perturb the measured lifetime. The ANDC (arbitrarily normalized decay curve) method [52] can account for the cascade effect analytically with the additional measurement of the time dependence of feeding transitions. However, using a laser removes the cascade problem by selectively exciting the state of interest. In all cases, a measurement of the beam velocity is required to transform the measured decay length to decay time.

The fast-beam laser technique has yielded the most precise measurement of any atomic lifetime [54]. These benchmark measurements on the first resonance lines of neutral Li and Na have an estimated accuracy of $\approx 0.1$–0.2% (limited principally by measurement of the beam velocity), and thus are useful in testing state-of-the-art ab initio atomic theory [54, 55]. Recently, in a measurement of the $6^2P_{3/2}$ lifetime in Cs [56], a more accurate method for the beam velocity determination was used in which the Doppler shift of the $6^2S_{1/2}$–$6^2P_{3/2}$ transition in the fast beam (relative to a stationary cell) was measured using interferometric methods. Measurement of the Doppler shift to 30 MHz provides the beam velocity to $\approx 1$ part in $10^4$.

Other methods of lifetime measurement have been demonstrated in a fast beam. Rapid Doppler-switching into resonance [57] and the use of a pulsed laser field [58] have been used in a collinear geometry. In addition, a pulsed laser has been used in the perpendicular geometry [59]. These techniques have yielded lifetimes for ions with accuracies of $\approx 1\%$.

### 15.3.4 Electron Spectrocopy

Electron spectroscopy in fast beams has been used primarily for the study of negative ions. Lifetimes [60] and energy levels of metastable states [61] of negative ions have been measured and the existence of barely bound negative ions [62] confirmed by fast beam spectroscopy. In addition, fast-beam measurements of the near-threshold behavior of photodetachment cross-sections have been used to study detachment dynamics tor simple systems, such as $H^-$ [63] and $He^-$ [64], in specialized situations, e.g., at relativistic velocity with high motional electric field ($\mathbf{v} \times \mathbf{B}$) [63]. While not all of these studies require energy analysis of the ejected electron, this additional capability allows the determination of electron affinities without resort to fitting the slowly rising edge of the

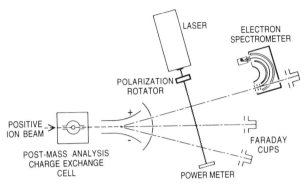

FIG. 6. Schematic of the crossed laser–ion beams apparatus used in photoelectron detachment spectroscopy [65].

threshold photodetachment cross section, $\sigma \propto \Delta E^{l+1/2}$, where $\Delta E$ is the energy above threshold and $l$ is the orbital angular momentum of the ejected electron.

Figure 6 shows a schematic of a crossed-beam photoelectron detachment apparatus [65]. Briefly, a fast beam of negative ions is formed by double charge exchange in Li vapor, with subsequent charge state separation from neutral and positively charged ions. Interaction with a high-power ($\approx 10^8$ W/cm$^2$), linearly polarized, energy-resolved photon beam from a flashlamp-pumped dye laser causes electron detachment. The energy of the electrons in the lab frame [given by Eq. (15.6)] is measured downstream by a 160° spherical-sector electron-energy analyzer. The forward and backward ejected peaks ($\varphi = 0°$ or 180°) result in a kinematic doubling of the detachment peaks for those electrons with c.m. velocity less than the ion beam velocity. The separation of these kinematically doubled peaks in the lab frame is, then, from Eq. (15.6), $\Delta E_{\mathrm{L}} = 4(m_{\mathrm{e}} E W / m_{\mathrm{i}})^{1/2}$, where $m_{\mathrm{e}}$ is the electron mass, $E$ is the ion beam energy, $W$ is the energy of the electron in the c.m. frame, and $m_{\mathrm{i}}$ is the mass of the ion. If photodetachment leaves the neutral in a different state, $W$ will change and result in another peak (or set of peaks) in the electron energy spectrum. If $E$ is unknown, the measured energy difference between two $\varphi = 0°$ peaks can be used to calculate the $W$ for the kinematically doubled peak, given $\Delta E_{\mathrm{L}}$ and the energy difference between the two states in the neutral, $E_{\mathrm{e}}$. The electron affinity of the atom is then given by $E_{\mathrm{a}} = E_{h\nu} - E_{\mathrm{e}} - W$, and very small electron affinities can be measured [62].

Measurement of angular distributions of photodetachment yields enables the determination of branching ratios for competing photodetachment channels [66]. For plane-polarized radiation within the electric dipole approximation and an independent electron model, the angular distribution of photodetached electrons is described by $1 + \beta P_2(\cos \theta)$, where $\beta$ is the asymmetry parameter, $P_2(\cos \theta)$ is the second-order Legendre polynomial, and $\theta$ is the angle between the laser polarization vector and the direction of electron ejection. In the apparatus in Figure 6, $\theta$ can be varied by simply rotating a $\lambda/2$ phase retarder (double Fresnel rhomb). The relative cross sections for detachment into the various exit channels are then determined by angular integration. When this is done under identical geometric conditions, the efficiency factors for the collection and detection of electrons from different exit channels can be cancelled, i.e., the ion beam energy can be varied in order to shift the different electron c.m. energies to the same energy in the laboratory frame. With the additional measurement of a well-known photodetachment cross section under the same kinematic conditions, e.g., $D^- \rightarrow D(^2S) + e$, an absolute scale for the relative cross-sections can be established.

## 14.4 Studies of Molecules

The motivation for the study of molecules using fast beams is much the same as that for atoms. Namely, the isolation and identification of reactive species, i.e., molecular ions, is often difficult. The fast beam provides a general method for the extraction and $m/q$ identification of the many species formed in an ion source, at the expense of a much reduced density. Unlike atoms, molecules can decay other than by radiation or ejection of an electron. In fact, the fluorescence quantum yield from a molecule is typically very small for all but the lowest excited states. Thus, other indirect methods for detection of resonant transitions are employed.

### 15.4.1 Photofragmentation

The study of molecular photofragmentation processes in fast beams has yielded detailed information on potential energy curves, bond energies, energy partitioning among dissociation fragments, and molecular constants for various electronic states [67]. Typically, two types of experimental information are available: the c.m. separation energies ($W$) at fixed photon energies, and the relative photodissociation cross sections as a function of photon energy. Information on potential curves [68], dissociation energies, and ionization potentials [69] is derived from detailed analysis of the various spectra. Optical–optical double resonance can be used to label states and simplify analysis [70].

A general-purpose apparatus which accomodates laser excitation in both collinear and crossed configurations and provides energy analysis of the charged fragments [8] is shown in Figure 7. The collinear geometry is only fully useful for "perpendicular" transitions, where photofragments are preferentially ejected perpendicular to the polarization axis of the laser and thus along the parent ion beam direction, or for dissociations with very little c.m. energy release. For "parallel" transitions, one must use the crossed laser–ion beam geometry.

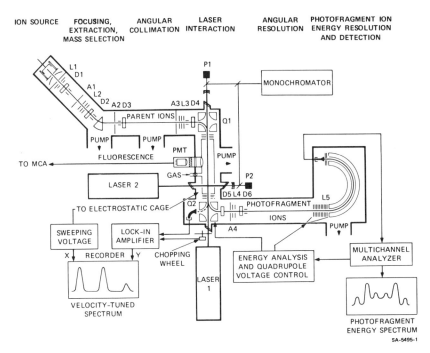

FIG. 7. Schematic of the multipurpose photofragmentation apparatus for molecular ions [8].

A comparison of the relative yields of the fragments under collinear and perpendicular excitation can be used to identify the nature of the bound–free transition. For example, in $Ar_2^+$, $^2\Pi_g \leftarrow {}^2\Sigma_u^+$ is a perpendicular transition, and thus detection of $Ar^+$ fragments is favored under collinear excitation, whereas $^2\Sigma_g^+ \leftarrow {}^2\Sigma_u^+$ is a parallel transition, and thus favored under perpendicular excitation [71–73]. The relative yield of the photofragments resulting from excitation to these two different electronic states changes by a factor of $\approx 6$ in changing excitation geometry [68].

Energy analysis of a charged photofragment is performed using a $180°$ electrostatic analyzer in combination with a deceleration system. The resolution, $\approx 1$ eV, in the lab frame is sufficient to distinguish parent ions dissociating from adjacent vibrational levels. By decelerating the photofragments before entering the electrostatic analyzer, its required resolution, $\Delta E/E$, is reduced. To scan the photofragment energy spectrum, the deceleration voltage is stepped synchronously with the potentials on the second quadrupole while the transmission energy of the electrostatic analyzer is kept fixed.

In addition to bound–free transitions, shape resonances can be studied. Inclusion of the centrifugal term in the interaction potential results in a barrier at

large internuclear separations, giving rise to quasi-bound levels (shape resonances) that lie above the dissociation limit. These shape resonances can be
selectively excited with a laser, and their study using photofragment spectroscopy allows direct determination of the energies of the resonance levels with
respect to the dissociation limit, the determination of their widths, and the
resolution of fine-structure effects [74, 75].

Moreover, bound–bound transitions can also be observed. For example,
vibration–rotation transitions may be observed using two lasers: the first to excite
a resonant transition from $(v'', N'')$ to $(v', N')$, and a second to drive a nonresonant transition from $(v', N')$ to a dissociative state [76]. Alternatively,
bound–bound transitions are observable if the upper state is predissociated, i.e.,
has a curve-crossing with a dissociative electronic state. Finally, indirect detection of vibration–rotation transitions is possible by charge-exchange methods.
Since the cross-section for various collisional processes may depend on vibrational state, a change in the neutralized (or ionized) fraction surviving passage
through a gas target is detectable when such a transition is made [77].

The major drawback to the photofragment spectroscopy techniques just
described, is that they cannot be applied to neutral species. A time- and
position-sensitive detector system has, however, been developed to study dissociative processes using fast neutral beams [78]. A schematic of the apparatus
is shown in Figure 8. Briefly, the dissociation event is induced at $t = 0$ by a laser

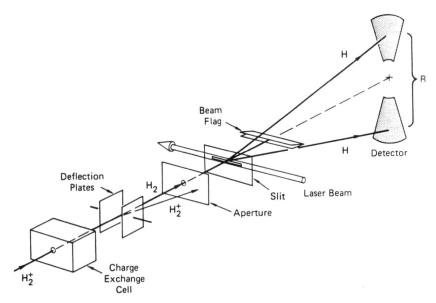

FIG. 8. Schematic of a photodissociation apparatus incorporating a time- and
position-sensitive detector for the study of neutral fragments [79].

(or other means), the photofragments separate downstream, and the position and time at which the two correlated particles strike the detector is recorded. Spatial resolution of 100 μm and time resolution of 400 ps is sufficient to measure vibrational and rotational structure. The method has the additional advantage that the solid angle for collection of the fragments is determined by the size of the detector, and thus much enhanced over that in the photofragment spectrometer in Figure 7. With the measurement of two quantities, $R$, the mutual separation and $\tau$, the flight time difference of the two fragments, the kinematics is completely determined. For unknown mass distribution between the fragments, it is necessary to measure the ratio of the radial displacements $R_1$ and $R_2$. The method has been used to study rovibrational energies and populations of molecules in fast beams [79], rovibrational product distributions from dissociative charge transfer reactions [80], and photodissociation [81].

## 15.4.2 Photodetachment Threshold Spectroscopy.

As for photofragmentation, photodetachment experiments are of basically two types: photodetachment threshold spectroscopy, where the photon energy is scanned while monitoring the onset of detachment products, or photoelectron spectroscopy of negative ions, where the energy of the ejected electron is measured while the photon energy is kept fixed. In principle, data from either method can be analyzed to yield electron affinities and molecular structural information. Threshold experiments [82] can yield much more precise values for electron affinities than electron spectroscopy experiments, because the latter are limited by the electron spectrometer resolution ($\approx 2$ meV). However, because of the lack of suitable sources, threshold spectroscopy with lasers is not applicable to very weakly bound species. Moreover, the usual bound–free transitions observed in photodetachment threshold spectroscopy in atoms yield integrated cross-sections as a function of photon energy, with multiple thresholds that can be difficult to interpret.

In the case of molecular negative ions, the breakdown of the Born–Oppenheimer approximation allows the coupling of vibrational and rotational degrees of freedom to electronic motion, permitting new spectroscopic approaches. Specifically, autodetachment spectroscopy [82] has shed light on propensity rules [83] for decay in molecular negative ions. This involves photoexcitation from the negative ion ground state to a quasi-bound excited state that lies above the detachment threshold and that may decay by electron ejection (autodetachment). The quasi-bound state may be a rovibrational state of the negative ion or, more generally, a dipole bound state. These occur in molecules with dipole moment greater that 2 debye, the electron being bound in the $1/r^2$ dipole potential. Depending on the lifetime of these quasi-bound states, very sharp structure may be observable in the photodetachment spectrum.

The apparatus required for negative ion autodetachment spectroscopy is essentially the same as that shown in Figure 7 for photofragment spectroscopy. A cw tunable laser is merged with a mass-selected, negative-ion beam in a collinear geometry. The residual parent beam is electrostatically separated with the second quadrupole, and the photodetached neutrals are easily detected by the secondary electrons emitted when they strike the glass window [84]. Since the detection of neutrals occurs after the second quadrupole, field-induced detachment may lead to the detection of transitions that are actually bound with respect to autodetachment. These can be distinguished by detecting slow photodetached electrons in addition to the neutrals [85]. Since the negative molecular ion may be moving quite slowly, threshold photoelectrons have very low kinetic energies in the lab frame, making their detection quite difficult. Nevertheless, selective detection of these slow electrons was demonstrated using a weak solenoidal guide field around the entire ion-beam/laser interaction region. This resulted in an enhanced detection sensitivity for threshold events by discriminating against processes which produce fast electrons [85].

In addition to the collinear cw fast beam methods, other complementary techniques are available for photodetachment threshold spectroscopy: pulsed photoelectron spectrometry, where a pulsed laser induces photodetachment and time-of-flight techniques are used to measure products [86]; photoelectron spectroscopy, with decelerated cw beams crossed with high-power, single-frequency UV radiation using a power buildup cavity [87]; zero–kinetic-energy photodetachment spectroscopy, where a pulsed laser induces photo-detachment and a delayed voltage pulse extracts the near-zero kinetic energy electrons for time-of-flight analysis [88]; and velocity-modulation spectroscopy in discharges [89].

## 15.5 Metrology

### 15.5.1 Tests of Special Relativity: Storage Ring Measurements

A basic result of special relativity is time dilation i.e., that a moving clock has its frequency modified by the factor $\gamma^{-1} = (1 - v^2/c^2)^{1/2}$, where $v$ is the velocity of the clock relative to the observer. By using an atom for a clock, one can measure the time dilation factor by comparing resonant frequencies for fast-moving and stationary atoms. The time dilation (also termed the transverse Doppler shift) factor is small compared to a very large first order Doppler shift [see Eqs. (15.8) and (15.9)], and two-photon absorption (TPA) with counter-propagating laser beams is required to eliminate the first-order shift. A verification of $\gamma$ at the 2.3 ppm level has been accomplished by comparing the frequency of two-photon absorption in a fast metastable neon beam ($E \approx 120$ keV, $\beta = 3.6 \times 10^{-3}$) with that in a slow beam [90, 91]. Tests of special relativity can

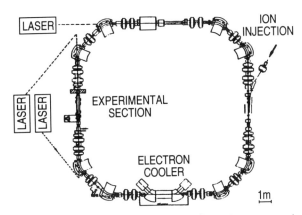

Fig. 9. Schematic of a laser spectroscopy experiment in a storage ring [96].

be done much more precisely if $\beta$ is increased, say by use of a storage ring [92]. A preliminary measurement using the $2^3S$ metastable state in $^7Li^+$ has yielded an accuracy of 32 ppm [93], but many ways to improve the measurement can be envisioned.

Storage rings such as that illustrated in Figure 9 promise to improve precision by vastly increasing the ion/radiation interaction time. Also, by repetitive interaction of the laser and ion beam (in successive passes of the ion beam), it is possible to use the techniques of laser cooling to compress the velocity spread of the ion beam to very low values [94]. This was first demonstrated using a 13.3 MeV $^7Li^+$ beam, where for metastable atoms in the $2^3S$ state, an initial temperature of 260 K was reduced to a longitudinal temperature of $<3$ K [95]. While this is higher than typically achieved in a directly extracted beam at lower energy, a longitudinal temperature of $<30$ mK has now been obtained with a 7.29 MeV $^9Be^+$ beam [96]. Cooling of ions in a storage-ring environment is greatly affected by the beam dynamics, i.e., a coupling of the longitudinal and transverse velocities through betatron motion and the Coulomb interaction. A number of auxiliary cooling methods were therefore required to reach this ultimate temperature. The possibility of observing coherent RF signals (Ramsey fringes) by successive interactions with the same RF cavity in a storage ring has also been investigated; it is found to be limited by magnetic field inhomogeneities [97].

## 15.6 Summary and Outlook

A wide variety of fields have benefited from fast beam laser spectroscopy, which allows study of reactive, rare, and unstable species under perturbation-free conditions. The additional benefits of a sample having a Doppler-compressed velocity distribution, with $m/q$ identification, and with a precisely controllable velocity, make high-resolution spectra much simpler to interpret than for, say, discharges. Moreover, the kinematic aspects of the fast beam allow one to observe fragmentation processes precisely and efficiently without the use of detectors which are specialized to a particular fragment.

Some recent developments will extend the capabilities of the technique. The advent of storage rings and the demonstration of laser cooling have opened the possibility of having beams with very narrow velocity distributions at relativistic speeds. This should lead to improved precision in tests of special relativity, and the possibility of observing crystallization in ion beams. The development of ion sources to produce usable currents of highly ionized atoms with appreciable metastable state content, and of cooled molecular ions with very little internal energy, will expand the number and variety of species that can be studied by this method. Atomic studies at higher ionization stages will provide ever more precise tests of QED and relativistic quantum mechanics. Molecular studies will characterize the trends in structural and dynamical behavior of systems as the size and composition is varied. Moreover, the possibility of selective charge exchange to shift the quantum state distribution to favor the state of interest provides increased sensitivity in specific cases, such as the study of retardation (Casimir) effects in Rydberg states in He. Finally, it is anticipated that the continuous evolution in laser technology will provide new probes in the deep UV, X-ray, and infrared regions.

# References

1. Neugart, R. (1987). Chapter 2 in *Progress in Atomic Spectroscopy*, Part D, H. J. Beyer and H. Kleinpoppen (eds.), p. 75. Plenum Press, New York and London.
2. Poulsen, O. (1983). In *Atomic Physics 8*, I. Lindgren, A. Rosen, and S. Svanberg (eds.), p. 485. Plenum Press, New York and London.
3. Andrä, H. J. (1979). Chapter 20 in *Progress in Atomic Spectroscopy*, Part B, W. Hanle and H. Kleinpoppen (eds.), p. 829. Plenum Press, New York and London.
4. Berry, H. G. (1977). *Rep. Prog. Phys.* **40**, 155.
5. Bashkin, S., ed. (1976). *Beam Foil Spectroscopy*. Springer-Verlag, Berlin.
6. Kaufman, S. L. (1976). *Opt. Comm.* **17**, 309.
7. Corney, A. (1977). *Atomic and Laser Spectroscopy*. Clarendon Press, Oxford.
8. Huber, B. A. *et al.* (1977). *Rev. Sci. Instrum.* **48**, 1306.
9. Geller, R. (1990). *Annual Reviews of Nuclear and Particle Science*, Vol. 40, p. 15. Annual Review Inc.
10. Leopole, D. G., Ho, J., and Lineberger, W. C. (1987). *J. Chem. Phys.* **86**, 1715.
11. Alton, G. D. *et al.* (1986). *Nucl. Instrum. Methods* **A244**, 142.
12. Heinemeier, J., and Hvelplund, P. (1978). *Nucl. Instrum. Methods* **148**, 65; *Nucl. Instrum. Methods* **148**, 425.
13. Meyer, F. W., Anderson, C. J., and Anderson, L. W. (1977). *Phys. Rev. A* **15**, 455.

14. Reynaud, C., Pommier, J., Tuan, V. N., and Barat, M. (1979). *Phys. Rev. Lett.* **43**, 570.
15. Anton, K. R. *et al.* (1978). *Phys. Rev. Lett.* **40**, 642.
16. Rapp, D., and Francis, W. E. (1962). *J. Chem. Phys.* **37**, 2631.
17. Ice, G. E., and Olson, R. E. (1975). *Phys. Rev. A* **11**, 111.
18. Buchinger, F. *et al.* (1982). *Nucl. Instrum. Methods* **202**, 159.
19. Pascale, J., Olson, R. E., and Reinhold, C. O. (1990). *Phys. Rev. A* **42**, 5305.
20. Arnold, E., Kühl, T., Otten, E. W., and von Reisky, L. (1982). *Phys. Lett.* **90A**, 399.
21. MacAdam, K. B., Gray, L. G., and Rolfes, R. G. (1990). *Phys. Rev. A* **42**, 5269.
22. Deck, F. J., Hessels, E. A., and Lundeen, S. R. (1993). *Phys. Rev. A* **48**, 4400.
23. Drake, G. W. F. (1988). *Can. J. Phys.* **66**, 586.
24. Johnson, W. R., and Sapirstein, J. (1992). *Phys. Rev. A* **46**, R2197.
25. Bayer, R. *et al.* (1979). *Z. Physik A* **292**, 329; Englert, M. *et al.* (1982). Appl. Phys. B **28**, 81.
26. Holt, R. A., Rosner, S. D., Gaily, T. D., and Adam, A. G. (1980). *Phys. Rev. A* **22**, 1563.
27. Riis, E., Berry, H. G., Poulsen, O., Lee, S. A., and Tang, S. Y. (1986). *Phys. Rev. A* **33**, 3023.
28. Scholl, T. J., Holt, R. A., and Rosner, S. D. (1989). *Phys. Rev. A* **39**, 1169; Scholl, T. J., Cameron, R., Rosner, S. D., Zhang, L., Holt, R. A., Sausonetti, C. J., and Gillaspy, J. D. (1993). *Phys. Rev. Lett.* **71**, 2188.
29. Dinneen, T. P., Mansour, N. B., Berry, H. G., Young, L., and Pardo, R. C. (1991). *Phys. Rev. Lett.* **66**, 2859.
30. Riis, E., Sinclair, A. G., Poulsen, O., Drake, G. W. F., Rowley, W. R. C., and Levick, A. P. (1994). *Phys. Rev. A* **49**, 207.
31. Greenlees, G. W., Clark, D. L., Kaufman, S. L., Lewis, D. A., Tonn, J. F., and Broadhurst, J. H. (1977). *Opt. Comm.* **23**, 236.
32. Keller, R. A., Bomse, D. S., and Cremers, D. A. (1981). *Laser Focus*, 75 October.
33. Fairbank, W. M., Jr. (1987). *Nucl. Instrum. Methods* **B29**, 407.
34. LaBelle, R. D., Fairbank, W. M., and Keller, R. A. (1989). *Phys. Rev. A* **40**, 5430.
35. Lewis, D. A., Tonn, J. F., Kaufman, S. L., and Greenlees, G. W. (1979). *Phys. Rev. A* **19**, 1580.
36. LaBelle, R. D., Hansen, C. S., Mankowski, M. M., and Fairbank, W. M. Preprint.
37. See, for example, Young, L., Childs, W. J., Dinneen, T., Kurtz, C., Berry, H. G., Engström, L., and Cheng, K. T. (1988). *Phys. Rev. A* **37**, 4213.
38. Kluge, H. J. (1979). Chapter 17 in *Progress in Atomic Spectroscopy*, Part B, W. Hanle and H. Kleinpoppen (eds.). Plenum Press, New York and London.
39. Rabi, I., Zacharias, J. R., Millman, S., and Kusch, P. (1938). *Phys. Rev.* **53**, 318.
40. Neumann, R., Träger, F., and zu Putlitz, G. (1987). Chapter 1 in *Progress in Atomic Spectroscopy*, Part D, H. J. Beyer and H. Kleinpoppen (eds.), p. 1. Plenum Press, New York and London.
41. Childs, W. J. (1992). *Phys. Rep.* **211**, 113.
42. Ramsey, N. F. (1956). *Molecular Beams*. Oxford University Press, New York.
43. Young, L., Kurtz, C. A., Beck, D. R., and Datta, D. (1993). *Phys. Rev. A* **48**, 173.
44. Borghs, G., De Bisschop, P., Odeurs, J., Silverans, R. E., and Van Hove, M. (1985). *Phys. Rev. A* **31**, 1434.
45. Scholl, T. J., Rosner, S. D., and Holt, R. A. (1987). *Phys. Rev. A* **35**, 1611.
46. Hessels, E. A., Arcuni, P. W., Deck, F. J., and Lundeen, S. R. (1992). *Phys. Rev. A* **46**, 2622.
47. Sen, A., Goodman, L. S., and Childs, W. J. (1988). *Rev. Sci. Inst.* **59**, 74.

48. Young, L., Dinneen, T. P., and Mansour, N. B. (1988). *Phys. Rev. A* **38**, 3812; Dinneen, T. P., Berrah Mansour, N., Kurtz, C. A., and Young, L. (1991). *Phys. Rev. A* **43**, 4824.
49. See, for example, Hemmer, P. R., Ontai, G. P., and Ezekiel, S. (1986). *J. Opt. Soc. Am. B* **3**, 219.
50. Lundeen, S. R., and Pipkin, F. M. (1986). *Metrologia* **22**, 9.
51. Andra, H. J., Gaupp, A., and Wittman, W. (1973). *Phys. Rev. Lett.* **31**, 501.
52. Curtis, L. J., Berry, H. G., and Bromander, J. (1971). *Phys. Letters* **34A**, 169.
53. Gaupp, A., Kuske, P., and Andrä, H. J. (1982). *Phys. Rev. A* **26**, 3351.
54. Guet, C., Blundell, S. A., and Johnson, W. R. (1990). *Phys. Lett. A* **143**, 384.
55. Froese Fischer, C. (1988). *Nucl. Instrum. Methods Phys. Res. B* **31**, 265.
56. Tanner, C. E., private communication.
57. Gaillard, M. L., Pegg, D. J., Bingham, C. R., Carter, H. K., Mlekodaj, R. L., and Cole, J. D. (1982). *Phys. Rev. A* **25**, 1975.
58. Poulsen, O., Andersen, T., Bentzen, S. M., and Nielsen, U. (1981). *Phys. Rev. A* **24**, 2523.
59. Ansbacher, W., Li, Y., and Pinnington, E. (1989). *Phys. Lett. A* **139**, 165; Gosselin, R. N., Pinnington, E. H., and Ansbacher, W. (1988). *Nucl. Instrum. Methods B* **31**, 305.
60. Bae, Y. K., and Peterson, J. R. (1984). *Phys. Rev. A* **30**, 2145.
61. Kvale, T. J., Alton, G. D., Compton, R. N., Pegg, D. J., and Thompson, J. S. (1985). *Rev. Lett.* **55**, 484.
62. Pegg, D. J., Thompson, J. S., Compton, R. N., and Alton, G. D. (1987). *Phys. Rev. Lett.* **59**, 2267.
63. Bryant, H. C., Mohagheghi, A., Stewart, J. E., Donahue, J. B., Quick, C. R., Reeder, R. A., Yuan, V., Hammer, C. R., Smith, W. W., Cohen, S., Reinhardt, W. P., and Overman, L. (1987). *Phys. Rev. Lett.* **58**, 2412.
64. Peterson, J. R., Bae, Y. K., and Huestis, D. L. (1985). *Phys. Rev. Lett.* **55**, 692.
65. Pegg, D. J., Thompson, J. S., Compton, R. N., and Alton, G. D. (1989). *Nucl. Instrum. Methods* **B40/41**, 221.
66. Pegg, D. J., Thompson, J. S., Dellwo, J., Compton, R. N., and Alton, G. D. (1990). *Phys. Rev. Lett.* **64**, 278.
67. Moseley, J. T., and Durup, J. (1981). *Annu. Rev. Phys. Chem.* **32**, 53.
68. Moseley, J. T., Saxon, R. P., Huber, B. A., Cosby, P. C., Abouaf, R., and Tadjeddine, M. (1977). *J. Chem. Phys.* **67**, 1659.
69. Helm, H., and Müller, R. (1983). *Phys. Rev. A* **27**, 2493.
70. Cosby, P. C., and Helm, H. (1982). *J. Chem. Phys.* **76**, 4770.
71. Zare, R. N., and Herschbach, D. R. (1963). *Proc. IEEE* **51**, 173.
72. Ozenne, J. B., Durup, J., Odom, R. W., Pernot, C., Tabché-Fouhaillé, A., and Tadjeddine, M. (1976). *Chem. Phys.* **16**, 75, and references therein.
73. van Asselt, N. P. F. B., Maas, J. G., and Los, J. (1976). *Chem. Phys.* **17**, 81.
74. Helm, H., Cosby, P. C., Graff, M. M., and Moseley, J. T. (1982). *Phys. Rev. A* **25**, 304.
75. Carrington, A., McNab, I. R., and Montgomerie, C. A. (1988). *Chem. Phys. Lett.* **149**, 326.
76. Carrington, A. (1986). *J. Chem. Soc. Faraday Trans. II* **82**, 1089.
77. Wing, W. H., Ruff, G. A., Lamb, W. E., and Spezeski, J. J. (1976). *Phys. Rev. Lett.* **36**, 1488.
78. deBruijin, D. P., and Los, J. (1982). *Rev. Sci. Instrum.* **53**, 1020.
79. Helm, H., and Cosby, P. C. (1987). *J. Chem. Phys.* **86**, 6813.

80. Walter, C. W., Cosby, P. C., and Peterson, J. R. (1993). *J. Chem. Phys.* **98**, 2860.
81. Cosby, P. C., and Helm, H. (1988). *Phys. Rev. Lett.* **61**, 298.
82. Lykke, K. R., Murray, K. K., Neumark, D. M., and Lineberger, W. C. (1988). *Phil. Trans. R. Soc. Lond. A* **324**, 179.
83. Berry, R. S. (1966). *J. Chem. Phys.* **45**, 1288.
84. Hefter, U., Mead, R. D., Schultz, P. A., and Lineberger, W. C. (1983). *Phys. Rev. A* **28**, 1429.
85. Mean, R. D., Lykke, K. R., Lineberger, W. C., Marks, J., and Brauman, J. I. (1984). *J. Chem. Phys.* **81**, 4883.
86. Johnson, M. A., and Lineberger W. C. (1988). Chapter XI In *Techniques for the Study of Ion Molecule Reactions*, J. M. Farrar and W. Saunders (eds.), p. 591. John Wiley and Sons, New York.
87. Ervin, K. M., and Lineberger, W. C. (1992). In *Advances in Gas Phase Ion Chemistry*, Vol. 1, p. 121. JAI Press Inc.
88. Müller-Dethlefs, K., Sander, M., and Schlag, E. (1984). *Chem. Phys. Lett.* **112**, 291; Sander, M., Chewter, L. A., and Müller-Dethlefs, K. (1987). *Phys. Rev. A* **36**, 4543.
89. Saykally, R. J. (1986). *Spectroscopy* **1**, 40.
90. McGowan, R. W., Giltner, D. M., Sternberg, S. J., and Lee, S. A. (1993). *Phys. Rev. Lett.* **70**, 251.
91. Riis, E., Andersen, L. U. A., Bjerre, N., Poulsen, O., Lee, S. A., and Hall, J. L. (1988). *Phys. Rev. Lett.* **60**, 81.
92. Habs, D. *et al.* (1989). *Nucl. Instrum. Methods* **B43**, 390.
93. Klein, R. *et al.* (1992). *Z. Phys. A* **342**, 455.
94. Javanainen, J., Kaivola, M., Nielsen, U., Poulsen, O., and Riis, E. (1985). *J. Opt. Soc. Am. B* **2**, 1768.
95. Schröder, S. *et al.* (1990). *Phys. Rev. Lett.* **64**, 2901.
96. Petrich, W., *et al.* (1993). *Phys. Rev. A* **48**, 2127.
97. Kristensen, M., Hangst, J. S., Jessen, P. S., Nielsen, J. S., Poulsen, O., and Shi, P. (1992). *Phys. Rev. A* **46**, 4100.

# 16. QUANTUM-BEAT, LEVEL-CROSSING, AND ANTICROSSING SPECTROSCOPY

T. F. Gallagher

Department of Physics
University of Virginia
Charlottesville, Virginia

## 16.1 Introduction

Quantum-beat, level-crossing, and anticrossing spectroscopy are inherently Doppler-free ways of obtaining precise spectroscopic information about excited states. Quantum-beat and level-crossing spectroscopy are coherent in the sense that it is necessary to create a coherent superposition of states to observe a signal. We begin by describing quantum-beat spectroscopy, and then show how it is related to the Hanle effect, the zero-field level crossing of Zeeman levels. While the Hanle effect was observed long ago, high-field level crossings were not observed until much more recently. In contrast to these methods, anticrossing spectroscopy is incoherent; only population differences are required to observe anticrossing signals, just as in RF spectroscopy.

In the sections which follow we describe quantum beats, level-crossing spectroscopy, and anticrossing spectroscopy. There exist several excellent reviews of these subjects to which the reader is referred for a more detailed description of the underlying theory [1–4].

## 16.2 Quantum Beats

### 16.2.1 Classical Picture

The point of quantum-beat spectroscopy is to measure the energy spacing between excited energy levels. To understand it we begin with the measurement of the excited-state Zeeman splitting of a very simple atomic system in which the ground state is a $^1S_0$ state and the excited state is a $^1P_1$ state, a spinless system. We first consider a simple classical picture. Imagine that the atom is in zero magnetic field and we excite it at time $t = 0$ with an $x$ polarized light pulse which is short compared to the lifetime of the excited state. After absorbing a photon, the excited atom has a charge distribution which oscillates along the $x$-axis. In

EXPERIMENTAL METHODS IN THE PHYSICAL SCIENCES
Vol. 29B

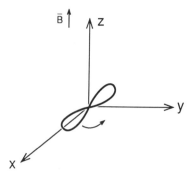

FIG. 1. The charge distribution of the electron cloud of an atom excited from a $^1S_0$ to a $^1P_1$ state by light polarized in the $x$ direction. In a magnetic field $B$ in the $z$ direction, the electron cloud rotates about the $z$ axis as shown by the arrow.

decaying, the atom reradiates light polarized along the $x$-axis, and the emitted light from a sample of atoms has a time-dependent intensity given by

$$I(t) = I_0 e^{-\Gamma t}, \tag{16.1}$$

where $1/\Gamma$ is the lifetime of the excited $P$ state and $I_0$ is the intensity at $t = 0$.

If we carry out the same excitation with the atoms in a nonzero $B$ field in the $z$ direction, the magnetic moment and the charge distribution of the excited atom rotate about the $z$-axis at the Larmor frequency $\omega_L$ after excitation; Figure 1 shows the rotating charge distribution. In this case the pulse is short compared to the inverse of the Larmor frequency as well. Both the polarization and the direction of the reradiated fluorescence change in time. The charge distribution and the polarization of the emitted light are aligned in the $x$ direction when $t = 0$, $\pi/\omega_L$, $2\pi/\omega_L$, ... , and in the $y$ direction when $t = \pi/2\omega_L$, $3\pi/2\omega_L$, ... . Similarly, no fluorescence is emitted in the $x$ direction when $t = 0$, $\pi/\omega_1$, $2\pi/\omega_L$, ... , and there is none emitted in the $y$ direction when $t = \pi/2\omega_L$, $3\pi/2\omega_L$, ... . As pointed out by Dodd and Series [2], the spatial dependence is roughly analogous to a rotating searchlight beam. In Figure 2, we show a graph of the $x$ and $y$ polarized fluorescence observed in the $z$ direction as a function of time after the excitation for $\omega_L \gg \Gamma$. The intensity oscillation is due to the rotation of the charge distribution and hence of the polarization of the fluorescence. Since the charge distribution is aligned along the $x$-axis twice in each rotation, the oscillation in the intensity is at twice the Larmor frequency. This frequency is the frequency difference between the quantum-mechanical $m = \pm 1$ states. By recording the time-resolved decay and measuring the beat frequency of the oscillations, we determine directly the energy separation between these two excited states. It is useful to note that if both the $x$ and $y$ polarized emissions in the $z$ direction are observed, only a simple decaying exponential is recorded. The polarization of the excitation and detection are critical to the observation of the beats.

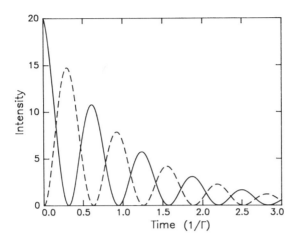

FIG. 2. Decay of the $x$ (———) and $y$ (− − −) polarized fluorescence observed in the $z$ direction subsequent to excitation at $t = 0$ by light polarized in the $x$ direction. Here, $\omega_L/\pi = 1.6\,\Gamma$.

## 16.2.2 Quantum Mechanical Description

To show that the oscillations observed in the fluorescence correspond to the excited-state energy splitting, we now consider a quantum-mechanical description of the Zeeman problem. In a magnetic field, the $^1P_1$ state energies are given by

$$W_{1,m} = W_p + \mu_B Bm, \tag{16.2}$$

where $m$, the magnetic quantum number, can take the values 0 and $\pm 1$. The $P$ eigenstates in the field are $\Psi_{1,m}(\mathbf{r}, t)$. Excitation at time $t = 0$ with light polarized in the $x$ direction leads to the excited-state wavefunction

$$\Psi_p(\mathbf{r}, t) = (\Psi_{1,1}(\mathbf{r}, t) + \Psi_{1,-1}(\mathbf{r}, t))/\sqrt{2}, \tag{16.3}$$

where

$$\Psi_{1,\pm 1}(\mathbf{r}, t) = \sqrt{\frac{3}{8\pi}} R(\mathbf{r}) \sin\theta\, e^{\pm i\phi} \cdot e^{-iW_{1,m} t/\hbar}. \tag{16.4}$$

Here $R(r)$ is a normalized $p$ radial function, and $r$, $\theta$, and $\phi$ are the conventional polar coordinates of the electron's position. Using Eq. (16.4), the wave function of Eq. (16.3) can be written as

$$\Psi_p(\mathbf{r}, t) = \sqrt{\frac{3}{4\pi}} R(r) \sin\vartheta \cos\left(\phi - \frac{\omega t}{2}\right) e^{-iW_p t}, \tag{16.5}$$

where $\omega = (W_1 - W_2)/\hbar = 2\mu_B B$. At $t = 0$, the wave function of Eq. (16.5) is aligned along the x-axis, and it rotates about the z-axis at the Larmor frequency $\mu_B B/\hbar$, just as in our simple classical picture.

The fluorescence radiated with x polarization is proportional to the square of the matrix element

$$X(t) = \int \Psi_s(\mathbf{r}, t)x\Psi_p(\mathbf{r}, t)d\mathbf{r}. \tag{16.6}$$

Using $x = r \sin \theta \cos \phi$ and $\psi_s(r, t) = R_s(r)/\sqrt{4\pi}$ yields

$$|X(t)|^2 = \frac{R_{sp}^2}{12} e^{-\Gamma t} \cos^2\left(\frac{\mu_B t}{\hbar}\right), \tag{16.7}$$

where $R_{sp}$ is the radial matrix element and the $e^{-\Gamma t}$ factor accounts for decay of the excited state. Using Eq. (16.7), we can write the intensity $I_x(t)$ of the x polarized fluorescence as

$$I_x(t) = \frac{I_0}{2}\left(1 + \cos\frac{2\mu_B B t}{\hbar}\right) e^{-\Gamma t}, \tag{16.8}$$

where $I_0$ is again the intensity at $t = 0$. The analogous expression for the intensity of the y polarized fluorescence has a minus sign in place of the plus sign in the parentheses of Eq. (16.8).

Equation (16.8) is the same result as the simple classical picture and describes Figure 2. Note that the oscillations in the detected fluorescence occur at the energy spacing of the excited state, a general property of quantum beats.

### 16.2.3 Generalizations

A quantum beat experiment has three parts: excitation, free evolution of the excited state, and detection. The natural description of the period of free evolution uses the energy eigenstates. However, the excitation and detection processes cannot be described using only one energy eigenstate. In the example we have considered thus far, the energy eigenstates are quantized along the z direction, while the natural descriptions of the x polarized excitation and detection use a set of states quantized along the x axis, which are not energy eigenstates. Consequently, excitation by a short pulse projects the ground state onto a coherent superposition of excited energy eigenstates. The amplitudes of the excited states are time-independent, but because of their different energies, their relative phases change with time. When they are projected onto a different set of final states by the emission of x polarized light, the projections vary in time, giving the observed beats.

In general, the detection of a quantum beat signal requires that several conditions be met. First, the exciting pulse must be short compared to the inverse of the frequency spacing of the excited states under study. Second, the exciting pulse must occur at a known time, so that we can observe the phase of oscillations in the detected signal relative to the excitation time. Third, the projection of the detection technique must be rapid compared to the inverse of the frequency spacing. Emission of a photon, for example, satisfies this criterion.

Excellent examples of quantum beats are the Na $nd$ fine structure beats observed by Haroche *et al.* [5]. In this case, the natural description of the optical excitation is in terms of the uncoupled $\ell m\, sm_s$ states, and a short optical pulse creates a coherent superposition of the coupled $\ell s\, jm_j$ energy eigenstates. Optical emission projects these energy eigenstates back onto the uncoupled $\ell m\, sm_s$ states, leading to the observed beats at the $nd$ fine-structure frequency in the detected fluorescence.

Specifically, Haroche *et al.*, excited Na atoms at low pressures, $10^{-6}$ torr, in a vapor cell with two 5 ns long laser pulses tuned to the $3s_{1/2} \rightarrow 3p_{3/2}$ and $3p_{3/2} \rightarrow nd_{3/2,\,5/2}$ transitions. Two of the choices of polarizations in the excitation and detection are shown in Figure 3. They detected the time-resolved polarized

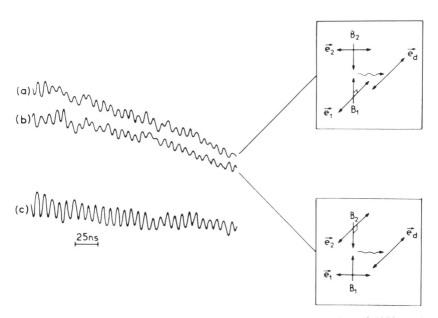

FIG. 3. Recording of fine-structure beats in the Na $9d$ level (averaging of 1000 runs) obtained by time-resolved fluorescence detection. Trace (a): signal obtained with configuration of polarizers $\vec{e}_1 \perp \vec{e}_2$, $\vec{e}_d \| \vec{e}_1$ as shown in the upper inset; trace (b) signal obtained with $\vec{e}_1 \perp \vec{e}_2$, $\vec{e}_d \| \vec{e}_2$ as shown in the inset; trace (c): result of subtracting trace (b) from trace (a). $\vec{e}_1$, $\vec{e}_2$, and $\vec{e}_d$ are the electric polarization vectors of laser beams 1 and 2 and the detected fluorescence. (From reference [5].)

*nd* → 3*p* fluorescence with a fast photomultiplier tube and transient digitizer to obtain the quantum beat signals shown in Figure 3. While the beat signals are clear, they do not have the 100% modulation shown in Figure 2, since the coherently excited states are not excited with equal amplitudes. Before leaving this case, it is useful to see how it meets the three requirements for observing beats. First, the exciting laser pulse length is shorter than 1/100 MHz. Second, it is straightforward to know when the laser pulse excites the atoms, because a small part of the laser beam can be detected with a fast photodiode. Third, the emission of individual photons automatically satisfies the third requirement, but it is also necessary to detect the resulting modulation in the fluorescence at 100 MHz. This frequency is near the frequency limit of most photomultipliers—although photodiodes are considerably faster—but is well below the bandwidth of present transient recorders.

Although optical detection is the most common approach, quantum beats need not be detected optically. Interesting examples are the quantum-beat measurements of the Na *nd* fine-structure intervals using field ionization detection by Leuchs and Walther [6] and Jeys *et al.* [7]. Using essentially the same optical excitation scheme as Haroche *et al.* [5], Jeys *et al.* [7] excited a coherent superposition of $nd_{3/2}$ and $nd_{5/2}$ states. The coherent superposition of fine-structure levels was allowed to evolve freely for a variable time of 0–10 µs, after which a rapidly rising small electric field pulse of 1 V/cm was applied to the atoms. The pulse projected the atoms from the coupled zero-field $\ell s\,jm_j$ fine structure states onto the uncoupled $\ell m\,sm_s$ states. The atoms were then exposed to a slowly rising field ionization pulse which ionized states of different *m* at different fields. By monitoring the signal corresponding to only one value of *m*, it was possible to observe the fine-structure beats with excellent signal-to-noise ratio. The projection by the field pulse is similar to the one produced by the emission of a photon. However, the field pulse only projects the $\ell s_j m_j$ states onto the $\ell m s m_s$ states if it is rapid. If it is slow, the evolution is adiabatic, and no beats are observed. Relative to fluorescence detection, this approach has two attractions. First, all excited atoms can be detected, not just those which fluoresce into a small solid angle. Second, at any delay time, all remaining atoms are detected. The signal is not spread over the entire radiative lifetime. A corollary to the previous statement is that is not possible to record the entire decay curve on one shot of the laser. Rather, the quantum beat curve is generated by slowly scanning the delay time of the field pulse over many shots of the laser.

## 16.2.4 Wave Packets

Wave packets are really just quantum beats of many levels, and the same general requirements exist for the observation of both. An illustrative example is the formation of a radial wave packet of Rydberg atoms [8–10]. At time *t* = 0, a short optical pulse is used to excite atoms from the ground state to a group of Rydberg states centered on principal quantum number $\bar{n}$. The exciting pulse projects the compact ground state onto a Rydberg wave packet propagating outward. The wave packet is reflected by the Coulomb potential at large radius and returns to the ionic core in the time of a classical orbit, $2\pi\bar{n}^3$, which is simply the inverse of the frequency spacing between the Rydberg levels. The highest probability of finding the electron oscillates between large and small radius. To observe the oscillation, it is possible to take advantage of the fact that a visible photon can only be absorbed by the Rydberg electron when it is near the ionic core. No photoionization by visible light occurs if the electron is far from the ionic core. Typically, such experiments are done with two short laser pulses to create and detect the wave packets. When the delay between the two pulses is scanned, a peak in the ionization occurs whenever the delay time is an integral multiple of the classical round-trip time.

Quantum-mechanically, if we use single-photon excitation from a ground $s$ state to the $np$ Rydberg series, we create a superposition of Rydberg states given by

$$\Psi_p(r, t) = \sum a_n \Psi_{np}(r, t), \qquad (16.9)$$

in which the coefficients $a_n$ are such that the wave functions all add constructively for $r \approx 0$ and $t = 0$ so that the probability is as shown in Figure 4a. How many $n$ states have nonnegligible values of $a_n$ is determined by the pulsewidth of the laser and its tuning. If the pulsewidth is 10% of the round-trip time, 10 states are excited. For $t > 0$, $|\Psi(r, t)|^2$ is not in general peaked near $r \approx 0$. For $t \approx \pi\bar{n}^3$, the probability of finding the electron is peaked at the outer turning point, as shown by Figure 4b. To the extent that the Rydberg states at $\bar{n}$ are equally spaced, at times $t = 2\pi\bar{n}^3 N$, where $N$ is an integer, the relative phases of the wave functions are the same as they are at $t = 0$, and $|\Psi(r, t)|$ is again peaked near $r \approx 0$. In fact, the $\Delta n$ intervals are not equal, and the returns of the wave packet to the origin corresponding to large $N$ are not observed. One

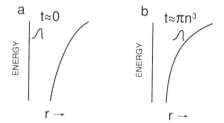

FIG. 4. Schematic drawing showing how the electron of a radial wave packet bounces back and forth in the Coulomb potential. (a) For $t \sim 0$, the electron is localized at $r \sim 0$. (b) For $t$ slightly greater than half the classical round-trip time, the electron has reflected from the large $r$ Coulomb potential and is returning to the ionic core.

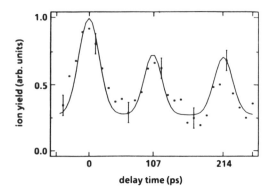

FIG. 5. Experimental and theoretical $K$ wave-packet signals (squares and solid curve, respectively). The signal shows a periodic enhancement with the period of the classical round-trip time (107 ps) for $\bar{n} = 89$. The width of the peaks indicate that about five states were significantly excited ($87 < n < 91$). (From reference [8].)

interesting difference between radial wave packets and the cases of quantum beats described earlier is that the phase evolution of the wave packet is purely radial, whereas in the quantum beats described earlier, it is purely angular.

A good example of wave packets is the work of Yeazell *et al.* [8]. They used 25 ps laser pulses of $\lambda = 571$ nm to excite $K$ atoms to $\bar{n} = 89$ Rydberg $ns$ and $nd$ states by two-photon excitation. With this pulse duration, they excited roughly five Rydberg states near $\bar{n} = 89$. A second 571 nm 1 photoionized the atoms after a variable delay, and the number of ions produced as a function of delay is shown in Figure 5. There are evident peaks in the number of ions observed at multiples of 107 ps, the classical orbit time. Equally interesting, it is evident that the modulation is not sinusoidal, like the beat signal of Figure 2, but is more sharply peaked. The sharpness comes from having a coherent superposition of more than two states. Just as two-level quantum-beat signals are analogous to the output of a laser with two modes, wave packet signals are analogous to the output of a mode-locked laser with many oscillating modes.

## 16.3 Hanle Effect and Level-Crossing Spectroscopy

If we return to the $^1S_0$–$^1P_1$ system in a $B$ field in the $z$ direction exposed to $x$ polarized exciting light, we can see how quantum beats are connected to the Hanle effect, the zero-field level crossing of the $m = \pm 1$ states. Imagine that we have a low-level cw light source illuminating the atoms. We can think of this source as a stream of photons, each of which excites an atom which later fluoresces. Since we cannot know when each photon excites an atom, we cannot determine the phase of any modulation in the fluorescence relative to the exciting photon. We can only measure the average or time-integrated value of the fluorescence. The first and third requirements for the detection of quantum beats are met, but the second is not, and we cannot observe quantum beats. However, we can observe a level-crossing signal. If the magnetic field is large so that $\omega_L \gg \Gamma'$, the atom rotates about the $z$-axis many times in its radiative lifetime. Consequently, the time integrated $x$ and $y$ polarized fluorescence emitted in the $z$ direction is almost the same, as is evident by examining Figure 2. On the other hand, as the magnetic field is reduced to zero, fewer rotations of the atom occur during the radiative lifetime, and when $\omega_L < \Gamma$, there is a large difference in the $x$ and $y$ polarized fluorescence. In this case, the atom radiates before the electron's orbit rotates through a large angle, and most of the fluorescence is $x$ polarized. When $B = 0$, the orbit does not rotate, and there is only $x$ polarized fluorescence. We can express these notions in a quantitative fashion by simply integrating the expression for $I_x(t)$ of Eq. (16.8) over time, yielding

$$\bar{I}_x = \frac{I_0}{T}\left[\frac{1}{\Gamma} + \frac{\Gamma}{\Gamma^2 + (2\mu_B B/\hbar)^2}\right], \tag{16.10}$$

where $T$ is the time between impinging photons. There is an analogous expression for $\bar{I}_y$ which has a minus sign in the brackets. Figure 6 is a graph of $\bar{I}_x$ and $\bar{I}_y$ vs. magnetic field $B$. The maximum in $\bar{I}_x$ and the minimum in $\bar{I}_y$ both occur at $B = 0$. Here the $m = \pm 1$ levels cross, no rotation of the excited state occurs, and the radiation must be polarized in the $x$ direction. The full width at half maximum of the increase in $\bar{I}_x$ and the decrease in $\bar{I}_y$ is $\Delta B = \Gamma\hbar/2\mu_B$. Rather small $B$ fields

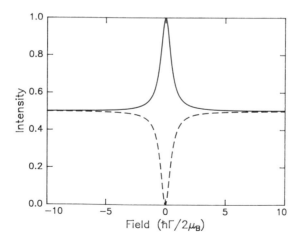

FIG. 6. Plots of the time-averaged $x$ and $y$ polarized fluorescence $\bar{I}_n$ (———) and $\bar{I}_y$ (– – –) vs. magnetic field $B$.

are required. A 10 ns lifetime, corresponding to $\Gamma = 10^8 \, s^{-1}$, leads to $\Delta B = 11$ gauss.

In the Hanle effect, the location of the signal, zero field, does not tell us anything we did not already know. However, there are high-field level crossings, the locations of which provide useful information. For example, we show in Figure 7 a general energy-level diagram of the Na $nd$ states as a function of magnetic field in the $z$ direction [11]. In order to show the levels on a universal graph, the magnetic field is plotted in terms of its ratio to the fine-structure interval $\delta W$. In Figure 7, the lowest field crossings between states differing in $m_j$ by two are circled. These level crossings are analogous to the level crossing of the $^1P_1 \, m = \pm 1$ levels at zero field. When a stream of $x$ polarized photons excites coherent superpositions of the states, away from the level crossing the $x$ polarization is reduced in the fluorescence detected in the $z$ direction because of the rapid evolution of the levels compared to the decay rate. However, when the field is set to the level-crossing field, the levels do not precess before decaying, and the fluorescence is largely $x$ polarized.

In the experiment of Fredricksson and Svanberg [11], Na atoms in a thermal beam were first excited from the ground $3s_{1/2}$ state to the $3p_{3/2}$ state with a powerful unpolarized Na resonance lamp. These atoms were further excited to the $nd_j$ states by an $x$ polarized cw blue dye laser propagating in the direction

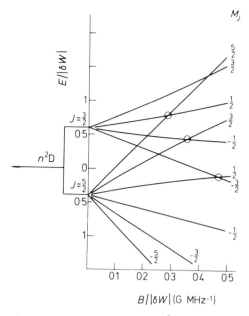

FIG. 7. Energy-level diagram for an inverted$^2$ $D$ doublet. Three $\Delta m_j = 2$ level crossings are indicated. The diagram is valid for all values of $|\delta W|$. (From reference [11].)

State                                                    Position (G)

9²D                                                      35·93

8²D                                                      50 21

7²D                                                      72·97

6²D                                                      110 8

5²D                                                      177·1

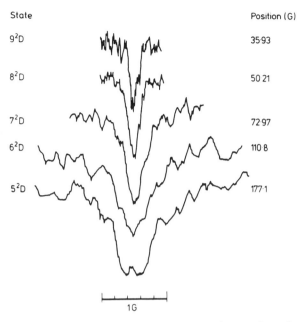

1G

FIG. 8. Experimental level-crossing curves for the Na 5d to 9d states. In each case the recording is made at the lowest-field level crossing of Figure 7. Each curve has been sampled for 1–2 h. (From reference [11].)

perpendicular to the magnetic field. The $y$ polarized $nd_j$–3p fluorescence was observed in the field direction as the magnetic field was slowly scanned through the region of the level crossing. In Figure 8 are shown the level-crossing signals recorded for the lowest $nd$ crossing of $n = 5$ to 9. As shown by Figure 8, the fluorescence drops sharply at the crossings. Also, with increasing $n$, the crossing occur at lower fields because of the decrease in the fine-structure interval with $n$, and the widths of the level-crossing signals also decrease because of the decreasing radiative decay rates. Both the fine-structure intervals and the radiative decay rates scale as $n^{-3}$.

## 16.4 Anticrossing Spectroscopy

Anticrossing spectroscopy, first discovered by Eck *et al.* [12], can most easily be understood by means of a simple example. Imagine two excited states A and B, of the same total azimuthal angular momentum, which would cross at the magnetic field $B_x$ if we ignored a small internal interaction between them, as shown by the broken lines of Figure 9.

A is coupled by an electric dipole transition to C, while B is similarly coupled to D. There is no coupling between A and D or B and C. As a concrete example,

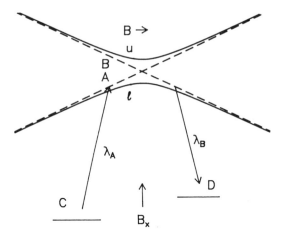

FIG. 9. Energy-level diagram for anticrossing spectroscopy with no coupling levels and A and B ($-$ $-$ $-$) exhibit linear energy shifts with field and cross at $B_x$. With the coupling $V$ included, the actual energy eigenstates $u$ and $\ell$ (————) do not cross.

A and C might be singlet and B and D triplet states. Including the interaction, a spin–orbit interaction for example, leads to the real energy eigenstates, $u$ and $\ell$, which do not cross but have an avoided crossing or anticrossing, as shown by the solid lines. The actual eigenfunctions are not the A and B state wavefunctions $\psi_A$ and $\psi_B$, but

$$\psi_u = \cos\theta\psi_A + \sin\theta\psi_B$$

and

$$\psi_\ell = -\sin\theta\psi_A + \cos\theta\psi_B, \qquad (16.11)$$

where

$$\tan 2\theta = \frac{2V}{(W_B - W_A)},$$

$$\sin 2\theta = \frac{2V}{\sqrt{4V^2 + (W_A - W_B)^2}}, \qquad (16.12)$$

$$\cos 2\theta = \frac{W_A - W_B}{\sqrt{4V^2 + (W_A - W_B)^2}},$$

and $0 < \theta < \pi/2$. Here $W_A$ and $W_B$ are the energies of A and B, ignoring their interaction, and $V$ is the interaction matrix element.

Far from the anticrossing, the eigenstates $\psi_{u,\ell}$ are equal to $\psi_A$ and $\psi_B$ to within a sign, and there is no observable effect of the coupling $V$. At the anticrossing, $\Psi_{u,\ell} = (\Psi_B \pm \Psi_A)/\sqrt{2}$, i.e., the wave functions of the eigenstates are 50–50 mixtures of A and B.

The mixed A–B character of the states at the anticrossing leads to the anticrossing signal. Imagine optically exciting the atoms with a source tuned to the C → A transition at $\lambda_A$, which is broad enough to cover the entire field tuning region. We detect light at $\lambda_B$ from the B → D transition. Away from the anticrossing, the energy eigenstates are the A and B states, and since we cannot excite the B state, we cannot observe the B → D emission at $\lambda_B$. At the anticrossing, the energy eigenstates are linear combinations of A and B, so they can both absorb the light at $\lambda_A$ and emit light at $\lambda_B$. As a result, when the field is scanned across the anticrossing, an increase in the B → D fluorescence is observed. The increase in the B → D fluorescence comes at the expense of the A → C fluorescence, which exhibits a corresponding decrease.

The experimental study of the He $nd$ states by Miller et al. [13] is a good example of anticrossing spectroscopy. As shown by the level diagram of Figure 10, there are three anticrossings of the singlet and triplet $nd$ states which occur at virtually the same magnetic field and lead to superimposed signals. The He atoms are excited by electron impact using a cw electron beam, which populates the $^1D_2$ states but not the $^3D_j$ states. When the emission at 4144 Å from the singlet $nd$ state to the $2p\ ^1P_1$ state is observed, a sharp decrease in the emission is observed at 15 kG because of the state mixing at the anticrossing. At the anticrossing, the same number of atoms is excited to the pair of mixed singlet–triplet states as is excited to the singlet state away from the avoided crossing. However, since both singlet and triplet lifetimes are comparable, at the anticrossing only half as many atoms decay to the $2p\ ^1P_1$ state. As shown by the lower trace of Figure 10, if the $nd\ ^3D_2\text{–}^3P_j$ emission at 3819 Å is observed, the signal rises sharply from zero at the anticrossing as expected. As can be seen from Figure 10, the crucial requirement is magnetic field homogeneity. The inhomogeneities must lead to energy inhomogeneities less than $V$, usually a less stringent requirement than for level-crossing spectroscopy.

As in the case of quantum beats, alternate means exist for detecting anticrossing signals. For example, Stoneman and Gallagher [14] detected anticrossings of $K\ ns$ states with $n-2$ Stark manifold states in electric fields. They excited $K$ atoms in an atomic beam with two 5 ns dye laser pulses from the $4s$ ground state to the $4p$ state and then to the $18s$ state in a static field near the anticrossing of the $18s$ state with the $n=16$ Stark states. Away from the anticrossing, the Stark states, which are composed of $\ell \geq 3$ states, are not excited. The atoms are exposed to room-temperature blackbody radiation for a period of 3 μs, after which they are exposed to a field ionization pulse to analyze the final states. The electric field is scanned through the anticrossing over many shots of the laser. Away from the anticrossing, the $18s$ atoms, but not those in Stark states, are driven primarily to the $18p$ state by the blackbody radiation. If

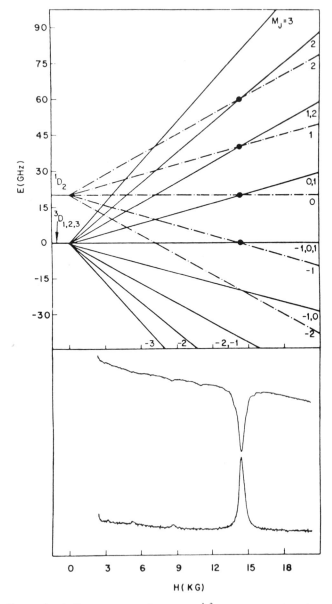

FIG. 10. Energy-level diagram (top) of the $n = 6^{1,3}D$ states as a function of m
field and actual spectra (bottom) of the $n = 6^{1,3}D$ anticrossing in He. Each of the s
traces was obtained in approximately 10 min running time. The top trace shc
decrease in light intensity of the $6^1D$ emission line at 4144 Å, while the lower trace
the corresponding increase in the $6^3D$ emission at 3819 Å. (From reference [13].)

the $18p$ field ionization signal is monitored as the field is scanned, it shows a 50% drop at the anticrossing. At the anticrossing, the same number of atoms is excited to the pair of mixed $18s - n = 16$ Stark states as away from the crossing. However, these two states have half the blackbody transition rate to the $18p$ state as does the pure $18s$ state, and the number of $18p$ atoms produced drops by half.

## References

1. Happer, W., and Gupta, R. (1978). In *Progress in Atomic Spectroscopy*, W. Hanle and H. Kleinpoppen (eds.). Plenum, New York.
2. Dodd, J. N., and Series, G. W. (1978). In *Progress in Atomic Spectroscopy*, W. Hanle and H. Kleinpoppen (eds.). Plenum, New York.
3. Beyer, H.-J., and Kleipopppen, H. (1978). In *Progress in Atomic Spectroscopy*, W. Hanle and H. Kleinpoppen (eds.). Plenum, New York.
4. Haroche, S. (1976). In *High Resolution Laser Spectroscopy*, K. Shimoda (ed.). Springer-Verlag, Berlin.
5. Haroche, S., Gross, M., and Silvermann, M. P. (1974). *Phys. Rev. Lett.* **33,** 1063.
6. Leuchs, G., and Walther, H. (1979). *Z. Physik A* **293,** 93.
7. Jeys, T. H., Smith, K. A., Dunning, F. B., and Stebbings, R. F. (1981). *Phys. Rev. A* **23,** 3065.
8. Yeazell, J. A., Mallalieu, M., Parker, J., and Stroud, C. R., Jr. (1989). *Phys. Rev. A* **40,** 5040.
9. ten Wolde, A., Noordam, L. D., Lagendijk, A., and van Linden van den Heuvell, H. B. (1988). *Phys. Rev. Lett.* **61,** 2099.
10. Meacher, D. R., Meyler, P. E., Hughes, I. G., and Ewart, P. (1991). *J. Phys. B* **24,** L63.
11. Fredricksson, K., and Svanberg, S. (1976). *J. Phys. B* **9,** 1237.
12. Eck, T. G., Foldy, L. L., and Weider, H. (1963). *Phys. Rev. Lett.* **10,** 239.
13. Miller, T. A., Freund, R. S., and Zegarski, B. R. (1975). *Phys. Rev. A* **11,** 753.
14. Stoneman, R. C., and Gallagher, T. F. (1985). *Phys. Rev. Lett.* **55,** 2567.

# 17. ATOM INTERFEROMETRY

Olivier H. Carnal

Holtronic Technologies S.A.
Marin, Switzerland

Jürgen Mlynek

Fakultät für Physik, Universität Konstanz
Konstanz, Germany

## 17.1 Introduction

Interferometry with massive particles such as neutrons and electrons [1] has already a long tradition and has been used to demonstrate purely quantum-mechanical effects, such as the Aharanov–Bohm effect and the sign reversal of spin-1/2 particles after a $2\pi$ rotation. It has also been used to measure phase shifts in the particle's wave function due to passage through solid-state material or electromagnetic fields. The logic next step in matter wave interferometry is the realization of interferometers for atoms, because atoms are composed of nucleons and electrons and form a more complicated structure, but are still so small that the interaction with the environment does not immediately destroy coherences.

Atom interferometers have excited much interest [2] because the atom's more complex internal structure allows to perform experiments unknown so far in matter wave interferometry. At the same time, the large atomic mass and the availability of very slow atoms to obtain long interaction times result in an increased sensitivity to gravitational forces and rotations, which could make atom interferometers a useful tool in metrology. The potential applications of atom interferometers can be divided into three subgroups:

1. Study of atomic properties such as electric polarizability, magnetic moment, scattering amplitudes, and atom–surface interactions.
2. Fundamental experiments to demonstrate quantum-mechanical effects such as the Aharonov–Casher effect [3], the Casimir–Polder force between a polarizable particle and a conducting plane [4], and phase diffusion after a spontaneous emission of a photon [5].
3. Use as highly sensitive gyroscope or accelerometer in metrology, navigation, or geology [6].

EXPERIMENTAL METHODS IN THE PHYSICAL SCIENCES
Vol. 29B

Because neutron and electron interferometers have been used with great success for many years, one may ask why it took so long to realize an atom interferometer (first successful demonstrations at the end of 1990). The main reasons for this delay are the following: First, the de Broglie wavelengths associated with atoms at room temperature are very small (typically 0.1–1 Å), and the velocity distribution in a thermal atomic beam is quite broad (width about equal to the mean velocity), which means that one has to accommodate the dispersion of beam–splitting devices. Additionally, since atoms do not pass through solid matter, novel technologies had to be developed to coherently split and recombine atomic waves.

The chapter is organized as follows: Some useful concepts in atom interferometry are outlined in Section 17.2; coherent beam splitters, the crucial elements in an atom interferometer, are presented in Section 17.3; and Section 17.4 is devoted to the different realizations of atom interferometers. In Section 17.5, we discuss further possible developments in the field of atom interferometry. The article concludes with an outlook.

## 17.2 Basic Concepts

Before talking about experimental techniques, we want to give a rough idea of the concept of atomic waves and present some basic formulas needed for the ensuing discussions. For thermal atoms, the velocities are small enough for the time-evolution of an atomic wave function to be described by the nonrelativistic Schrödinger equation [7]. For time-independent external potentials $V(\vec{r})$ and free atoms with total energy $E$ and mass $m$, the solution for the atomic wave function $\Psi(\vec{r})$ is given by a linear superposition of plane waves:

$$\Psi(\vec{r}) = A \cdot e^{i\vec{k}\vec{r}}. \tag{17.1}$$

The de Broglie wavelength of this matter wave is $\lambda_{dB} = 2\pi/k$, with

$$k = |\vec{k}(\vec{r})|\sqrt{(2m/\hbar^2)(E - V(\vec{r}))}, \quad (\hbar = h/2\pi, \ h \text{ being Planck's constant}). \tag{17.2}$$

For nonrelativistic particles, the group velocity $v_g = \partial\omega/\partial k = 1/\hbar \cdot \partial E/\partial k = \hbar k/m$ is identical to the velocity of the classical particle $v$, so that one obtains the important de Broglie relation between classical motion and particle wavelength:

$$\lambda_{dB} = h/mv. \tag{17.3}$$

For a helium atom with a velocity $v = 1000$ m/s, the corresponding de Broglie wavelength $\lambda_{dB}$ is roughly 1 Å and decreases for heavier and faster atoms. We would like to point out that the wavelength of an atom is normally smaller than

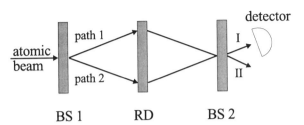

FIG. 1. Basic elements forming an atom interferometer. Atoms originating from either a beam source or a trap are coherently split in the first beam splitter BS 1, redirected again in a reflecting device RD, which could also be a beam splitter, and finally overlapped in the second beam splitter BS 2. In this geometry, the atoms reach the detector D on two distinct paths and can experience different phase shifts along their path, leading to a variation of the output intensities in beams I and II. The splitting must not necessarily be in the transverse direction, but can also occur longitudinally.

the diameter of the electronic orbits inside the atom itself, unless very slow atoms are used. Equation (17.3) shows also that a velocity spread $\Delta v$ in a beam directly leads to an uncertainty in atomic wavelength $\Delta \lambda_{dB} = h/m \cdot \Delta v/v^2$. This effect can be minimized by compressing the velocity distribution with laser forces [8, 9].

An interferometer can be decomposed into the following elements (see Figure 1): An incoming beam of atoms, usually described by an incoherent super-position of plane waves with different $k$-vectors, is split into two output beams in the first beam splitter either longitudinally or transversely (along or perpendicularly to the atomic beam), so that one has access to one of the two beams without affecting the other. What is important is that this beam splitter preserves the phase of the atomic wave function, so that there is a well-defined phase relation between the atomic wave functions in path 1 and path 2. The two beams have to be overlapped again with beam deflectors and are finally recombined. This can be done—e.g., with another beam splitter—such that a part of beam 1 and a part of beam 2 leave the interferometer in the same direction, forming the two output beams I and II. What happens after recombination? Let us assume that the atomic wave function on path 1 has accumulated a phase $\phi_1$ which is both a function of path length $L_1$ and effective refractive index $n(\vec{r}) = |\vec{k}(\vec{r})|/k_0$ along this path:

$$\phi_1 = \int_{path 1} \vec{k}(\vec{r}) \cdot d\vec{r}, \tag{17.4}$$

with $|\vec{k}(\vec{r})|$ given by Eq. (17.2). The phase shift along path 2 is calculated analogously. Things are getting more complicated if non-Hamiltonian (dissipative) and non-adiabatic processes occur, e.g., after spontaneous emission of a photon in the interferometer or non-adiabatic passage through a quickly turned-on optical field. But to compute the effect of slowly varying external fields on the atomic wave function, this scalar treatment is valid.

If both the beam separation and recombination are perfect and the intensities in both arms of the interferometer are equal, the total wave function at the output is given by

$$\Psi_{tot} = A\,[\exp(i\phi_1) \pm \exp(i\phi_2)], \tag{17.5}$$

depending on whether atoms are detected in output port I or II.

The intensity recorded at the output is then a direct measure of the acquired phase difference between the two paths $\Delta\phi = \phi_1 - \phi_2$, and therefore is sensitive to path and potential-energy differences:

$$I_{I,\,II} \propto 1/2 \cdot (1 \pm \cos(\Delta\phi)). \tag{17.6}$$

As one can see from Eq. (17.6), noise in the output intensity leads to an uncertainty in the determination of the induced phase shifts. Because the particle flux in an atom interferometer is normally very low and the probability of finding $N$ atoms in an integration time $t$ is given by Poisson statistics, the fluctuations in the intensity measurement can be considerable. The contribution of this noise to the phase uncertainty is given by [10]

$$\Delta\phi_{min} = 1/\sqrt{N} = 1/\sqrt{\beta t}, \tag{17.7}$$

with $\beta$ being the flux of atoms through the interferometer. An important consideration in designing an atom interferometer is therefore not only a short atomic wavelength and large interaction region, but also a high flux of atoms.

Supersonic atomic beams or atoms launched from an atom trap are ideal candidates for this purpose. These atomic sources are described in detail in other chapters of this book. Supersonic sources have the advantage of producing a very monoenergetic and intense atomic beam with little divergence and velocities around 100–1000 m/s. They are preferably used when the wavelength dispersion is a problem. Atoms in a trap have velocities below 1 m/s and can be stored at very high densities, but they spread out in space immediately after release from the trap. Trap sources are used whenever long wavelengths or interaction times are required.

## 17.3 Beam Splitters for Atoms: A Short Review

Beam splitters can be divided roughly into two big subgroups: those which only act on the external degrees of freedom, without changing the internal state of the atom leaving the beam splitter; and those relying on a coupling of internal and external states of motion. If the internal states are different in the two beams forming the interferometer, one could call it, in analogy to classical optics, a polarizing beam splitter.

Beam splitters for atoms were already realized as long as 65 years ago, when atoms were diffracted from cleaved ionic crystals [11]. In these experiments, the electronic states of an atom are slightly shifted when approaching the atomic surface. This leads to a phase shift in the atomic wave function and, as a result of the crystal periodicity, to a phase grating for atoms. The various diffracted beams reflected off the surface have a well-defined phase relation and are largely separated (diffraction angles of 0.1–1 radian possible) and could therefore be used in an interferometer.

Instead of using periodic crystal structures given by nature, recent developments in microtechnology allow the fabrication of free-standing gratings with a period $d$ below 100 nm [12] so that atoms passing through the structure are diffracted into beams separated in momentum space by

$$\Delta p = h/d. \tag{17.8}$$

Since $d$ is much larger than the period in a solid-state crystal, the diffraction angles given by $\Delta p/p$, with $p$ the atomic momentum, are therefore quite small. Another disadvantage is the fact that such an amplitude grating has a low efficiency because at most 10% of the incoming intensity is deflected into the first diffraction order. This is mostly compensated by its universal use with any kind of atomic species and its small sensitivity to the surface roughness of the structure. Up to now, diffraction by a free-standing grating has been demonstrated by two groups with diffraction angles of about $10^{-4}$ rad [13, 14].

The interaction of an atom having a permanent magnetic or induced electric moment with static electric or magnetic fields leads to another class of beam splitters. In the Stern–Gerlach effect [15], the energy of atomic states with different magnetic moments is altered in a magnetic field, and the presence of a strong field gradient leads to a splitting of the different magnetic sublevels. Analogously, the interaction of an induced electric dipole moment with an electric field gradient can be used as a beam splitter for atoms. Reasonable field gradients can already lead to considerable deflection angles of about $10^{-2}$ rad. These devices produce outgoing beams in orthogonal internal states. The experimental requirements to build, e.g., a Stern–Gerlach interferometer are rather stringent if the beams have to be well separated in space [16]. An experimentally easier approach uses the splitting of an atomic beam along the beam axis with a longitudinal field gradient [17]. There, the splitting occurs in time but not in space, which restricts the applications of such a device.

The near-resonant interaction of an atom with a laser field has led to a variety of powerful beam splitters. All of them are based on the coupling between the atom's electric dipole moment and the light field, but are considerably different from each other so that they deserve independent discussions.

Let us first consider the effect of a standing light field (frequency $\omega$) aligned perpendicular to an atomic beam of two-level atoms (with $|g\rangle$ the ground and $|e\rangle$

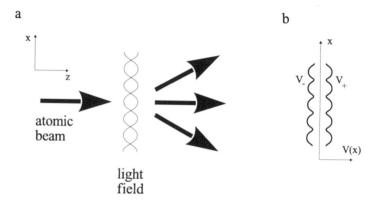

FIG. 2. Diffraction of atoms crossing a standing wave perpendicular to the atomic trajectory. (a) Experimental setup with atoms propagating along $z$ and light intensity variations along $x$. (b) Corresponding potential energies $V_\pm(x)$ of the dressed states in the optical field as given by Eq. (17.9).

the excited state of the atom and a transition frequency $\omega_0$) (see Figure 2a). The dipole interaction in the optical field leads to a coupling of the two levels and to two new energy eigenvalues $V_\pm$ as a function of the detuning from resonance $\Delta = \omega - \omega_0$ and the Rabi frequency $\omega_R = \langle g | \vec{d} | e \rangle \cdot \vec{E}/\hbar$ ($\vec{d}$: atomic dipole moment, $\vec{E}$: electric field) [18]:

$$V_\pm(x) = \frac{\hbar}{2} \cdot [-\Delta \pm \sqrt{\omega_R^2(x) + \Delta^2}], \qquad (17.9)$$

where $x$ is the transverse position in the light field. For a standing wave, $V_\pm(x)$ is shown in the inset of Figure 2b. As the atomic $k$-vector changes with potential energy, the interaction with this transverse periodic modulation corresponds to a diffraction from a phase grating. The phase shift $\phi(x)$ is given by Eq. (17.4).

The first convincing experimental demonstration of diffraction of atoms from a standing light wave was achieved in 1983 by a group at MIT, Cambridge, Massachusetts, using a beam of sodium atoms [19], and improved results with a large laser detuning to ensure adiabatic evolution into one eigenstate were obtained in 1986 [20]. The same group also observed a closely related effect, Bragg scattering, by passing atoms through a nonfocused laser beam with a large diameter [21].

Instead of having a two-level atom pass through a standing light wave such that it "sees" many nodes and antinodes of the field, one can decrease the atomic beam's transverse extension so much that it is smaller than the period of the standing wave. If the atoms cross the field at the position in between field maxima or minima, the atoms see a transverse field gradient and are deflected to either the right or the left, depending on whether they are in the eigenstate $\Psi_+$

or $\Psi_-$, corresponding to $V_+$ or $V_-$. If the laser is tuned to exact resonance ($\Delta = 0$) the passage through the field happens nonadiabatically, and the atomic wave function initially in the ground state is put into a linear superposition of equal components of $\Psi_+$, and $\Psi_-$. As a consequence, the beam is coherently split into two beams with a momentum transfer which mainly depends on laser power. Because of its strong analogy to the Stern–Gerlach experiment with magnetic moments, this effect is called the *optical* Stern–Gerlach effect, and the experiment has been performed by our group in Konstanz with metastable helium atoms [22].

Very recently, our group has also demonstrated a beam-splitting scheme that uses diffraction from a standing wave, but with an additional magnetic field applied parallel to the laser beam [23]. In this so-called magneto-optical beam splitter, the magnetic field is chosen such that the Rabi frequency in the optical electric field is twice the Rabi frequency of the magnetic moment. This resonance condition ensures that once an atom has absorbed one photon from one direction it continues to absorb from that direction and emit into the other. Therefore, this scheme leads to a clearly two-peaked diffraction pattern with larger deflection angles than obtained in a normal standing-wave diffraction. This setup requires an atom with at least two magnetic sublevels in either the ground or excited state.

Coherent population trapping has been used to cool atoms below the recoil limit given by $\Delta p = \hbar k$ and is based on the presence of an atomic dark state which does not interact with the electromagnetic field. Researchers at the ENS in Paris used a three-level system in metastable helium to demonstrate the effect [24], and the final transverse momentum distribution showed two peaks at $p = \pm \hbar k$, corresponding to the dark state $|\Psi\rangle = 1/\sqrt{2} \cdot \{|m_J = -1, p = -\hbar k\rangle - |m_J = +1, p = +\hbar k\rangle\}$, $m_J$ being the magnetic quantum number in the ground state and $p$ the transverse momentum. Since this represents a coherent superposition of two distinct momentum eigenstates, the two output beams could be used in an interferometer. However, the setup is very sensitive to external magnetic fields, and the initial atomic beam has to be collimated rather tightly because the capture range into that dark state is very small.

At the end of this section, we consider a beam splitter very similar to the one just described which relies on the transfer of a single photon recoil to the atom. The concept is the excitation of a two-level atom in a coherent superposition of ground and excited levels by applying a resonant radiation pulse with either a chopped laser or a short interaction time with a stationary field such that the Bloch vector rotates by an angle $\pi/2$ or $\pi$ (a so-called $\pi/2$- or $\pi$-pulse) [25]. The interaction length determines the probability amplitudes of being in the ground or excited state. As the excited state has absorbed the photon momentum, a momentum splitting between ground and excited level occurs: $|\Psi\rangle = a_g |g, p = p_0\rangle + a_e |e, p = p_0 + \hbar k\rangle$, thus forming a beam splitter for atoms. A larger momentum transfer of $2\hbar k$ can be achieved by using a Raman transition between hyperfine components of the ground state [26].

## 17.4 Realizations of Atom Interferometers

As in many other fields of fundamental research, there has been quite an animated, almost philosophical discussion about what defines a "real" atom interferometer. As the quantum-mechanical theory is based on linear super-position of complex-valued wavefunctions, basically every physical effect in the microscopic world can be explained by the interference of wave functions of different eigenstates. As a consequence, the whole discussion relies on the subtle difference between the terms "interference" and "interferometer." The inter-ferometers presented below are nowadays generally accepted as atom inter-ferometers and are mainly listed chronologically.

Although many different beam-splitting techniques exist, only a few have been actually employed to realize an atom interferometer. Up to now, diffraction from microstructures (see 17.4.1 and 17.4.2), single photon recoil in optical fields (see 17.4.3. and 17.4.4), and the longitudinal Stern–Gerlach and Stark effects (see 17.4.5), respectively, have been used.

### 17.4.1 Young's Double Slit

Probably the simplest configuration for an atom interferometer is Young's double slit (named after T. Young, who demonstrated the wave behavior of light with a similar setup in 1807). It consists of an entrance slit (or small source), a double slit, and a detector, which is either a slit, a microchannel plate or a grating (see Figure 3a). This setup was first demonstrated by our group with a supersonic beam of helium atoms ($\lambda_{dB} = 1$ Å) and a slit separation of $d = 8$ μm [27]. It was used shortly afterwards by a group in Tokyo with ultracold neon atoms ($\lambda_{dB} \sim 20$ nm at the double slit) released from a trap and a 6 μm slit separation to demonstrate a phase shift due to the Stark effect [28].

The entrance slit acts as a wavefront splitting device and creates a spatial coherence between the two apertures of the double slit. These two slits act as two coherent atomic sources, and this leads to an interference pattern in the plane of detection (at distance $L$) with interference fringes whose maxima are separated by $\Delta x = L\lambda_{dB}/d$. This interference pattern can be detected by measuring the atomic flux through a single slit as a function of the slit displacement, or by introducing a grating with the periodicity of the interference pattern. The whole pattern can also be imagined simultaneously with a microchannel plate, as demonstrated by the group in Tokyo. The interference pattern observed in our experiment is shown in Figure 3b, and the period of the fringe pattern of about 8 μm is in good agreement with the formula given earlier.

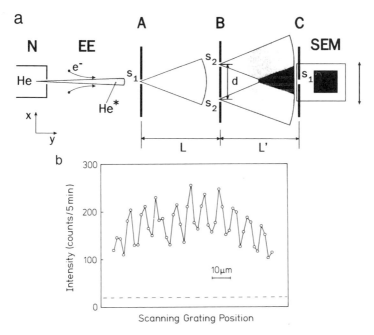

FIG. 3. Double slit interferometer. (a) Schematic setup of the three elements. A: entrance slit or small source, B: double slit, and C: detector with high spatial resolution (slit widths $s_1 = 2$ μm, $s_2 = 1$ μm). (b) Experimental results of the authors showing the atomic intensity as a function of position of a gold grating in the detector plane with a period equal to the one of the atomic interference pattern [27], where BG denotes the background noise of the electron multiplier detector.

The advantages of this setup are its conceptual simplicity and the use of only a few mechanical structures. Its use as a measuring tool is quite restricted, because the beam separation cannot be increased without considerable loss in intensity (both the width of the entrance and the double slit have to be decreased inversely proportional to the beam separation $d$), and the atomic flux is therefore usually very small. The number of fringes in the detection plane is furthermore limited by the width of the velocity distribution in the beam. This interferometer type may be useful whenever high spatial confinement of an interferometer arm is required (given by the slit widths which are normally 1–2 μm), e.g., in probing the short-ranged Casimir potential at micrometer distances from a conducting surface.

## 17.4.2 Three-Grating Interferometer

An alternative scheme for an interferometer based on diffraction from micro-structures is the three-grating arrangement used by a group at MIT [29]. As the name reveals, this interferometer mainly consists of three transmission gratings and two collimating slits (see Figure 4a). The first grating coherently splits the atomic beam, and the second redirects the split beams such that they recombine at the third grating. For simplicity, diffraction orders not used in the interferometer are not shown in the figure. Since the period of the interference pattern at the position of the third grating is independent of the atomic wavelength and is only given by the grating period $d$, this setup is achromatic, which is very useful, considering the often broad velocity distribution in an atomic beam.

The experiment performed at MIT used sodium atoms ($\lambda_{dB} = 0.16$ Å) seeded in an argon beam and silicon nitride gratings with $d = 400$ nm in a first demonstration and 200 nm in improved later version. This leads to a maximum beam separation of 27 μm and 54 μm, respectively. With the atomic beam tightly collimated by two entrance slits, this allows insertion of a thin metallic foil between the two well-separated paths for the introduction of independence phase shifts. Interference fringes were recorded as a function of the transverse displace-

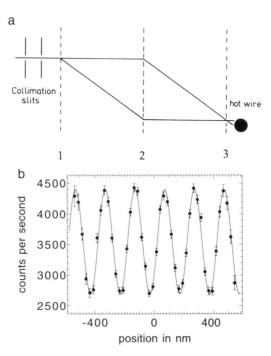

Fig. 4. Three-grating interferometer. (a) Schematic setup. The first grating splits a collimated beam, and a second identical grating diffracts both beams again, such that by moving the third grating along the $x$-axis, the interference pattern is recovered. (b) Experimental data of the intensity distribution using the ionization of sodium atoms on a hot wire detector (with kind permission of C. Ekstrom).

ment of the last grating (as shown in Figure 4b) by measuring the atomic flux on a hot wire detector placed behind the last grating. A crucial part of the experiment is the quality of the gratings, because the grating distortions must be small compared to the grating period over the whole area of the grating. As in the double-slit interferometer, the alignment of the three diffraction structures has to be done very carefully in order not to reduce fringe visibility.

Experiments with this interferometer have been performed recently to measure the polarizability of the sodium ground state and the phase shift in a background gas, and to demonstrate interference revivals due to precession of different magnetic sublevels in a magnetic field [30].

The interferometer presented in Section 17.4.2 is the only one which has actually employed the physical separation of the two interferometer arms, allowing elements to be inserted in between. Although beam separations do also occur in interferometers based on light interactions (as presented in 17.4.3 and 17.4.4), this fact is not crucial in these experiments and has not yet been exploited.

## 17.4.3 Ramsey Interferometer

Almost at the same time as the mechanical interferometers, two interferometers based on optical beam splitters were demonstrated. One of them uses the Ramsey excitation geometry with four spatially separated traveling waves (see Figure 5a). This technique had been applied in ultrahigh-resolution spectroscopy for many years, but it was pointed out only in 1989 that the Ramsey fringes can be interpreted in terms of atomic interference [31] and that the device could measure phase differences between two paths by analyzing the shift of the interference fringes in frequency.

The setup is as follows: A resonant traveling laser wave excites the incoming state $|g, 0\rangle$ (atom in ground state and zero transverse momentum) into a linear superposition of ground and excited state: $\Psi = a_g |g, 0\rangle + a_e |e, +\hbar k\rangle$. The same happens in the other three interaction zones, so that there are $2^4 = 16$ beams emerging from the last interaction zone, but only four of them are superpositions of beams which were separated in space previously (beams I–IV in Figure 5a). Each trapezoid can be interpreted as a distinct Mach–Zehnder interferometer, but since the two interferometers are in different internal states between the second and third laser fields, either of the recoil components can be suppressed by deflecting or depopulating the atoms with another laser beam [32, 33]. The beam intensity of output port I can be measured by its fluorescence, and the fringes in the signal are recorded as a function of laser detuning (and not position as in 7.4.1. and 7.4.2) (see Figure 5b).

Required for this type of interferometer are a long-lived atomic state whose lifetime is longer than the time-of-flight through the interferometer (with the

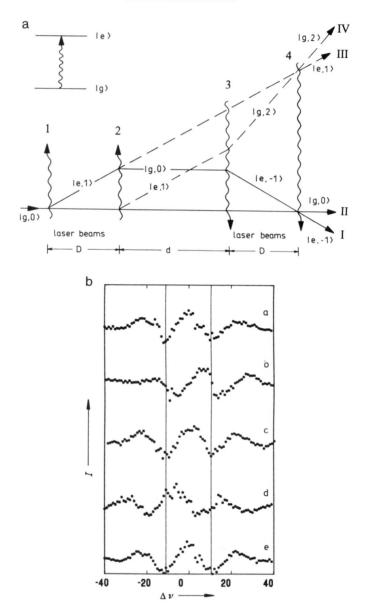

FIG. 5. Ramsey interferometer. (a) Experimental setup. An incoming atom in state $|g, 0\rangle$ is split and recombined by momentum transfer due to absorptions and emissions of photons in the four traveling waves. The solid and dotted trapezoids correspond to two distinct interferometers. (b) Interference patterns in the fluorescence from beam I as a function of laser detuning. Traces a, c, e: no rotation, $\Omega = 0$, b: $\Omega = -0.09\ \mathrm{s}^{-1}$, and d: $\Omega = 0.09\ \mathrm{s}^{-1}$ (with kind permission of F. Riehle).

consequent need to build an ultrastable laser), as is available in calcium and magnesium, and four laser beams of equal and precisely controlled intensity.

Researchers in Braunschweig and Bonn, Germany, have employed such a Ramsey interferometer to measure a rotation of the beam apparatus [32] and phase shifts due to the different interaction of ground and excited state with static or optical fields (static and AC Stark shifts) [33, 34]. This is facilitated by the fact that the two arms of the interferometer are in different internal states between laser beam 1 and 2 or 3 and 4.

### 17.4.4 Light-Pulse Interferometer with Raman Transitions

A very similar idea to the one just described is pursued in an interferometer using stimulated Raman transitions to coherently split and recombine an atomic beam [26]. This technique combines the advantages of a long lifetime of a magnetic sublevel and the large photon recoil associated with an optical transition. Instead of having four stationary beams, a group at Stanford University has developed a technique in which a beam of sodium atoms, launched from an atomic trap, is coherently split and recombined with three pulses of two counterpropagating, circularly polarized laser fields with frequencies $\omega_1$ and $\omega_2$ and $\vec{k}$-vectors $\vec{k}_1 = -\vec{k}_2 = \vec{k}$, respectively. The two frequencies are chosen such that the difference frequency is close to the hyperfine splitting of the ground state: $\omega_1 - \omega_2 \approx \omega_{hfs}$ (see Figure 6a) and the atom's ground levels are coherently coupled. By choosing the pulse duration such that the ground-state

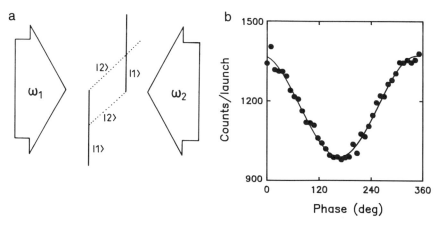

FIG. 6. Light-pulse interferometer. (a) Principle of the interferometer using stimulated Raman transitions between levels $|1\rangle$ and $|2\rangle$ and a pulse sequence $\pi/2–\pi–\pi/2$ of the two counterpropagating laser beams with frequencies $\omega_1$ and $\omega_2$. (b) Innterference fringe in the number of atoms in state $|2\rangle$ leaving the interferometer as a function of the phase of the final $\pi/2$-pulse for $T = 50$ ms (with kind permission of M. Kasevich and S. Chu).

coherence rotates by first an angle $\pi/2$, then $\pi$, and finally $\pi/2$ again, the atom originally in state $|1, \vec{p} = \vec{p}_0\rangle$ is first transferred to a linear combination of $|1, \vec{p}_0\rangle$ and $|2, \vec{p}_0 + 2\hbar\vec{k}\rangle$, then redirected in the second pulse and recombined in the last pulse, so that the transition amplitude to state $|2, \vec{p}_0 + 2\hbar k\rangle$ for an atom initially in state $|1, \vec{p}_0\rangle$ is given by

$$a_{2, \vec{p}_0 + 2\hbar\vec{k}} = \tfrac{1}{2}[e^{-i\Delta\phi} + 1], \tag{17.10}$$

where the phase $\Delta\phi$ is zero when $\omega_1$ and $\omega_2$ are time-independent. In the case of free-falling atoms in a gravitational field (acceleration $\vec{g}$), however, the frequencies have to be swept in time, giving rise to a phase shift $\Delta\phi = \vec{k} \cdot \vec{g}T^2$, with $T$ the time between two pulses. A measurement of the number of atoms detected in state 2 by two-photon resonant photoionization gives directly $\Delta\phi$ via Eq. (17.10). One interference fringe in the number of atoms was observed as a function of the phase of the final $\pi/2$-pulse and is shown in Figure 6b. The fringe visibility makes a measurement of the earth's gravitational acceleration $g$ with a resolution of $3 \times 10^{-8} g$ possible, approaching the resolution of "falling cube" gravimeters [35].

This interferometer can be operated in two configurations, with the two laser beams directed either collinearly with or perpendicularly to the atomic beam. The combination of Raman transitions with Ramsey spectroscopy has been successfully used to measure the ratio of $\hbar$ to the mass of cesium [36], which could be useful to determine the fine-structure constant $\alpha$.

When operated with laser-cooled atoms, the two beams in the interferometer can be separated by almost 1 cm, a macroscopic distance. This type of interferometer seems to use to be one of the most promising, since it provides a large separation and can be operated in either a longitudinal or transverse configuration. The experimental realization, however, is very difficult because it requires, e.g., a magneto-optical atomic trap, a highly stable RF oscillator, and a sophisticated detection method.

### 17.4.5 Longitudinal Stern–Gerlach and Stark Interferometers

The idea of using the Stern–Gerlach effect to coherently split and recombine an atomic beam (see Section 17.3) has been around for a long time, but it was pointed out that an experiment would be very difficult if the atomic beam was split in space.

To bypass these difficulties, a group in Villetaneuse, France, split an atomic beam longitunally. This is easier to accomplish, with the drawback that one has no direct access to the two beams [17]. The setup used by the French group is shown in Figure 7a. A partially polarized beam of metastable hydrogen atoms in the state $2s_{1/2}$, $F = 1(\lambda_{dB} = 0.4 \text{ Å})$ is prepared in a linear superposition of magnetic sublevels by a non-adiabatic passage through a magnetic field perpendicular to the atomic beam with a gradient along the beam. The different

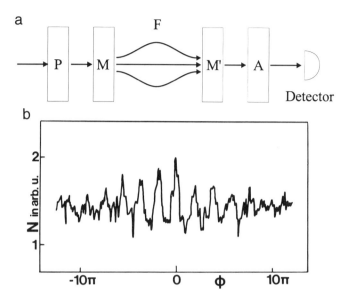

FIG. 7. Longitudinal Stern–Gerlach interferometer. (a) Experimental setup. Hydrogen atoms in state $2s_{1/2}$ are magnetically polarized in P and the different magnetic sublevels mixed in region M. The three sublevels $m_F = 0$, $\pm 1$ experience different potential energies in region F before being remixed in M′ and analyzed in A. (b) Observed interference pattern in the intensity of Lyman-$\alpha$ photons behind the analyzer as a function of the calculated phase shift $\Phi$ in region F caused by changing the current through the magnet coil (with kind permission of J. Baudon).

magnetic sublevels enter a constant magnetic-field region F and after 10 cm are mixed or recombined again in a region identical to the one used to coherently split the beam. Finally, an analyzing magnetic field selects a particular magnetic polarization, whose intensity is then measured by detecting Lyman-$\alpha$ photons emitted in the decay of the $2p_{1/2}$ state to the ground state (the Lamb–Retherford method [37]). A typical interference pattern is shown in Figure 7b as a function of the phase shift accumulated in region F.

Interference fringes are obtained in the beam intensity by changing the magnetic field strength $B_F$ and arise from the different potentials experienced by the magnetic sublevels in region F. The longitudinal Stern–Gerlach interferometer has been applied to demonstrate the effect of topological phases on the atomic wavefunction for a non-adiabatic cyclic evolution [38].

This interferometric device, based on the longitudinal splitting of the atomic wave packet in a static magnetic field, has been widely recognized as an interferometer, whereas a very similar setup based on the longitudinal Stark splitting in an electric field had already been demonstrated 15 years ago in the former Soviet Union without being accepted as an interferometer [39]. For

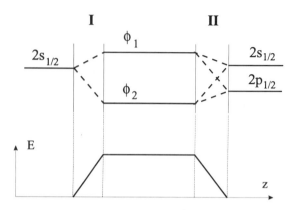

FIG. 8. Experimental setup of the longitudinal Stark interferometer. A beam of hydrogen atoms in state $2s_{1/2}$ is both split in time and recombined by electric field gradients along the beam direction. The potential energy difference of the eigenstates $\phi_1$ and $\phi_2$ in the intermediate region leads to an interference pattern in the number of atoms leaving the interferometer in state $2p_{1/2}$ [39].

completeness, let us briefly outline the principle of operation of this experiment (see Figure 8). A very fast beam of hydrogen atoms in the $2s_{1/2}$ state is produced by charge exchange of 20 keV protons emerging from an ion source. Coherent superpositions of $2s_{1/2}$ and $2p_{1/2}$ states are produced in the mixing region I by nonadiabatic passage through an electrode, creating a strong gradient in the electric field. In the subsequent region of constant field $E$, the atom is in a linear superposition of the two eigenstates $\phi_1$ and $\phi_2$ which depend on $E$. These states are mixed again in region II, and the intensity of $2p_{1/2}$ states in the beam is detected through the emission of Lyman-$\alpha$ photons. This setup allowed the determination of the Lamb shift and the hyperfine splitting in hydrogen.

## 17.5 Proposals for Atom Interferometers

Because there is no fundamental strategy to develop new types of inter-ferometers with improved performance, the selection of proposals presented in this chapter is, of course, very subjective and far from complete. It should mainly give an idea of other concepts of interferometers that are actively investigated in various laboratories around the world.

The double-slit setup presented in Section 17.4.1 is conceptually easy and works with few elements, but suffers from severe intensity problems when the beam separation is increased. A technique often employed in electron inter-ferometry [40] could overcome this problem. Instead of using only the diffraction from the slit apertures to recombine the two beams, a phase shift is introduced at the double slit which redirects the phase fonts much more efficiently. This technique is called biprism and is based on a constant phase gradient across both paths of the interferometer. A constant phase shift is equivalent to a classical force on the atom, since an atomic plane wave is bent by a fixed angle. Via Eq. (17.4), the appropriate phase shift is obtained by, e.g., applying an optical field perpendicular to the atomic motion, which creates a potential $U(x) \propto |x|/x_0$, with $x$ the distance from the center of the double slit and $x_0$ a distance matched to the beam separation. This potential can be obtained in a large-period standing wave with a node between the two slits (see Figure 9). With this setup, the slit structures can be made much wider or even be omitted, increasing the intensity considerably.

Another proposal relies on the optical Lloyd's mirror geometry (see Figure 10), a device based on the division of wavefronts. Atoms emerging from a small aperture can arrive at the detector plane either on a direct trajectory or via reflection off an evanescent wave, thus creating an atomic interference pattern in the detector plane. An optical evanescent wave is an electromagnetic wave decaying rapidly with an exponential behavior and is produced by totally reflecting a laser beam at a glass–air interface. This gradient in light intensity acts as a repulsive potential for atoms if the light frequency is tuned above the atomic resonance transition ($\Delta > 0$; see Section 17.3) [41]. This setup could be used, for example, to probe the solid-state surface at the place of the evanescent field, because the minimum distance of the atoms to the surface is determined by the detuning and intensity of the light field. Since the interaction with the surface

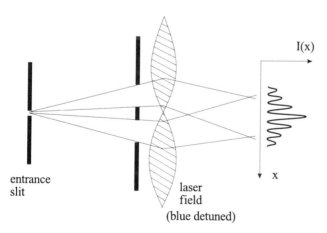

FIG. 9. Schematic setup of an interferometer similar to a biprism in electron interferometry. Atoms are coherently split by a small entrance slit and subsequently pass two wide slits. The two beams are redirected in a blue detuned standing laser field, whose nodes are adjusted such that the atoms experience a constant field gradient with opposite sign in both slits.

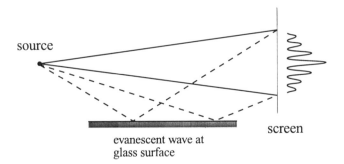

FIG. 10. Proposal for an interferometer based on Lloyd's mirror geometry in classical optics. Atoms emerging from a small source can reach the detector either on a direct path or via reflection off an evanescent wave. This light field is created by total reflection at a glass–vacuum interface.

may destroy coherences between the direct and reflected beams, the disappearance of the interference fringes with decreased light intensity leads to information about the surface.

Large beam separations are the main goal of further developments in atom interferometers. The nicest way to build a reliable and stable interferometer with large beam separations would be a setup analogous to the perfect silicon crystal interferometer with neutrons: a crystal interferometer using the diffraction from cleaved surfaces. One problem is the fact that atoms do not penetrate solid materials, which means that such a beam splitter has to be operated in reflection and not in transmission, as is the case with neutrons. The following setup has been proposed by Toennies [42] (see Figure 11); if once successfully demonstrated it would revolutionize atom interferometry. The interferometer is constructed out of a perfect crystal of, e.g., lithium fluoride and has three planes standing out of a ground plate. An incoming beam of helium atoms would be diffracted at plane 2 and the first diffraction orders redirected at plane 1. Finally, after two further diffractions, the two beams would be recombined and would leave the crystal through a hole in plane 3. With that, beam separations in the $10^{-2}$ m range with an interferometer only a few centimeters long could become possible. The main problem is the requirement that all surfaces have to be flat within an atomic wavelength across the atomic beam, in order not to lose coherence. Additionally, these surfaces have to be kept clean over a sufficiently long period of time.

## 17.6 Summary and Outlook

Atom interferometry is a very young field in physics and has evolved rapidly since the first demonstrations in 1990. With many different kinds of inter-

## Perfect LiF Crystal

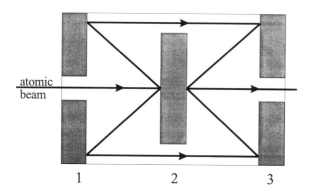

atomic
beam

1                    2                    3

Fig. 11. Proposal for a perfect crystal interferometer for atoms. Helium atoms in the ground state enter the interferometer through a hole in plane 1 and are diffracted on plane 2 of a perfect LiF crystal. Via four diffractions on lanes 2, then 1, 3, and 2 again, the atoms are split and recombined, leading to interference in the output beam [42].

ferometers now available, the first experiments making use of this novel tool have lately been performed with applications in metrology and in atomic physics.

In future, the research will focus both on the development of new types of interferometers with larger beam separations and special geometries to allow investigations of, e.g., surface or atomic properties, and on the application of atom interferometers to test the principles of quantum mechanics or general relativity, which requires an extreme resolution and accuracy. Parallel to that, intensities and velocity distributions of atomic sources have to be improved by laser cooling or related methods to minimize statistical uncertainties in the phase measurement.

When looking back at the developments in neutron interferometry, one hopes that finally one type of interferometer will be as compact, robust, and reliable as the perfect crystal interferometer with neutrons, and that it will be the size of a small suitcase. This would enormously boost the field of atom interferometry.

## References

1. Badurek, G., Rauch, H., and Zeilinger, A., eds. (1988). Special issue of *Physica B* (Vol. 151), *Matter Wave Interferometry*, Proc. of the Int. Workshop on Matter Wave Interferometry.
2. Levy, B. (1991). "Atoms Are the New Wave in Interferometers," *Physics Today*, July, p. 17; Mlynek, J., Balykin, and Meystre, P., eds. (1992). Special issue of *Appl. Phys. B* (Vol. 54, May 1992), *Optics and Interferometry with Atoms*; Sigel, M., Adams,

C. S., and Mlynek, J. (1992). *Atom Optics*. Int. School of Physics, Varenna; Pritchard, D. E. (1992). In *Atom Interferometers, Proc. of the 13th Int. Conference on Atomic Physics, Munich, Germany*, T. W. Hänsch and H. Walter, eds.

3. Aharonov, Y., and Casher, A. (1984). *Phys. Rev. Lett.* **53**, 319.
4. See, e.g., Boyer, T. H. (1970). *Ann. of Physics* **56**, 474.
5. Sleator, T., Carnal, O., Pfau, T., Faulstich, A., Takuma, H., and Mlynek, J. (1991). "Atom Interferometry with Mechanical Structures," in *Proc. of the 10th Int. Conference on Laser Spectroscopy, Font-Romeu, France*, M. Ducloy *et al.* (eds.).
6. Clauser, J. F. (1988). *Physica B* **151**, 262.
7. See, e.g., Sears, V. (1989). *Neutron Optics*. Oxford University Press, Oxford.
8. See, e.g., Chu, S., and Wieman, C., eds. (1989). Special issue of *J. Opt. Soc. Am. B* (Vol. 6), *Laser Cooling and Trapping of Atoms*; Hulet, R. G., in this volume.
9. Faulstich, A., Schnetz, A., Sigel, M., Sleator, T., Carnal, O., Balykin, V., Takuma, H., and Mlynek, J. (1992). *Europhys. Lett.* **17**, 393.
10. Scully, M. O., and Dowling, J. P. *Phys. Rev. A*, in press.
11. Estermann, I., and Stern, O. (1930). *Zeits. f. Physik* **61**, 95.
12. Ekstrom, C. R., Keith, D. W., and Pritchard, D. E. (1992). *Appl. Phys. B* **54**, 369.
13. Keith, D. W., Schattenburg, M. L., Smith, H. I., and Pritchard, D. E. (1988). *Phys. Rev. Lett.* **61**, 1580.
14. Carnal, O., Faulstich, A., and Mlynek, J. (1991). *Appl. Phys. B* **53**, 88.
15. Gerlach, W., and Stern, O. (1922). *Z. Phys.* **8**, 110.
16. Schwinger, J., Scully, M. O., and Englert, B.-G. (1988). *Z. Phys. D* **10**, 135.
17. Robert, J., Miniatura, Ch., Le Boiteux, S., Reinhardt, J., Bocvarski, V., and Baudon, J. (1991). *Europhys. Lett.* **16**, 29.
18. Dalibard, J., and Cohen-Tannoudji, C. (1985). *J. Opt. Soc. Am. B* **2**, 1707.
19. Moskowitz, P. E., Gould, P. L., Atlas, S. R., and Pritchard, D. E. (1983). *Phys. Rev. Lett.* **51**, 370.
20. Gould, P. L., Ruff, G. A., and Pritchard, D. E. (1986). *Phys. Rev. Lett.* **56**, 827.
21. Martin, P. J., Oldaker, B. G., Miklich, A. H., and Pritchard, D. E. (1988). *Phys. Rev. Lett.* **60**, 515.
22. Sleator, T., Pfau, T., Balykin, V., Carnal, O., and Mlynek, J. (1992). *Phys. Rev. Lett.* **68**, 1996.
23. Pfau, T., Kurtsiefer, C., Adams, C. S., Sigel, M., and Mlynek, J. (1993). *Phys. Rev. Lett.* **71**, 3427.
24. Aspect, A., Arimondo, E., Kaiser, R., Vansteenkiste, N., and Cohen-Tannoudji, C. (1988). *Phys. Rev. Lett.* **61**, 826.
25. See, e.g., Allen, L., and Eberly, J. H. (1975). *Optical Resonance and Two-Level Atoms*. Wiley, New York.
26. Kasevich, M., and Chu, S. (1991). *Phys. Rev. Lett.* **67**, 181.
27. Carnal, O., and Mlynek, J. (1991). *Phys. Rev. Lett.* **66**, 2689.
28. Shimizu, F., Shimizu, K., and Takuma, H. (1992). *Jpn. J. Appl. Phys.* **31**, 46, and *Phys. Rev. A* **46**, R17.
29. Keith, D., Ekstrom, C. R., Turchette, Q. A., and Pritchard, D. E. (1991). *Phys. Rev. Lett.* **66**, 2693.
30. Pritchard, D. E. Private communication.
31. Bordé, C. J. (1989). *Phys. Lett. A* **140**, 10.
32. Riehle, F., Kisters, T., Witte, A., Helmcke, J., and Bordé, C. J. (1991). *Phys. Rev. Lett.* **67**, 177.
33. Sterr, U., Sengstock, K., Müller, J. H., Bettermann, D., and Ertmer, W. (1992). *Appl. Phys. B* **54**, 341.

34. Rieger, V., Sengstock, K., Sterr, U., Möller, J. H., and Ertmer, W. (1993). *Opt. Comm.* **99,** 172.
35. Faller, J. E., and Marson, I. (1988). *Metrologia* **25,** 49.
36. Weiss, D. S., Young, B. C., and Chu, S. (1993). *Phys. Rev. Lett.* **70,** 2706.
37. Lamb, W. E., Jr., and Retherford, R. C. (1952). *Phys. Rev.* **86,** 1014.
38. Miniatura, Ch., Robert, J., Gorceix, O., Lorent, V., Le Boiteux, S., Reinhardt, J., and Baudon, J. (1992). *Phys. Rev. Lett.* **69,** 261.
39. Sokolov, Yu. L., and Yakolvlev, V. P. (1982). *Zh. Eksp. Teor. Fiz.* **83,** 15; *Sov. Phys. JETP* **56,** 7.
40. Möllenstedt, G., and Düker, H. (1956). *Zeits. f. Phys.* **145,** 377.
41. Cook, R. J., and Hill, R. K. (1982). *Opt. Comm.* **43,** 258.
42. Toennies, J. P., Max-Planck-Institut für Strömungsforschung, Göttingen, Germany. Private communication.

# 18. CLASSICAL SPECTROSCOPY

## J. Bland-Hawthorn[1]

Department of Space Physics and Astronomy
Rice University, Houston, Texas

## G. Cecil

Department of Physics and Astronomy
University of North Carolina, Chapel Hill, North Carolina

## 18.1 Introduction

This chapter is a basic review of the fundamental principles of modern spectroscopy, designed to provide a research student access to a widespread field. We begin by describing the basic properties of spectrometers and emphasize the relative merits of certain techniques. Attention is given to recent developments in monochromators, in particular, narrowband and tunable filters. Extensive historical reviews are to be found elsewhere [1, 2]. While spectroscopic techniques continue to evolve, they rely on either dispersion (refractive prisms) or multibeam interference (diffractive gratings, interferometers). Prisms are of great historical importance [3–5], but their performance is now surpassed by transmissive and reflective gratings [6–8]. At the close of the 19th century, three important developments occurred. The curved grating removed the need for auxiliary optics, thereby extending spectroscopy into the far ultraviolet and infrared wavelengths [7]. This was soon followed by the development of the Fourier transform [9] and Fabry–Perot [10] interferometers that have many uses today. Good general discussions can be found in references [11–13]. More specialized treatments are as follows: prisms [11, 14, 15], gratings [2, 16], Fabry–Perot [17, 18] and Fourier transform interferometers [19, 20].

## 18.2 Basic Principles

The most useful figure of merit of a spectrometer is the product of the resolving power ($\mathcal{R}$) and the throughput ($\mathcal{T}$). The *throughput* is defined as

[1]Current address: Anglo-Australian Observatory, P.O. Box 296, Epping, NSW 2121.

EXPERIMENTAL METHODS IN THE PHYSICAL SCIENCES
Vol. 29B

$\mathcal{T} = A \cdot \Omega$, where $A$ is the normal area of the beam and $\Omega$ is the solid angle subtended by the source. The *resolving power* is defined as $\mathcal{R} = \lambda/\delta\lambda$ where $\delta\lambda$ is known as the *spectral purity*, or the smallest measurable wavelength difference at a given wavelength $\lambda$. In a properly matched optical system, the throughput. or equivalently, the flux through the spectrometer, depends ultimately on the entrance aperture and the area of the dispersive element. The theoretical limit to the resolving power is set by the characteristic dimension of the spectrometer (e.g., prism base). In practice, it is often advantageous to accept a lower value of $\mathcal{R}$, for example, by widening the entrance slit to allow more light to enter the spectrometer. Indeed, if an observed spectral line is not diffraction-limited, the flux through the spectrometer is inversely related to the resolving power of the system. Thus, it makes sense to compare the relative merits of spectrometers at the same effective resolving power. If we match the area of the dispersing element for each technique—prism, grating, Fabry–Perot etalon, Fourier transform beam-splitter—we find that the acceptance solid angles of the latter techniques have a major throughput (Jacquinot) advantage over the others.

### 18.2.1 Throughput Advantage

Jacquinot [21] demonstrated the relative merit of prisms, gratings and Fabry–Perot etalons on the basis of throughput. The simplest spectrometer comprises a *collimator* optic to equalize the optical path lengths of each ray between the entrance aperture and the *disperser*, and a *camera* optic to reverse the action of the collimator by imaging the dispersed light onto the *detector*. If the solid angle subtended by the object, or *source*, at the distance of the collimator is $d\Omega$, and the projected area of the collimator is $dA$, the radiant flux falling on the collimator from a source of brightness $B$ is given by $dF = B \cdot dA \cdot d\Omega$. The quantity $dA \cdot d\Omega = dF/B$ is simply the throughput or the *étendue* of the system. In a properly matched (lossless) optical system, the étendue is a constant everywhere along the optical path, in which case the brightness of the source is equal to the brightness of the image. This is easily seen for a simple lens with focal length $f$. A small element $dA$ at the focal (or object) distance subtending a solid angle $d\Omega$ is brought to a focus at the image plane distance $f'$ with area $dA'$ subtending an angle $d\Omega'$. The linear magnification, both horizontally and vertically, is $f'/f$. Therefore $dA'/dA = (f'/f)^2$. However, this is compensated exactly by the change in $d\Omega'/d\Omega$ such that $dA \cdot d\Omega = dA' \cdot d\Omega'$. Note that the $f$/ratio of the optic sets the solid angle of the system. Therefore, it is important to match all optical elements in series such that the output of one element fills the aperture of the next. On the basis of throughput alone, the Fabry–Perot is superior to grating instruments, which in turn have superiority over prisms. However, in practice, Fabry–Perots are normally used for high-resolution observations over a narrow wavelength interval. A better comparison is with the Fourier transform spectrometer (FTS) which, for the same bandwidth, has a much higher throughput than all slit-aperture spectrometers in the same configuration.

## 18.2.2 Resolving Power

The resolution of a spectrometer is set by the bandwidth limit imposed by the dispersing element. An elegant demonstration using Fourier optics is quoted by Gray [16]. When a beam of light passes through an aperture of diameter $L$, in the far field approximation, a Fraunhofer diffraction pattern arises.[2] The width of the central intensity spike is proportional to $\lambda/L$, which is roughly $\delta\lambda$ for most resolution (e.g., Rayleigh, Sparrow) criteria. Because the dispersing element defines a finite baseline or aperture, the ultimate instrumental resolution is set by the diffraction limit. This has the simple consequence that the highest spectroscopic resolutions have generally been obtained with large spectrometers.

The theoretical value of $\Re$ is rarely achieved in practice, not least because of optical and mechanical defects within the instrument. The width of the *instrumental profile*, i.e., the response of the spectrometer to a monochromatic input, must be matched carefully to the size of each detector element (or *pixel*) and the sampling interval defined by the angular dispersion. Gray [16] shows that an observed spectrum arises from the product of three functions, $S_\lambda = B_\lambda \cdot III_\lambda \cdot O_\lambda$, where $O_\lambda$ is the original spectrum, $B_\lambda$ is a rectangle function that defines the baseline of the dispersing element, and $III$, is a Shah function that describes the regular sampling. Because the Fourier transform of the observed spectrum is the convolution of the transform of the individual functions, discrete sampling causes the transform of the original spectrum to replicate with a periodicity $1/\delta x$, where $\delta x$ is the sampling interval. If the original spectrum is undersampled by the spectrometer, the adjacent orders of the transformed spectra overlap and cannot be disentangled uniquely. This problem of *aliasing* can be avoided by *Nyquist sampling*, i.e., sampling the source spectrum at twice the frequency of the highest Fourier component that you wish to study.

## 18.2.3 Detector Constraints

Most contemporary spectrometers employ panoramic electronic detectors because they are highly sensitive (more than 80% of the incident photons can be detected in many cases), provide digital output, have a linear response over a dynamic range of $10^5$, and record photons across a two-dimensional field. Several corporations now routinely manufacture low-noise, large-area charge-coupled devices (CCDs): in the optical, $2048^2$ 15 μm and $4096^2$ 7.5 μm pixel arrays are now available; in the infrared, arrays of $1024^2$ pixels have been fabricated. The photosensitive area is currently limited to the 10 cm diameter of

---

[2]The radial diffraction pattern is a $sinc^2$ function, where $sinc\ y = \sin y/y$. Recall that the sinc function is the Fourier transform of the rectangle ("top hat") function.

a silicon wafer, but suitably designed CCDs can be edge-butted to form large mosaics. A full description of the limitations and capabilities of modern CCDs can be found in reference [22]. At optical wavelengths, the largest detector formats are 35 cm photographic plates. The Kodak TechPan emulsions have a quantum efficiency close to 5% and an effective resolution element of roughly 5 μm.

To minimize the thermal and collisional excitation of electrons (which contribute a "dark current"), arrays must be cooled to below 30 K for InSb infrared arrays (which therefore require inconvenient cryogens) or to below 210 K for modern optical CCDs (obtained using thermoelectric coolers). Both require water-free atmospheric chambers to prevent frost, which restricts access to the focal plane. Uncorrelated noise sources combine in quadrature, so the effective signal-to-noise ratio counted at each pixel is

$$S/N \approx \frac{s_v \Delta_v t}{\sqrt{s_v \Delta_v t + 2([b_v \omega \Delta_v + d]t + \sigma_R^2)}}, \qquad (18.1)$$

for which $\sigma_R$ is the read noise (generated in the output amplifier, in rms electrons); $s_v \Delta_v$, $b_v \Delta_v$, and $d$ are the number of electrons per second generated by the object, background, and dark current, respectively, in a frequency interval $\Delta_v$; $t$ is the exposure time (seconds); and $\omega$ is the solid angle subtended by each pixel. The factor of two in the denominator assumes that the corresponding noise sources are measured on separate exposures, then subtracted from the data frame. Read noise on modern optical CCDs is a few electrons rms; infrared arrays are at least 10 times noisier, but their performance can be improved with nondestructive, multiple read-out. In practice, the $S/N$ ratio will be smaller than predicted by Eq. (18.1) because multiplicative ("flat field") and additive ("bias offset") gains must be determined empirically and applied to each detector element [23]. Except at very low and high flux levels, the CCD is a linear detector so these calibration steps are often quite successful.

## 18.2.4 Multiplex Advantage

In recent decades, the sense of what constitutes a multiplex (Fellgett) advantage has evolved. The traditional meaning arises from single-element or row-element detectors, which used to prevail at infrared wavelengths. With a single-element detector, a two-dimensional image (either spatial–spatial or spatial–spectral) was made by scanning at many positions over a regular grid. The same image is more easily obtained with a one-dimensional detector array after aligning one axis of the image with the detector and then shifting the detector in discrete stages along the other axis. For a single-element detector, Fellgett [24] realized that there is an important advantage to be gained by recording more than one spectral increment (channel) simultaneously if the signal detection is limited by detector (background) noise. If the receiver observes in sequence $n$ spectral channels dispersed by a prism or grating for a total exposure time of $\tau$, the $S/N$ ratio within each channel is proportional to $\sqrt{\tau/n}$. Thus, a Fourier transform device (see Section 18.6), which observes all $n$ spectral channels for the entire duration, has a *spectral* multiplex advantage of $\sqrt{n}$ compared with conventional slit spectrometers. Thus, a multiplex advantage makes more efficient use of the available light.

In principle, all spectroscopic techniques can achieve a multiplex advantage with the use of a multiplexed or coded *aperture mask*. Spectrometers with cylindrical symmetry (e.g., Fabry–Perots) use circular masks, while slit aperture devices (e.g., gratings) use rectangular masks. Figure 1 illustrates the special case of Hadamard coded apertures. A cyclic mask $H_{ij}$ with $n$ rows is placed at the entrance slit; a particular column can be aligned with the slit by sliding the mask. In this way, the dispersing element can be illuminated through each of the mask columns in turn. A one-dimensional detector array is aligned with the spectral dispersion so that each spectral channel receives all of the modulated signal through the slit at a discrete frequency. For each spectral channel, the modulated signal $M_i = H_{ij}O_j$ can be used to derive the original signal $O_j$ at each position along the slit. This assumes that the (square) Hadamard array $H_{ij}$ can be inverted and that the influence of systematic errors can be minimized. The multiplex advantage is roughly $\sqrt{n}/2$ if the mask columns have approximately half the holes open.

With the advent of large-format, low-noise detectors in many wavebands, the conventional definition of the multiplex advantage is only significant for the highest spectral resolutions. It has become difficult to generalize under what conditions a multiplex advantage might prevail [11]. For certain applications, it can be inefficient to match the acceptance solid angle of the spectrometer to the

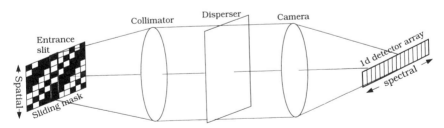

FIG. 1. A rudimentary Hadamard spectrometer and one-dimensional detector array. The only light to pass through the vertical entrance slit is through the mask holes aligned with the slit. After sliding each column over the slit, it is possible to reconstruct the spectrum at each spatial position along the slit even though the detector does not extend in this direction.

imaging detector. As an example, optical fibers in astronomy—after major improvements in their blue to near-infrared response—have afforded a significant *spatial* multiplex advantage in recent years [25, 26]. The light from discrete sources over a sparse field can be collected with individual fibers, which are then aligned along a slit. The multiplex advantage, when compared with conventional slit spectrometers, is simply the number of fibers if the signal attenuation along the fiber length is neglible.

### 18.2.5 Optics Design

All spectrometers make use of the constructive interference of light by dividing the wavefront (e.g., gratings), by dividing the amplitude of the wavefront (e.g., Fabry–Perots), or by decomposing the wavefront into orthogonal polarization components (e.g. Lyots). The design of a spectrometer is in essence a competition between a dispersing element that deviates light into different angles according to wavelength, and optics that focus the light at the detector with minimum aberration. Two lenses in the optical path are normally sufficient to counteract the dominant aberrations [12]. In most cases, the dispersing element must be illuminated with parallel (collimated) light. Hence, the acceptance angle (i.e., $f$/ratio) of the collimating optic must be the same as that of the primary concentrating optics to transfer all the light. If the collimated beam is too small, the Jacquinot advantage of the spectrometer is reduced [27, 28]. If the collimated beam is too large, light is lost from the optical system.

It is important to realize that all optical systems lose light at surfaces along the optical path. On occasion, the scattered light will simply leave the system. More often, the stray light finds its way back into the optical path to be imaged at the detector as a spurious "ghost" signal that can be difficult to distinguish from real signals. Scattered light can also dramatically increase the background signal at the detector, thereby reducing contrast and setting a limit on the sensitivity that can be reached within a given exposure time. The manner in which this happens is specific to the instrument. In later sections, we describe a few of the ghost families that occur within gratings and Fabry–Perot spectrometers. Antireflective coatings (including newly developed coatings whose index of refraction increases smoothly through their thickness), aperture stops, baffles, and ingenious optical designs are part of the arsenal to combat these anomalies.

In later sections, we describe diffraction gratings (18.4), interference filters (18.5.1), Fabry–Perot (18.5.2) and Fourier Transform spectrometers (18.6). We also discuss recent technological developments for selecting a bandpass whose central wavelength can be tuned over a wide spectral range. The most promising of these developments are those which utilise anisotropic media, particularly the Lyot filter (18.5.3) and the acoustooptic filter (18.5.4).

## 18.3 Prisms

Refractive prisms are no longer in common use as primary dispersers in slit spectrometers, although they are frequently used as cross dispersers in high-order spectroscopy and they play an important role in immersion gratings. However, prisms highlight some of the basic principles discussed in the previous section. Their operation is fully specified by Snell's law of refraction and the dispersive properties of the medium. A light ray incident on the face of a prism with refractive index $n_\lambda$ deviates by an angle $\delta = i - r$, where $i$ is the angle to the face normal, and $r$ is the refracted angle. If we take the external medium to be air, then $\sin i = n_\lambda \sin r$. For a restricted wavelength range, the Hartmann dispersion formula, $n_\lambda = a + b(\lambda - c)^{-1}$, provides a good approximation to many materials in the Schott glass catalog. The constants $a$, $b$, and $c$ depend on the material.

Figure 2a illustrates the angle convention for a ray passing through a prism. We note that $\delta_1 = i_1 - r_1$, $\delta_2 = i_2 - r_2$, and the prism apex angle, $\alpha = r_1 + r_2$, such that the deviation is given by $\theta = \delta_1 + \delta_2 = i_1 + i_2 - \alpha$, or equivalently

$$\theta(\alpha, \lambda, i_1) = i_1 - \alpha + \sin^{-1}(n_\lambda \sin[\alpha - \sin^{-1}(n_\lambda^{-1} \sin i_1)]). \quad (18.2)$$

We can approximate the angular dispersion $d\theta/d\lambda$ by finding the gradient between two discrete wavelengths $\lambda_0$ and $\lambda_1$. If we plot $(\theta(\alpha, \lambda_1, i_1) - \theta(\alpha, \lambda_0, i_1))/(\lambda_1 - \lambda_0))$ over a range of $i_1$ for a flint prism, say, we find that it increases with $\alpha$ to a theoretical maximum at which point $\alpha \approx 74°$ and $i_1 = 90°$ [14]. This corresponds to a ray at glancing incidence on the first face followed by a symmetric passage or. equivalently, minimum deviation through the prism. Alternatively, we can substitute $i_1$ with $r_1$ and differentiate to find

$$\frac{\partial}{\partial r_1} \theta(\alpha, \lambda, r_1) = \frac{n \cos r_1}{\sqrt{1 - n^2 \sin^2 r_1}} - \frac{n \cos(\alpha - r_1)}{\sqrt{1 - n \sin^2(\alpha - r_1)}}. \quad (18.3)$$

Thus, minimum deviation occurs when $r_1 = r_2 = \frac{1}{2}\alpha$, which corresponds to a symmetric passage through the prism. At minimum deviation, we let $i = i_1 = i_2$ and $r = r_1 = r_2$. After substituting $\theta = 2i - r$ and $\alpha = 2r$, one finds

$$\frac{d\theta}{d\lambda} = \frac{2 \sin \frac{1}{2}\alpha}{\sqrt{1 - n^2 \sin^2 \frac{1}{2}\alpha}} \frac{dn}{d\lambda}. \quad (18.4)$$

The maximum apex angle is not used in practice because most of the light is reflected at the first surface. Manufacturers cut prisms to apex angles (e.g., $30°$, $45°$, $60°$) that minimize wasted glass. The $60°$ prism is a good compromise for which the angular dispersion is approximately $n \, dn/d\lambda$. This function rises rapidly to large and small wavelength cut-offs [11, 21]. Thus, prisms disperse most efficiently at their absorption limits and therefore find use from ultraviolet to infrared wavelengths (100 nm–60 µm). A major disadvantage of prisms is the strongly nonlinear angular dispersion. If the collimated beam is matched to the

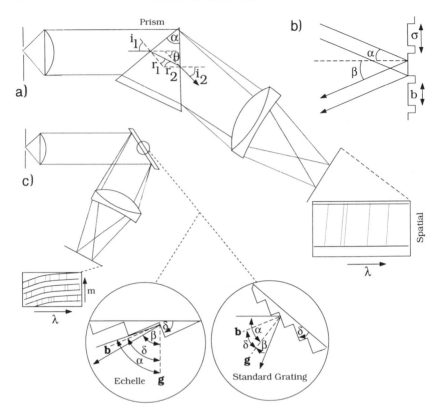

FIG. 2. (a) A rudimentary prism spectrometer. (b) Angle and groove definitions for a conventional, reflective diffraction grating. (c) A blazed reflection-grating spectrometer. The standard grating and echelle arrangements are shown as insets, where the dashed-line vectors **g** and **b** are the normals to the grating and blaze facets, respectively.

prism, it follows from Eq. (18.4) that $\mathcal{R} = B \, d\theta/d\lambda = L \, dn/d\lambda$, where $L$ is the length of the prism base and $B$ is the beam diameter. At grazing incidence, $B = 0$ and therefore $\mathcal{R} = 0$. Finally, for a large 10 cm glass prism, $dn/d\lambda \approx 10^{-4} \, \text{nm}^{-1}$, and $\mathcal{R} \approx 10^4$, which is roughly the practical limit of prism spectrometers.

## 18.4 Diffraction Gratings

If a plane wave of wavelength $\lambda$ is incident at an angle $\alpha$ to the perpendicular of a periodic grating with groove spacing $\sigma$, the outbound reflected beams at an angle $\beta$ interfere constructively according to the *grating equation*

$$\sigma(\sin \alpha + \sin \beta) = m\lambda, \tag{18.5}$$

where $m$ is the order of interference and the angle conventions are illustrated in Figure 2b. It is common practice to make $\alpha > \beta$ to minimize scattered light. The angular dispersion follows by differentiating this equation:

$$\frac{d\beta}{d\lambda} = \frac{m}{\sigma \cos \beta}. \tag{18.6}$$

In many applications $\beta > 30°$, which causes the angular dispersion to be slightly nonlinear, and therefore the wavelength scale at the detector plane must be calibrated from a reference spectrum. Standard ruled gratings have $\sigma = 1/600$ to $1/1200$ mm and are normally used in relatively low order ($<5$). The longest wavelength accessible with a grating is $2\sigma$, so gratings in the infrared tend to be coarsely ruled. For a grating length $L$, the maximum resolving power increases as $\mathcal{R} = Lm/\sigma$ and can exceed $10^5$. The spectral purity, $\delta\lambda$, depends on the collimator focal length, $f_{coll}$, such that

$$\delta\lambda = -\cos \alpha \frac{w}{f_{coll}} \frac{\sigma}{m}, \tag{18.7}$$

where $w$ is the width of a detector element.

Fourier optics demonstrates that the flux distribution in the focal plane is the Fourier power spectrum of the transmission (or reflection) function over the collimated beam. The grating acts as a filter that blocks all spatial frequencies except those associated with its groove frequency and the spatial frequency content of each groove [16]. The number of grooves and the maximum path-length difference is set by the grating length $L$. The rectangle function that defines each grating groove produces a broad diffraction envelope (called the *blaze function*) that modulates the flux of each spectral order, and whose width increases as the groove narrows (see Figure 3). If we ignore the entrance slit, the wide rectangle function that defines the grating produces a high-frequency diffraction envelope (or *cluster*) at each spectral order which is modulated by the broad diffraction envelope. The width of each cluster decreases as the overall length of the grating increases. The maximum intensity of the secondary peak in a cluster is a fraction $N^{-2}$ of the central peak, where $N (= L/\sigma)$ is the total number of grooves in the grating. Note that the discontinuities in optical path length produced by the light-blocking strips between the slits or the steps between the mirrored facets provide the high spatial frequencies that are essential for shaping the profile by interference.

## 18.4.1 Grating Fabrication

Plane reflective gratings are constructed by ruling with a diamond tool either an aluminum or a gold coating on a low-expansion glass substrate with a diamond saw. To reduce costs, many epoxy-resin replicas are made from each

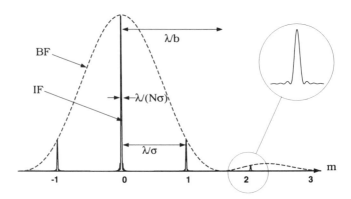

FIG. 3. Instrumental response of a diffraction grating with blaze angle $\delta = 0$. BF is the sinc$^2$ blaze function and IF is the interference response to a monochromatic source. When $\sigma = b$, only the $m = 0$ order passes flux because the other peaks coincide with the BF zeros.

glass master. Wear on the tool and changes in the operating environment during the grooving process limit the maximum ruled area to about $35 \times 45$ cm, but gratings can be mosaiced in much the same way as silicon chips (Section 18.2.3). It takes several weeks of cutting at 10 grooves min$^{-1}$ to fabricate a grating, during which no distance should drift by more than 20 nm. The actual resolution attained with a grating is smaller than calculated due to slow variations in the ruling density that can degrade coherence, broaden the wings of the instrumental profile, and produce spurious spectral-line and continuum features at the detector [29, 30]. Periodic deviations in the engraved facets from flaws in the ruling engine produce periodic spectral features near bright spectral lines (Rowland ghosts). When one or more periodic deviations are present, interference of the ghost diffraction patterns can generate spurious lines far from the original spectral feature (Lyman ghosts).

Holographic gratings do not suffer from these mechanical constraints and therefore can be fabricated with much higher groove densities (3500–6000 grooves mm$^{-1}$). They are formed by imaging the interference pattern of a laser-fed Fizeau, Michelson (see Section 18.6), or Twyman Green interferometer onto a glass plate that has been covered with photoresistive emulsion. The unexposed areas are etched away in an acid bath. and the resulting sinusoidal surface undulations form a grating. The profile of each facet must be squared to maintain high efficiency [31]. Another advantage of these gratings is that the astigmatism of an off-axis spectrograph can be compensated for by shaping the wavefronts of the interfering beams. However, all gratings produce at least two kinds of stray light [32]: a Lorentzian component predicted by diffraction theory, and a Rayleigh-scattered component due to microscopic surface defects.

## 18.4.2 Influence of the Entrance Slit

The entrance aperture serves a dual function in slit spectrometers. First, it restricts flux to a particular region of the source and ensures that the flux is dispersed onto a uniform and low-level background that is imaged by the detector. Second, the recorded spectral lines are inverted images of the slit width, convolved by the grating profile and further broadened by aberrations within the optics. The choice of a slit width is a compromise between reduction of the source intensity and the intrinsic resolution of the grating. To avoid light loss by diffraction, the width cannot be reduced below about five times the operating wavelength. In practice, it is hard to make an adjustable slit narrower than 10 μm that maintains parallelism.

The slit may actually be an aperture mask with many separate openings (slitlets) distributed across the focal plane, but arranged so that spectra do not overlap at the detector. Even more flexibility is possible by positioning many optical fibers at widely distributed points in the focal plane [25, 26]. Such "fiber feeds" have been attached to existing spectrometers to greatly increase their efficiency for certain projects. One end of the fibers either can be attached to precut holes in a custom mask or can be moved to arbitrary positions by a robot arm. The other ends can then be lined up along the spectrograph slit so that the spectra do not overlap whatever the fiber position. It is common practice to reserve some of the fibers for direct observations of the background signal. If the source does not occupy the entire length of a slitlet, aperture masks produce better background sampling. This is because each fiber has a slightly different response: A fiber that sees only the background does not exactly match an adjacent fiber that is illuminated by the combined source and background signal. Fibers are also used as "image scramblers" when it is important to minimize spectral artifacts that are introduced as a point source moves around in a large slit.

The collimator $f$/ratio is fixed to that of the primary collecting optics for full illumination, while its diameter must be that of the grating to maintain the grating resolution. The ratio of the camera and the collimator focal lengths, $f_{cam}/f_{coll}$, demagnifies the scale along the slit. The camera focal length is set to reduce the slit image width $W$ down to the width of two detector elements $2w$ to satisfy the sampling theorem (Section 18.2.2), where

$$w = -\frac{\cos\alpha \, f_{cam}}{\cos\beta \, f_{coll}} W. \tag{18.8}$$

The *anamorphic factor* $\cos\alpha/\cos\beta$ arises from the different beam sizes seen by the camera and the collimator. It is common for this factor to be less than unity to maximize spectral resolution, in which case the grating is more face-on to the camera than to the collimator. The camera diameter is also matched to the grating length, which now fixes the $f$/ratio. Speeds faster than $f/2$ usually require catadioptric (Schmidt) cameras, which have fairly inaccessible focal planes, although dioptric cameras are preferred if the detector obscures a significant fraction of the beam. Solid-glass Schmidt cameras are often used for fast systems because the higher index of refraction allows the $f$/ratio to increase by the same factor.

In many applications, the input scale is imposed by the source. This is particularly the case for large-aperture ($>3$ m diameter) astronomical telescopes. The main optics produce an image scale at the entrance slit of $206,265/(\rho D_{pr})$ (which is typically $\leq 10$ angular seconds of arc mm$^{-1}$), where $D_{pr}$ (in millimeters) and $\rho$ are the diameter of the primary optic and its $f$/number, respectively. At even the best sites, images are blurred by turbulence at 8–10 km altitude to angular diameters of $\gtrsim 0.6$ seconds of arc, which forces both slits and fibers to be $\gtrsim 100$ μm wide in order to pass most of the light. Thus, for high values of $\mathcal{R}$, a large, high-density grating is required to compensate for the wide slit. This in turn requires large-diameter optics and a long focal length for the collimator. In addition, because CCD pixels are rarely larger than 25 μm, the camera $f$/ratio must be about four times faster than the collimator to demagnify the image. Cassegrain beams commonly have $\rho = 7$, which implies very fast, complex, and expensive camera optics. New generation 8–10 m telescopes will use low-order adaptive optics to partially correct the distorted stellar wavefront for atmospheric blurring. This will reduce the core of the stellar image to $\approx 0.2$ seconds of arc in diameter, allowing a smaller slit, grating, and optics with minimal light loss. A recent development has been to segment the telescope mirror (pupil image) or the focal plane image with microlens arrays. These generate many sub-images which are brought to focus onto a fiber bundle prior to the slit allowing a much simpler camera design.

### 18.4.3 Transmission Grating

Transmission gratings are used mostly in conjunction with a prism ("grism") or a lens ("grens"), either in contact or air-spaced. Transmission gratings are commonly used as slitless systems which are only practical for spectroscopy of point sources on a weak background signal. Because no auxiliary optics are required, they can often be incorporated into an existing optical train to provide some wavelength selection while maintaining high throughput. The aberrations associated with a noncollimated beam (principally coma) can be minimized with suitable grating or detector tilts if the linear dispersion is comparatively low. Note that the spectra are dispersed across the field of view. In slitless systems, the spectra are superposed on the variable background signal, and will overlap if the field is crowded with sources.

## 18.4.4 Blazed Grating

As the intergroove separation $\sigma - b$ is decreased (see Figure 2b), more of the light is concentrated into the nondispersed $m = 0$ order (Figure 3). Time delays can be introduced across the grating to shift the peak of the diffraction envelope to an angle that corresponds to dispersive nonzero orders. Such *blazing* is easy to do on a reflective grating by grooving at an angle $\delta$ to its normal. Most high-efficiency gratings are reflective and have peak efficiencies approaching 80%. The change in effective width of the grooves broadens the grating diffraction peak asymmetrically, with a more abrupt decline in efficiency on the shorter-wavelength side. Once again, the blaze function has the form $\mathrm{sinc}^2 \gamma$, for which

$$\gamma = \frac{\pi\sigma\cos\delta}{\lambda} [\sin(\beta - \delta) + \sin(\alpha - \delta)]. \tag{18.9}$$

In terms of the quoted blaze wavelength $\lambda_0$ and spectral order $m$, $\gamma = m\pi(\lambda_0 - \lambda)/\lambda$, and $\gamma = 1$ when $\alpha = \beta = \delta$. The wavelength of the blaze peak is $\lambda_b = 2\sigma(\sin\alpha)/m$. The blaze curve drops to 40% of its peak value at the wavelengths $\lambda_\pm = m\lambda_0/(m \mp \frac{1}{2})$. Grating manufacturers quote the blaze wavelength for $m = 1$ and $\alpha = \beta$ (Littrow configuration). When the required exit angle $\beta_b \neq \alpha$, the blaze is shifted (typically by 10%) to the shorter wavelength

$$\lambda_b = \lambda_0 \cos \tfrac{1}{2}(\alpha - \beta_b). \tag{18.10}$$

At high order, the wavelength range spanned by the blaze curve is small. When there is a need for large wavelength coverage with moderate resolution ($\mathcal{R} \approx 10^{4-5}$), the gratings are operated in the *echelle* mode (Section 18.4.5). Off-the-shelf gratings are available with the blaze peak at one of several strong spectral lines. Because the blaze function is quite strongly peaked at low orders, the ability to manipulate the orientation of plane gratings accurately is an important design goal for an efficient spectrometer. This capability is nontrivial to provide within the sealed cryostat of an infrared spectrometer. In the infrared, it is also hard to eliminate unwanted higher orders that can extend into the visual.

Because the grating constitutes a series of stepped mirrors, the blaze efficiency depends strongly on the polarization state [16], and strong discontinuities with amplitudes $\gtrsim 20\%$ (called Wood's anomalies) are often present close to the edges of the blaze distribution $\lambda_\pm$. Modern astronomical telescopes usually mount spectrometers at the Nasmuth foci, which are fed by a 45° tertiary mirror. Reflection off this mirror induces or alters the polarization of the incident light, and the varying polarization angle after the further reflection off the grating greatly complicates spectrophotometric calibrations.

## 18.4.5 Echelle Grating

While echelles have a low density of grooves ($\approx 80$ mm$^{-1}$), they are operated at a large spectral order $m \approx 200$ and high tilt (hence large anamorphic demagnification) to maintain a large path difference. The adjacent orders overlap, so they are cross-dispersed by directing the light through a prism or secondary grating that is oriented perpendicular to the echelle. In this way the different spectral orders are distributed as curved arcs across a two-dimensional detector array. The main advantage of an echelle is that most of the detector area is used to record spectra (several orders $m$ simultaneously; see Figure 2c). The large tilt of an echelle means that a rectangular mosaic of gratings is necessary for full illumination by a large collimated beam. All of the separate glass substrates must have identically low coefficients of thermal expansion (e.g., Schott Zerodur or Corning ULE glass) to ensure a high $\mathcal{R}$ at the end of the several-week–long fabrication cycle of the gratings.

Spatial information along the entrance slit is limited by the small separation between spectral orders. During data reduction, the different orders are extracted, then overlapped, to yield one spectrum that can span the entire sensitivity range of the detector while maintaining $\mathcal{R} \approx 10^{4-5}$. The blaze peaks when $m\lambda = 2\sigma \sin \alpha$, which corresponds to many wavelengths with adjacent spectral orders because $m$ is large. The blaze curves of the different orders introduce undulations in the combined spectrum with large amplitude, which must be removed by recording the known continuous spectrum of a calibration source with the same setup. The variations in $S/N$ ratio along the spectrum can complicate the data analysis.

## 18.4.6 Curved Grating

Curved gratings are used to avoid focus degradation from differential chromatic dispersion across a large wavelength range, when one is operating in a wavelength region where lenses are ineffective and mirror reflectances are low, or when it is desirable to make the spectrometer as compact as possible. The grating surface is figured to act as a collimator, a camera, or both. The gratings are one-of-a-kind and are therefore expensive, with limited flexibility in the choice of camera focal length to alter the projected slit widths and spectral coverage. In addition, because the grating is an off-axis mirror, it introduces astigmatism unless an auxiliary mirror is introduced. The magnitude of the astigmatism increases with the length of the rulings, and the entrance slit must be aligned precisely with the ruling direction to avoid degrading the spectrograph focus. Further discussion of grating mounts and spectrograph aberrations can be found in references [11] and [12].

### 18.4.7 Immersion Grating

While transmission gratings (Section 18.4.3) have been largely superseded by other grating types (e.g. Section 18.4.4), prisms have found important uses in the context of immersion gratings [33]. Like the transmission grating, the ruling is placed on the downstream face of the prism, but now the collimated beam is reflected and refracted along a direction close to the original path. The principle here is that the resolving power of the grating is increased by the refractive index of the medium [34]. Immersion gratings are particularly useful at infrared wavelengths where high index materials are readily available. Anamorphic immersion gratings [35] increase the resolving power still further by utilizing highly wedged prisms to enhance the anamorphic factor of the collimated beam (Section 18.4.2).

## 18.5 Multiple-Beam Interferometers

### 18.5.1 Interference Filter

These monochromators allow a narrow spectral band-pass to be isolated. The principle relies on a dielectric spacer sandwiched between two transmitting layers (single cavity). The transmitting layers are commonly fused silica in the ultra-violet, glass or quartz in the optical, and water-free silica in the infrared. Between the spacer and the glass, surface coatings are deposited by evaporation which partly transmit and reflect an incident ray. Each internally reflected ray shares a fixed phase relationship to all the other internally reflected rays. For a wavelength $\lambda$ to be transmitted, it must satisfy the condition for constructive coherence such that, in the $m$th order,

$$m\lambda = 2\mu l \cos \theta_R, \qquad (18.11)$$

where $\theta_R$ is the refracted angle within the optical spacer. The optical gap $\mu l$ is the product of the thickness $l$ and refractive index $\mu$ of the spacer. An interference filter is normally manufactured at low order so that neighboring orders spanning very different wavelength ranges can also be used [36]. Additional cavities. while expensive, can be added to decrease the band-pass or to make the filter response more rectangular in shape. Either the glass material or an absorptive broadband coating is normally sufficient to block neighboring orders.

Filter manufacturers normally provide data sheets that describe operation at room temperature and in a collimated beam. If the filter is used in a converging beam, the band-pass broadens asymmetrically and the peak transmission shifts to shorter wavelengths. In the collimated beam, the peak transmission shifts to smaller wavelengths by an amount which depends on the off-axis angle. In either beam, the wavelength response of the interference filter can be shifted slightly (tuned) to shorter wavelengths with a small tilt of the filter to the optical axis. If

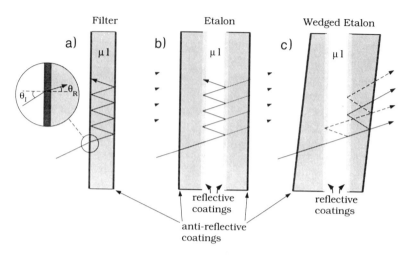

FIG. 4. (a) Interference filter: the internal structure is not shown. (b) Fabry–Perot etalon. (c) Wedged Fabry–Perot etalon.

$\lambda_I$ is the wavelength of a light ray incident at an angle $\theta_I$ (Figure 4a), then from Snell's Law and Eq. (18.11), it follows that

$$\left(\frac{\lambda_I}{\lambda_N}\right)^2 = \left[1 - \frac{\sin^2\theta_I}{\mu}\right],\tag{18.12}$$

for which $\lambda_N$ is the wavelength transmitted at normal incidence. To shift the band-pass to longer wavelengths, one must increase the filter temperature; typically, one can achieve $0.2\ \text{Å}\ \text{K}^{-1}$.

A more versatile approach in constructing narrowband filters is to use dielectric, multilayer thin-film coatings. A highly readable account of optical interference coatings is provided by Baumeister and Pincus [37]. One of the most successful of these is the quarter-wave stack, in which alternate layers of high and low refractive index media are used. Through judicious combinations of refractive index and layer thickness, it is possible to select almost any desired bandwidth, reflectance, and transmittance [38]. However, filters with band-passes narrower than $\lambda/100$ are difficult and costly to manufacture.

In the next section, we discuss scanning Fabry–Perot etalons which, for the purposes of this review, use air gap spacers and are routinely operated at low and high order. There exists another class of interference filters that is essentially a single-cavity Fabry–Perot with a solid dielectric spacer. These *etalon filters* employ a transparent piezoelectric spacer, e.g., lithium niobate, whose thickness and, to a lesser extent, refractive index can be modified by a voltage applied to both faces. Once again, tilt and temperature can be used to fine-tune the band-pass if it is important to keep the piezoelectric voltages modest. High-quality spacers with thicknesses less than a few hundred microns are difficult to manufacture, so etalon filters are normally operated at high orders of interference.

Any air-glass interface reflects about 4% of the incident light. This can be significantly reduced (1% or better) by the application of an antireflective coating. In its simplest form, this constitutes a single $\lambda/4$ layer of, say, $MgF_2$ whose refractive index is close to $\sqrt{\mu}$ (Figure 4a). A multilayer "V-coat" with alternating layers of $TiO_2$ and $MgF_2$ dielectric coatings can reach 0.25% reflectivity at specific wavelengths. Many advances in spectroscopic techniques in recent years have arisen from the refinement of multilayer coatings. The most recent developments have used rare earth oxides and very thin (5 Å) metallic layers. However, coating performance is currently limited by the availability of pure transparent dielectrics with high refractive indices.

### 18.5.2 Fabry–Perot Spectrometer

There exists a wide class of multiple-reflection interferometers [36]. With the exception of interference filters, the Fabry–Perot remains the most popular in this class. Fabry–Perots are not true monochromators in the sense of interference or acousto-optic filters (Section 18.5.3) and, indeed, require an auxiliary low resolution monochromator to block out neighboring orders. Figure 5 shows the simple construction of a Fabry–Perot spectrometer: An etalon $d$ is placed in a collimated beam between a collimator $c$ and a camera lens $e$. The internal structure of the etalon is shown in Figure 4. The image plane detector $g$ resides within a cooled detector housing; light passes through a window $f$. An interference filter $b$ is normally placed close to the focal plane a or in the collimated beam (Figure 5b). The etalon comprises two plates of glass kept parallel over a small separation $l$ (Figure 4b), where the inner surfaces are mirrors coated with reflectivity $\Re$. The transmission of the etalon to a monochromatic source $\lambda$ is given by the Airy function

$$\mathcal{A} = \left(1 + \frac{4\Re}{(1 - \Re)^2} \sin^2(2\pi\mu l \cos\theta/\lambda)\right)^{-1}, \tag{18.13}$$

where $\theta$ is the off-axis angle of the incoming ray and $\mu l$ is the optical gap. The peaks in transmission occur at $m\lambda = 2\mu l \cos\theta$, where $m$ is the order of constructive interference. From this equation, it is clear that $\lambda$ can be scanned physically in a given order by changing $\theta$ (tilt scanning), $\mu$ (pressure scanning), or $l$ (gap scanning). Both tilt and pressure scanning suffer from serious drawbacks which limit their dynamic range [39]. With the advent of servocontrolled, capacitance micrometry [40, 41], the performance of gap scanning etalons surpasses that of other techniques. These employ piezoelectric transducers, which undergo dimensional changes in an applied electric field or develop an electric

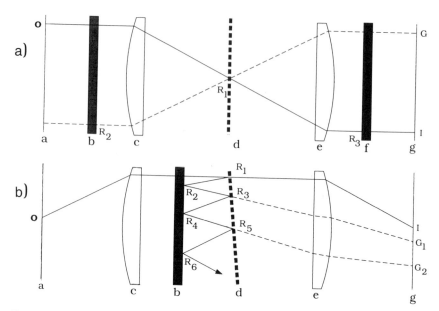

FIG. 5. Ghost families arising from internal reflections within a Fabry–Perot spectrometer (a) Diametric ghosts. Rays from the object O form an inverted image I and an out-of-focus image at $R_3$. The reflection at $R_1$ produces an out-of-focus image at $R_2$. The images at $R_2$ and $R_3$ appear as a ghost image G at the detector. (b) Exponential ghosts. The images at $R_2$ and $R_4$ appear as ghost images $G_1$ and $G_2$, respectively.

field when strained mechanically. In practice, the etalon plates are built to a characteristic separation or *zero-point gap* about which they move through small physical displacements. The scan range is limited by the dynamic range of the transducers to roughly $\pm 5$ μm. For an arbitrary etalon spacing in the range of a few microns to 30 mm, it is now possible to maintain plate parallelism to an accuracy of $\lambda/200$ while continuously scanning over several adjacent orders about the zero-point gap [42].

The resolution of the Fabry–Perot is set primarily by the reflective coating that is applied to the etalon plates. The reflective finesse is defined as $N_R = \pi \sqrt{\mathfrak{R}}/(1 - \mathfrak{R})$. Ideally, this constant defines the number of band-pass widths (spectral elements) over which an etalon may be tuned without over-lapping of orders. In other words, the *finesse* is the ratio of the interorder spacing (*free spectral range*) and the spectral purity ($\Delta\lambda/\delta\lambda$). The etalon is fully specified once the free spectral range and spectral purity are decided [43, 44] because this sets the order of interference ($m = \lambda/\Delta\lambda$), the resolving power ($\mathfrak{R} = \lambda/\delta\lambda = mN_E$), and the zero-point gap ($l_0 = \lambda^2/(2\mu\Delta\lambda)$).

However, what is measured by the instrument is the *effective finesse* $N_E$, which is always less than or roughly equal to the theoretical value of $N_R$. There are two effects that serve to degrade the theoretical resolution. If the beam passing through the etalon is not fully collimated, the instrumental function broadens and its response shifts to smaller wavelengths. This is analogous to what happens to an interference filter in a converging beam [45]. The induced profile degradation is measured by the *aperture finesse*, or $N_A = 2\pi/(m\Omega)$, where $\Omega$ is the solid angle set by the $f$/ratio of the incoming rays. The profile degradation is negligible in beams slower than $f$/15. Another important source of degradation arises from defects in the flatness of the etalon plates. The *defect finesse* is given by $N_D = \lambda/(2\,\delta l)$, where $\delta l$ is the rms amplitude of the micro-defects. In practice, the defect finesse will also include terms for large-scale bowing and drifts in parallelism [39]. In essence, to realize an effective finesse of 50 requires that the spacing and parallelism be maintained to $\lambda/100$. In summary, the effective finesse is set by the characteristics of the etalon surfaces, for which we can write $1/N_E^2 = 1/N_R^2 + 1/N_A^2 + 1/N_D^2$.

There are two primary methods for extracting spectral information from imaging Fabry–Perot spectrometers. The *interferogram* or *areal method* measures the perturbations in the radii of the interference rings to determine spectral differences across the field of observation. Roesler [46] shows that if the radius $r$ of the $k$th ring has been perturbed by $\delta$, then the corresponding fraction of a free spectral range is $(r_{k+\delta}^2 - r_k^2)/(r_{k+1}^2 - r_k^2)$. While this method is of historical importance, it is now more convenient to assemble the image frames into a three-dimensional data stack and to form spectra along one axis. In this *spectral method*, the etalon is scanned to obtain a sequence of narrowband images taken over a fixed grid of etalon spacings. As the gap is scanned, each pixel of the detector maps the convolution of the Airy function and the filtered spectrum at that point [43, 44].

It is possible to radically alter the resolving power and free spectral range by scanning etalons in series simultaneously [46]. In principle, all possible combinations of a mixture of 3–5 high- and low-finesse etalons can be used to mimic the effect of a tunable filter. While it is still necessary to use a broadband filter to block unwanted orders, only a handful are needed to cover the full optical spectrum. However, to use a Fabry–Perot as a tunable filter without the necessary phase correction requires that we restrict the observations to the Jacquinot central spot. This is defined as the field about the optical axis within which the peak wavelength variation with field angle does not exceed $\sqrt{2}$ of the etalon band-pass [21]. A good discussion of Fabry–Perot based tunable filters is given by reference [39]. The basic principle is to use the etalon in low order so that widely different wavelength regions are accessible by using the adjacent orders. At optical and near-infrared wavelengths, Eq. (18.11) indicates that to reach the lowest orders requires gap spacings of only a few microns. A major obstacle, however, is that special techniques are now needed to deposit nonlaminar dielectric coatings [38]. There have been recent advances in maintaining $\lambda/200$ parallelism over plate separations of 30 mm down to 1.8 μm [39], allowing for free spectral ranges of 0.05–50 nm at optical wavelengths. Presently, techniques are under investigation for purging incompressible dust grains from the etalon air gap. Thus, in principle, it should be possible to operate several etalons at low order, both separately and in tandem, to simulate a variable band-pass tunable filter over the visible or near-infrared spectrum. However, the design, manufacture, and stability of extended bandwidth coatings remain important obstacles for broadband tunable filters.

Even a minimal Fabry–Perot arrangement can have eight or more optically flat surfaces. At some level, all of these surfaces interact separately to generate spurious reflections. The periodic behavior of the etalon requires that we use a narrowband filter somewhere in the optical path. Typically, the narrowband filter is placed in the converging beam before the collimator or after the camera lens (Figure 5a). The filter introduces ghost reflections within the Fabry–Perot optics (Figure 5b). The pattern of ghosts imaged at the detector is different in both arrangements, as illustrated in Figure 5. The dominant reflections are mostly deflected out of the beam by tilting the etalon through a small angle with respect to the optical axis. A more difficult problem arises from the optical blanks which form the basis of the etalon. These can act as internally reflecting cavities (cf. Figure 4a) that generate a high-order Airy pattern at the detector [18, 43]. Traditionally, the outer surfaces have been wedge-shaped to deflect this spurious signal out of the beam (Figure 4c). Even curved lens surfaces occasionally produce "halation" around point source images which may require experiment-ing with both biconvex and planoconvex lenses when designing a focal reducer.

### 18.5.3 Birefringent Filter

The underlying principle of birefringent filters is that light originating in a single polarization state can be made to interfere with itself [47]. The Michelson interferometer (Section 18.6) achieves interference by splitting the input beam and sending the rays along different path lengths before recombining them. By analogy, an optically anisotropic, birefringent medium can be used to produce a relative delay between ordinary and extraordinary rays aligned along the fast and slow axes of the crystal. (A birefringent medium has two different refractive indices, dependending on the plane of light propagation through the medium.) Title and collaborators have discussed at length the relative merits of different types of birefringent filters [see references in 47]. The filters are characterized by a series of perfect polarizers (Lyot filter [48, 49]), partial polarizers, or only an entrance and an exit polarizer (Solc filter [50]). The highly anisotropic off-axis behavior of uniaxial crystals give birefringent filters a major advantage. Their solid acceptance angle is one to two orders of magnitude larger than is possible with interference filters (Section 18.5.1), although this is partly offset by half the light being lost at the entrance polarizer. To our knowledge, there has been no attempt to construct a birefringent filter with a polarizing beam-splitter, rather than an entrance polarizer, in order to recover the lost light.

The Lyot filter is conceptually the easiest to understand. The entrance polarizer is oriented 45° to the fast and slow axes so that the linearly polarized, ordinary, and extraordinary rays have equal intensity. The time delay through a crystal of thickness $d$ of one ray with respect to the other is simply $d \, \Delta\mu/c$, where $\Delta\mu$ is the difference in refractive index between the fast and the slow axes. The combined beam emerging from the exit polarizer shows intensity variations described by $I^2 \cos(2\pi \, d \, \Delta\mu/\lambda)$, where $I$ is the wave amplitude. As originally illustrated by Lyot [49], we can isolate an arbitrarily narrow spectral band-pass by placing a number of birefringent crystals in sequence where each element is half the thickness of the preceding crystal. This also requires the use of a polarizer between each crystal so that the exit polarizer for any element serves as the entrance polarizer for the next. The resolution of the instrument is dictated by the thickness of the thinnest element.

The instrumental profile for a Lyot filter with $s$ elements is

$$\mathscr{L} = \frac{1}{4^s} \frac{\sin^2(2^s \pi \, d \, \Delta\mu/\lambda)}{\sin^2(\pi \, d \, \Delta\mu/\lambda)}, \qquad (18.14)$$

where $d$ is now the thickness of the thinnest crystal element. By analogy with the Fabry–Perot (Section 18.5.2), if $\lambda_0$ is the wavelength of the peak transmission, the filter bandwidth ($= 0.88\lambda_0^2/(2^s d \, \Delta\mu)$), the free spectral range ($= \lambda_0^2/(d \, \Delta\mu)$), and the effective finesse ($= 1.13 \, 2^s$) are easily derived.

It should be noted that $\lambda_0$ can be tuned over a wide spectral range by rotating the crystal elements. But to retain the transmissions in phase requires that each crystal element be rotated about the optical axis by half the angle of the preceding thicker crystal. The NASA Goddard Space Flight Center has recently produced a Lyot filter utilizing eight quartz retarders with a 13-cm entrance window. The retarders, each of which are sandwiched with half-wave and quarter-wave plates in addition to the polarizers, are rotated independently with stepping motors under computer control. They achieve a band-pass of 4–8 Å, tuneable over the optical wavelength range (3500–7000 Å).

## 18.5.4 Acousto-optic Filter

In 1969, Harris and Wallace introduced a new type of electronically tunable filter that makes use of collinear acousto-optic diffraction in an optically anisotropic medium [51]. Acousto-optic tunable filters (AOTF) are formed by bonding piezoelectric transducers such as lithium niobate to an anisotropic birefringent medium. The medium has traditionally been a crystal, but polymers have been developed recently with variable and controllable birefringence. When the transducers are excited at frequencies in the range 10–250 MHz, the ultrasonic waves vibrate the crystal lattice to form a moving phase pattern that acts as a diffraction grating. A related approach is to use liquid crystals made of nematic (anisotropic) molecules. When a transverse electric field is applied, the molecules align parallel to the field because of their positive dielectric anisotropy and form a uniaxial birefringent layer. In acousto-optic filters, the incident light Bragg-scatters off the moving pattern from one polarization state into its orthogonal state. The birefringent interaction permits a large angular aperture which is unattainable with isotropic Bragg diffraction. If the crystal is thick enough and the driving power high enough (several watts for each square centimeter of aperture), only a limited band of optical frequencies in the incident light is cumulatively diffracted for a given acoustic frequency. The wave vector of the diffracted output beam is the vector sum of the input wave vector and the acoustical wave vector; out-of-band frequencies remain undeviated. Crystals such as $TeO_2$, quartz, and $MgF_2$ are highly transmitting and have an efficient acousto-optic response in different wavelength regimes from 200 nm to 5 μm. For currently attained crystal homogeneity and thickness, $\mathfrak{R}$ can reach $10^4$, and out-of-band light is rejected down to a level of about $10^{-5}$.

In the more useful noncollinear filters [52, 53], the acoustic and optical wave vectors differ in such a way that the phase differences introduced by variations in the angle of incidence can be approximately compensated by the different refractive indices of the ordinary and extraordinary rays $n_o$ and $n_e$, respectively. For an extraordinary polarized incident beam, $n_i = ((\cos \theta_i/n_o)^2 + (\sin \theta_i/n_e)^2)^{-1/2}$. If the incident angle is $\theta_i$, the vacuum central wavelength is $\lambda_0$, and the acoustic wavelength is $\Lambda$, the diffracted angle $\theta_d$ depends only weakly on $\lambda$ and is given by solving

$$\left(\frac{\lambda_0}{\Lambda}\right)^2 = n_o^2 + n_i^2 - 2n_o n_i \cos[\theta_i - \theta_d(\lambda)]. \tag{18.15}$$

If $\tan \theta_d = (n_o/n_e)^2 \tan \theta_i$ (the noncritical phase matching configuration), and the interaction length is short enough, the acceptance angle for diffraction-limited imaging can increase to almost 28° (i.e., an $f/2$ beam) to provide throughput comparable to that of a Fabry–Perot. However, $\mathfrak{R}$ is then limited by the small attainable density of the acoustic driving power. The diffracted ordinary and extraordinary rays emerge displaced to either side of the undiffracted out-of-band light by angles of 1–10° (Figure 6). The undeviated out-of-band light is removed with an aperture mask in the collimated beam. It is particularly difficult at infrared wavelengths to keep this light from scattering back into the optical path, thereby increasing the background. While the corresponding $f$/numbers of 50 to 5 are too slow for imaging spectroscopy of low surface-brightness sources, for brighter objects these filters simultaneously deliver the two orthogonally polarized images with more than 65% efficiency and so are reliable and compact imaging spectro*polarimeters* [54]. In beams as fast as $f/2$, the incident light can be sent through a linear polarizer in front of the filter, and a crossed polarizer removes one of the alleviated beams after passage through the crystal for a net throughput in excess of 30%.

The instrumental profile of an acousto-optic filter of length $L$ that operates at a central wavelength $\lambda_0$ in a collimated beam is like that of a grating,

$$T = T_0 \, \text{sinc}^2 \frac{\Delta k L}{2\pi}, \qquad (18.16)$$

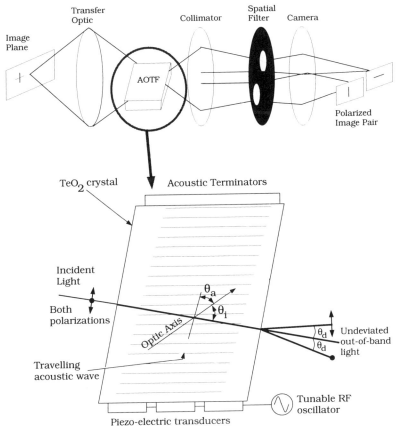

FIG. 6. A crystalline, noncollinear acousto-optic tunable filter. $\theta_i$ is the angle of incidence relative to the optic axis, $\theta_a$ the angle between the acoustic wave and optic axis, and $\theta_d(\lambda)$ the diffraction angle. The beams separate when they exit the filter.

with FWHM

$$\Delta\lambda = \frac{0.9\,\pi\lambda_0^2}{Ld\,\sin^2\theta_i} = \frac{2\pi n}{\Delta k},\tag{18.17}$$

where $L$ is the interaction length, $n$ is one index of refraction, $\Delta k$ is the mismatch between optical and acoustic wave vectors, and $\theta_i$ is the incident angle relative to the crystal optical axis. The dispersive term, $d(\lambda)$, becomes very large near a band edge. The band-pass can be altered rapidly across a large wavelength range merely by tuning the power and frequency of the acoustic wave, to form a composite band-pass shape with widely separated, broad multiple peaks. Such spectral multiplexing across several selected pass bands is a unique capability of acousto-optic filters.

These filters can be used alone as moderate-resolution imaging spectrometers, even in the ultraviolet and infrared, where it is extremely difficult to produce interference filters with bandwidths narrower than $\lambda/100$ which also have high transmission and good off-band rejection. Alternatively, they can be used in a collimated beam as the order-sorting filter of a Fabry–Perot etalon. The spatial resolution can be as good as 15 μm, well matched to the size of CCD detector elements. Crystal performance does not appear to deteriorate with age, unlike multilayer coatings used with interference filters and Fabry–Perot etalons.

Current disadvantages include their expense, long fabrication time, and small size (<25 mm square for crystals that are uniform enough for good imaging) relative to interference filters. restrictions that are likely to be lifted as the commercial market develops. Another concern that is particularly acute in the infrared is the power dissipated and heat generated during their operation. Nonetheless, an acousto-optic filter today is often much cheaper than a comparably performing Fabry–Perot and control electronics, and also avoids the complications of selecting among the multiple spectral orders that characterize etalons.

## 18.6 Two-Beam Interferometers

All spectroscopic techniques rely ultimately on the interference of beams that traverse different optical paths to form a signal. The prism uses essentially an infinite number of beams, whereas the grating uses a finite number of beams set by the number of grooves. The Fabry–Perot uses a smaller number of beams set by the instrumental finesse. Bell [19] notes that as the number of beams decreases, the throughput (and therefore efficiency) of the spectrograph increases. Because at least two beams are required for interference, Bell concludes that two-beam interferometers are the ultimate in spectrometers. The two most commonly used Fourier transform devices divide either the wavefront (lamellar grating interferometer) or the wave amplitude (Michelson interferometer). The efficiency of the latter is ≲50% whereas the former technique can approach 100%. Lamellar gratings are discussed in reference [13].

A simple two-beam Michelson interferometer is shown in Figure 7 and forms the basis of the Fourier transform spectrometer. The collimated beam is split into two beams at the front surface of the beam-splitter. These beams then undergo different path lengths by reflections off separate mirrors before being imaged by the camera lens at the detector. The device shown in Figure 7 uses only 50% of the available light. It is possible to recover this light but the layout is involved [55, 56]. For all systems, the output signal is a function of path difference between the mirrors. At zero path difference (or arm displacement), the waves for all frequencies interact coherently. As the movable mirror is scanned, each input wavelength generates a series of transmission maxima. Commercially available devices usually allow the mirror to be scanned continuously at constant speed, or to be stepped at equal increments. At a sufficiently large arm displacement, the beams lose their mutual coherence.

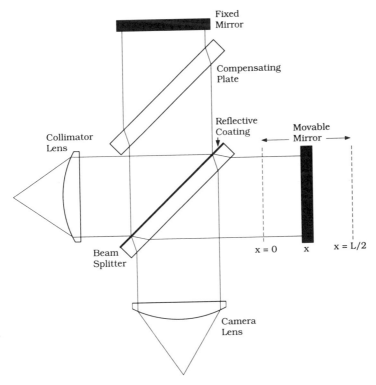

FIG. 7. Two-beam Michelson interferometer (Fourier transform spectrometer).

The spectrometer is scanned from zero path length ($x = y = 0$) to a maximum path length $y = L$ set by twice the maximum mirror spacing ($x = L/2$). The superposition of two coherent beams with amplitude $b_1$ and $b_2$ in complex notation is $b_1 + b_2 e^{i2\pi vy}$, where $y$ is the total path difference and $v$ is the wavenumber. If the light rays have the same intensity, the combined intensity is $2b^2(1 + \cos 2\pi vy)$, or equivalently, $4b^2 \cos^2 \pi vy$, where $b = b_1 = b_2$. The combined beams generate a series of intensity *fringes* at the detector. If it was possible to scan over an infinite mirror spacing at infinitesimally small spacings of the mirror, the superposition would be represented by an ideal Fourier transform pair, such that

$$b(y) = \int_{-\infty}^{\infty} B(v)(1 + \cos 2\pi vy)\, dv, \qquad (18.18)$$

$$B(v) = \int_{-\infty}^{\infty} b(y)(1 + \cos 2\pi vy)\, dy, \qquad (18.19)$$

where $b(y)$ is the output signal as a function of path length $y$ and $B(v)$ is the spectrum we wish to determine. $B(v)$ and $b(y)$ are both undefined for $v < 0$ and $y < 0$: We include the negative limits for convenience. Notice that

$$b(y) - \tfrac{1}{2}b(0) = \int_{-\infty}^{\infty} B(v) \cos 2\pi vy\, dv, \qquad (18.20)$$

$$B(v) = \int_{-\infty}^{\infty} [b(y) - \tfrac{1}{2}b(0)] \cos 2\pi vy\, dy. \qquad (18.21)$$

The quantity $b(y) - \tfrac{1}{2}b(0)$ is usually referred to as the *interferogram*, although this term is sometimes used for $b(y)$. The spectrum $B(v)$ is normally computed using widely available fast Fourier transform methods.

It is clear that the output signal is sinusoidal about some mean continuum value. If the wavefronts have different intensities, the depth of the modulation decreases and the mean continuum level increases. This is undesirable because the background continuum constitutes a source of noise. The contrast of the fringe amplitude with respect to the background is known as the *fringe visibility*. It is particularly important to ensure that all rays undergo the same optical path; otherwise, the output signal is asymmetric with respect to zero path length. In Figure 7, the beam that reflects off the movable mirror passes through the beam-splitter three times. The compensating plate ensures that the beam that reflects off the fixed mirror undergoes the same optical path.

Efficient beam-splitters are crucial to the operation of an FTS device. These often comprise dielectric sheets, wire grids, or multilayer dielectric coatings on substrates (Section 18.5.1), depending on the wavelength of operation. Particular care must be taken over the internally reflecting rays. Typically, the primary transmitted ray dominates and the secondary transmitted ray can be neglected. However, the primary and secondary *reflected* rays (cf. Figure 4a) are comparable in intensity. To maximize peak efficiency, this requires a beam-splitter which maintains constructive interference between the rays over as much of the wavelength range as possible.

In practice the ideal Fourier transform pair is not realized. The finite maximum baseline $y = L$ set by the maximum mirror displacement ultimately limits the instrumental resolution, although under certain circumstances the size of the source can impose a stronger constraint [19]. The response of most interferograms declines with higher wavenumbers. At some intermediate value of $y$, the signal-to-noise ratio may fall to unacceptably low levels, in which case the effective resolution may be somewhat lower than the theoretical value. The FTS is scanned at discrete sampling intervals (mirror spacings), which causes the computed spectrum to replicate at wavenumber intervals that are inversely proportional to the sampling interval (Section 18.2.2). If the baseline $L/2$ is sufficiently large to resolve all details within the spectrum, the replicated spectra will not overlap. Thus, the integrals in Eqs. (18.18) and (18.19) should be replaced by summations over finite limits.

The optical alignment of an FTS is particularly involved [19]. Traditionally, this is done by using, say, a He–Ne laser along the optical axis and aligning the beam-splitter with the movable and fixed mirrors separately and then together. The mirrors must be aligned and sufficiently flat so as to introduce no wavefront errors greater than $\lambda/8$ [13]. This is a major optomechanical challenge over the large arm displacements ($\lesssim 1$ m) of a two-beam interferometer. In some respects, the Fabry–Perot constraint of $\lambda/2N$ is easier to meet because the plates are optically contacted at a physical spacing of. say, 1 mm, and then scanned through a few orders about this spacing, a total mechanical distance of only a few microns. There are many ways to introduce phase errors into the computed spectrum. This is particularly so if the stepping does not start at the precise position for the zero path difference.

While attention has been given to the difficulties of two-beam interferometry, the disadvantages are few when compared with other spectrometers. In particular, the instrumental profile can take any form after mathematical filtering. The interferogram of a single wavelength is given by $b(y) = B(v_0) \cos 2\pi v_0 y$. The expected spectrum is then

$$S(v) = B(v_0) \sum_{y=-L}^{y=L} \cos 2\pi v_0 y \cos 2\pi v y \, \delta y. \qquad (18.22)$$

It is straightforward to show that this reduces to

$$S(v) = B(v_0)L \left( \text{sinc } 2\pi(v_0 - v)L + \text{sinc } 2\pi(v_0 + v)L \right). \qquad (18.23)$$

Because the second term is negligible ($<1\%$) at optical and infrared wavelengths, the basic instrumental profile is a sinc function. This has the highest resolution possible for a baseline of $L$.

At the expense of resolution, the convolution theorem [57] allows one to modify digitally the instrumental response, i.e., for an arbitrary digital filter $f(y)$, we can write

$$S(v) = \mathrm{FT}[f(y)b(y)] = F(v) * B(v) \tag{18.24}$$

where $*$ denotes convolution, FT is the Fourier transform operator, and $F(v) = \mathrm{FT}[f(y)]$. A common reason to modify the instrumental response is to reduce the side lobes of the sinc function, a process known as *apodization* [57]. A list of apodizing functions is given in reference [19]. In particular, if the interferogram is convolved with a triangle function, $f(y) = 1 - |y|/L$, the instrumental profile now takes the form of a $\mathrm{sinc}^2$ function which is the response of both the grating (Section 18.4) and the acousto-optic filter (Section 18.5.4).

## Acknowledgements

We thank P. D. Atherton (Queensgate Instruments), S. C. Markham (MediMedia), J. O'Byrne (University of Sydney), P. L. Shopbell (Rice University and Caltech), and Lady Anne Thorne (Blackett Laboratory) for critical readings of an early manuscript.

## References

1. Williams, D. (1976). *Meth. Expt. Phys.* Vol. 13, Chapter 1. Academic Press, New York.
2. Stroke, G. W. (1967). *Ency. Phys.* **29**, 426.
3. Newton, I. (1703). *Optiks*.
4. Melvill, T. (1752). *Physical and Literary Essays*.
5. Fraunhofer, J. (1817). *Ann. Phys.* **56**, 264.
6. Fraunhofer, J. (1823). *Ann. Phys.* **74**, 337.
7. Rowland, H. A. (1882). *Phil. Mag.* **13**, 469.
8. Lord Rayleigh (1874). *Phil. Mag. XLVII*, **81**, 193.
9. Michelson, A. A. (1887). *Phil. Mag.* **24**, 463.
10. Fabry, Ch., and Perot, A. (1897). *Ann. Chim. Phys.* **12**, 459.
11. Thorne, A. P. (1988). *Spectrophysics*. Chapman and Hall.
12. Schroeder, D. J. (1987). *Astronomical Optics*. Academic Press.
13. Meaburn, J. (1976). *Detection and Spectrophotometry of Faint Light*. Reidel, Holland.
14. Kitchin, C. R. (1991). *Astrophysical Techniques*. Adam Hilger.
15. Klinkenberg, P. F. A. (1976). *Meth. Expt. Phys.* **13A**, 253.
16. Gray, D. F. (1992). *The Observation and Analysis of Stellar Photospheres*. Cambridge University Press.
17. Vaughan, J. M. (1989). *The Fabry–Perot Interferometer: History, Theory, Practice and Applications*. Adam Hilger.

18. Hernandez, G. (1986). *Fabry–Perot Interferometers*. Cambridge University Press.
19. Bell, R. J. (1972). *Introductory Fourier Transform Spectroscopy*. Academic Press.
20. Steel, W. H. (1983). *Interferometry*. Cambridge University Press.
21. Jacquinot, P. (1954). *J. Opt. Soc. Amer.* **44**, 761.
22. Janesick, J., and Elliot, T. (1992). In *Astronomical CCD Observing and Reduction Techniques, ASP Conf. Ser.* **23**, 1.
23. Newberry, M. V. (1991). *PASP* **103**, 122.
24. Fellgett, P. (1951). Thesis, Cambridge University.
25. Barden, S. C., ed. (1988). *Fiber Optics in Astronomy I, ASP Conf. Ser.,* Vol. 3.
26. Gray, P. M., ed. (1993). *Fiber Optics in Astronomy II, ASP Conf. Ser.,* Vol. 37.
27. Jacquinot, and Dufour, Ch. (1950). *J. Phys. (Paris)* **11**, 427.
28. Jacquinot, P. (1960). *Rep. Prog. Phys.* **23**, 267.
29. Breckinridge, J. B. (1971). *Appl. Opt.* **10**, 286.
30. Calatroni, J. A. E., and Garavaglia, M. (1973). *Appl. Opt.,* **12**, 2298; erratum (1974). *Appl. Opt.* **13**, 1009.
31. Hutley, M. C. (1982). *Diffraction Gratings*. Academic Press, New York.
32. Woods, T. N., Wrigley, R. T., Rottman, G. J., and Haring, R. E. (1994). *Appl. Opt.* **33**, 4273.
33. Longhurst, R. S. (1973). *Geometrical and Physical Optics*. Longman, London.
34. Dekker, H. (1987). In *Instrumentation for Ground-based Optical Astronomy* (L. B. Robinson, Ed.), p. 183, Springer-Verlag, New York.
35. Wynne, C. G. (1991). *Mon. Not. R. Astron. Soc.* **250**, 796.
36. Malacara, D. (1988). In *Methods of Experimental Physics*, Vol. 26, Chapter 1. Academic Press, New York.
37. Baumeister, P., and Pincus, G. (1970). *Sci. Am.* **223:6**, 59.
38. Trauger, J. T. (1954). *Appl. Opt.* **15**, 2998.
39. Atherton, P. D., and Reay, N. K. (1981). *Mon. Not. R. Astron. Soc.*
40. Jones, R. V., and Richards, J. C. S. (1973). *J. Phys. E.: Sci. Instrum.* **6**, 589.
41. Hicks, T. R., Reay, N. K., and Scaddan, R. J. (1974). *J. Phys. E.: Sci. Instrum.* **7**, 27.
42. Atherton, P. D., Reay, N. K., Ring, J., and Hicks, T. R. (1981). *Opt. Eng.* **20**, 806.
43. Bland, J., and Tully, R. B. (1989). *Astron. J.* **98**, 723.
44. Atherton, P. D., Taylor, K., Pike, C. D., Harmer, C. F. W., Parker, N. M., and Hook, R. N. (1983). *Mon. Not. R. Astron. Soc.* **201**, 661.
45. Lissberger, P. H., and Wilcock, W. L. (1959). *J. Opt. Soc. Amer.* **49**, 126.
46. Roesler, F. L. (1974). In *Methods of Experimental Physics*, Vol. 12A. Academic Press, New York.
47. Title, A. M., and Rosenberg, W. J. (1981). *Opt. Eng.* **20**, 815.
48. Öhman, Y. (1938). *Nature* **141**, 157.
49. Lyot, B. (1944). *Ann. d'Ap.* **7**, 31.
50. Solc, I. (1965). *J. Opt. Soc. Amer.* **55**, 621.
51. Harris, S. E., and Wallace, R. W. (1969). *J. Opt. Soc. Amer.* **59**, 744.
52. Chang, I. C. (1981). *Opt. Eng.* **20**, 824.
53. Suhre, D. R., Gottlieb, M., Taylor, L. H., and Melamed, N. T. (1992). *Opt. Eng.* **31**, 2118.
54. Hayden-Smith, W., and Smith, R. M. (1990). *Exp. Astronomy* **2**, 329.
55. Connes, P., and Michel, G. (1975). *Appl. Opt.* **14**, 2067.
56. Larson, H. P., and Fink, U. (1975). *Appl. Opt.* **14**, 2085.
57. Bracewell, R. N. (1986). *The Fourier Transform and Its Applications*. McGraw-Hill.

# 19. PULSED LASER RAMAN SPECTROSCOPY OF DYNAMIC SYSTEMS

Carter Kittrell

Department of Chemistry and Rice Quantum Institute
Rice University, Houston, Texas

## 19.1 Introduction

Although the Raman effect was first observed in 1928 [1, 2], it is only recently that the remarkably diverse utility of Raman spectroscopy has been realized. Application of laser technology began a renaissance in the field [3] because, for example, a focused argon ion laser beam provides orders of magnitude greater irradiance than can a spectrally filtered mercury lamp. New developments in detectors, filters, and spectrometers have greatly improved sensitivity and suppression of background. Raman spectroscopy is used in a wide variety of applications, primarily for analytical purposes and as a probe of molecular and crystal structure. For these applications, much effort is made to ensure the sample remains unchanged by exposure to the intense laser light [4, 5].

More recently, an increasing number of researchers are trying to use spontaneous Raman spectroscopy to study transient behavior of systems formed by intentionally exciting or disrupting the molecular bonds in the sample. Transient processes may be examined by using an intense laser pulse to photolyze the sample preceding the Raman probe pulse, or an input wavelength may be selected which photolyzes the sample while the Raman spectrum is collected. Although it is inherently difficult to maintain high concentrations of transient species, the time evolution may still be probed by taking advantage of the essentially "instantaneous" response of Raman scattering (RS). Time scales ranging from many nanoseconds down to a few femtoseconds are now being studied. Spontaneous Raman scattering has been used to monitor *product formation* from a chemical reaction, to observe and characterize *excitation and relaxation processes*, to monitor *transient* species, and even to determine the nuclear motion in the first few femtoseconds of a direct *photodissociation* process.

Increasingly sensitive and specialized instrumentation is being developed for RS, especially systems utilizing multiwavelength and continuously tunable pulsed ultraviolet lasers. Tunability allows the targeting of particular electronic bands. It is very helpful for following the time evolution of photodissociating species, and for obtaining Raman excitation profiles (REPs). It also helps in

393

identifying and avoiding spurious fluorescence features. While much of the instrumentation for analytical RS is standardized, the ultraviolet Raman laser spectrometer systems used for dynamic studies are generally custom designed and use many different laser systems ranging from continuous wave (cw) to subpicosecond sources. Tunable ultraviolet–visible pulsed laser systems having pulse widths ≈10 ns are most frequently used [6]. The design of pulsed laser Raman spectrometer systems is described in some detail, and references to other types of laser systems are provided. In many cases, recent references are cited in preference to those which originally introduced a new idea, so that subsequent developments may be readily seen.

## 19.2 Raman Scattering

The theory of Raman scattering has been discussed in a number of books [7–9]. Here the essential features of the theory are discussed to provide a guide. Normal or nonresonant Raman scattering (RS), which was first studied by Raman and is still the most widely used, utilizes excitation wavelengths far from any absorption band. It is an inelastic scattering process, in which the incident photon gives up (or gains) some energy to the target molecule. The energy gained by the molecule, usually as vibrational excitation, is equal to that lost by the photon. The absence of dissipative processes during the "instantaneous" scattering typically leads to sharp spectral features, even for complex molecules, although the corresponding vibrational motions can be quite complicated for larger polyatomics [10, 11].

RS may viewed as excitation from the ground electronic state to a virtual level followed by return to a different level of the ground state (Figure 1a). This simple picture must be viewed with caution, as the scattering is a simultaneous, not sequential, process. The traditional method for determining the scattering intensities to various final vibrational levels uses the approach based on the Kramers–Heisenberg–Dirac (KHD) sum-over-states [12] and a vibronic coupling picture with a four-term A, B, C [13], and D summation [14]. When the laser frequency is tuned near or on an absorption band, resonance Raman scattering (RRS) results (Figure 1b), The scattering intensity increases, sometimes by orders of magnitude [15, 16], and there is selective enhancement of some vibrational modes. Overtone progressions may appear; these have been studied in many simple and complex systems, including benzene [17], carotenoids [18], metal oxides, halides [19, 20], and complexes [21]. During the scattering process, the nuclei experience forces associated with the excited-state potential energy surface (PES) and move from their equilibrium positions, giving rise to the observed RRS. Overtones are especially prominent when the motion on the excited PES corresponds to a particular vibrational mode of the ground state.

This includes the beginning of bond-breaking motion which leads to photo-dissociation. The intensities of features associated with different vibrational modes provide information about the excited PES.

The sum-over-states method has been extensively applied to determine excited state properties [13, 14, 22–24]. Large displacements of the excited potential energy minimum give rise to pronounced overtone activity, so the observed overtones can be used to elicit motion on the PES [19]. However, summation over all vibrational levels of the appropriate excited states can become intractable to computation when there are a large number of normal modes [12, 25, 26]. There are two additional classes of theory that can elucidate excited-state behavior. The transform method [27], including a many-body approach [25], originated with time-correlator theory [28, 29] and was independently developed from sum-over-states methods [30]. It works well for complex molecules [26] even when there is only a limited amount of usable Raman data, as the absorption profile contributes to calculating a Raman excitation profile (REP) [31]. It has been extended to include non-Condon vibronic coupling that utilizes overtone/fundamental intensity ratios rather than absolute scattering cross sections [32]. The inverse transform method is the complement as the

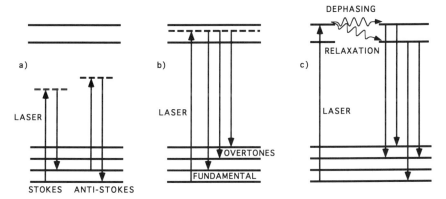

FIG. 1. Raman and resonance Raman scattering. (a) Normal Raman scattering, showing a Stokes shift to lower frequency originating from the lowest vibrational level, and an anti-Stokes shift to higher frequency originating from a vibrationally excited level. The energy lost (gained) from the laser photon precisely matches the energy gained (lost) by the molecule in vibrational excitation (de-excitation) of the ground electronic state. (b) Resonance Raman scattering (RRS). The laser is tuned to near-resonance with a level in the upper electronic state. The same energy balance holds as for (a), because there are no dissipative processes during the scattering. A series of overtones may be observed. (c) Laser-induced fluorescence (LIF). The photon is absorbed, forming an excited state that may undergo dephasing and/or relaxation. The emitted photon energy corresponds to the energy difference between the upper and lower levels.

absorption band is calculated from an experimental REP [33, 34]. Time-dependent wave-packet dynamics using semiclassical methods were developed by Heller and co-workers [35, 36], and pure quantum mechanical methods were later applied [37–39]. Numerical methods are generally needed for large-amplitude motion. The equivalence of time-independent and time-dependent methods has been experimentally demonstrated using $Br_2$ as an example [40]; comparisons have also been made between the wave-packet and transform/time-correlator methods [41].

## 19.2.1 Molecular Dynamics and Wave-Packet Propagation

Overtones provide a clear signal of wave-packet propagation [42] on the excited PES (see Figure 2). Observation of long RS overtone progressions in ozone and methyl iodide led to application of the wave-packet propagation theory to Raman spectroscopy of molecules undergoing direct photodissociation [43]. When methyl iodide is excited by 250 nm radiation, the molecule undergoes a transition from the bound ground-state surface to an excited surface which is repulsive for the C–I bond. The C and I nuclei immediately begin to separate on the time scale of a few femtoseconds. As the bond stretches, there is a finite probability that the incident photon will be inelastically scattered and the molecule returned to the ground state. Because the bond is stretched, some of the nuclei are displaced relative to their ground-state equilibrium positions, and the molecule is left in a vibrationally excited level. The energy of the RRS photon is equal to the energy of the incident photon minus the vibrational energy deposited in the molecule. The more the bond is stretched, the greater the frequency shift of the scattered photon. The resulting series of sharp Raman emission lines corresponding to increasing quanta of energy deposited in a particular vibrational mode is known as an overtone progression. The motion of the nuclei is not observed in real time, but rather as scattered photon energy loss. Analysis of the frequency-dependent data, however, provides the evolution of the breaking bond on a femtosecond time scale. Rapid motion on the excited PES leads to highly enhanced overtone progressions. A caveat is associated with this intuitive picture: Raman scattering is a coherent process, so all three steps, excitation, propagation, and emission, happen simultaneously. The experiment does not have the means to time-resolve the emission; the total time-integrated emission into each level is recorded. Although the time-dependent model presents a rather different physical picture than the time-independent sum-over-states KHD theory, they have been shown to be equivalent [40].

The method of analysis just discussed is quite general. Virtually all molecules show similar behavior: When moved from a ground PES to an excited PES, the nuclei are no longer at equilibrium and move along the gradients of the excited PES. Some examples are $H_2O$ [44], $C_2H_4$ [45], $CH_3I$ [46, 47]. The wave-packet propagation approach can even elicit molecular dynamics in fairly complex molecules, such as twisting about double bonds in excited *trans*-1,3,5-hexatriene [48]. Interference by other excited electronic states can be detected by "de-enhancement" due to PES crossings [49] and by polarization studies in $CH_3I$ [50]; the PES curve-crossing hypothesis in reference [51] is disputed in reference [52] on the basis of selection rules. The solvent environment may alter the photodissociation dynamics, as shown for $I_3^-$ [53].

Motion on the excited PES may be stimulated whether or not the state is dissociative. Indeed, "if there is a change in the molecular geometry in the excited electronic state that mirrors a normal mode of the ground-state molecule,

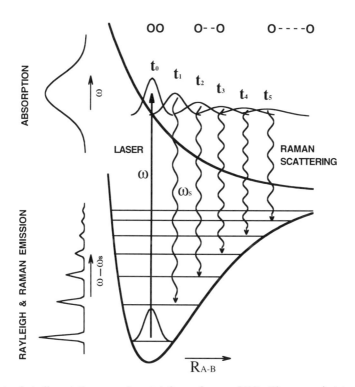

FIG. 2. A photodissociation experiment (after reference [43]). The ground-state wave function is promoted to the repulsive excited surface, where the wave packet evolves and spreads on a femtosecond time scale along the reaction coordinate. Only a small portion of the ground-state wave-packet amplitude is promoted at any instant (see text). The development of phase structure is not shown. The dotted line connecting the round symbols at the top indicates the stretching of the bond linking the two parts, A and B, of the molecule which is dissociating. The absorption spectrum is uncertainty-broadened. The small fraction of the molecules that scatter to various vibrationally excited levels in the ground state give rise to the sharp Raman signals observed.

then that normal mode will be active in the resonance Raman spectrum" [54]. The largest displacements have the highest overtone intensities, and a large damping factor diminishes overtone intensities [55]. For a bound excited state, RRS provides information on the early stages of movement of the nuclei, as illustrated in Figure 3a. In predissociation, the initial motion may or may not be along the bond-breaking coordinate. The wave packet may split with partial reflection and transmission when encountering a barrier on the PES, as suggested in Figure 3b, or it may proceed mostly to a different exit channel [56]. While direct photodissociation generally takes place in a time short compared to a corresponding period of the vibrational mode, indirect photodissociation processes take longer [57], and rovibrational structure has been observed using REPs in predissociating $NH_3$ [58] and $CH_3I$ [59]. RRS may also be used to probe dynamics on the ground-state PES [42].

Vibrational motions that are not excited by the move to the new PES generally give weak Raman features, although off-resonance interference from a nearby electronic state may alter this rule [50]. Long overtone progressions may or may not be observed for large chromophores, depending in part on whether or not the initial motion of the nuclei becomes quickly distributed among many vibrational modes when projected back onto the ground-state PES. Sum rules are used to

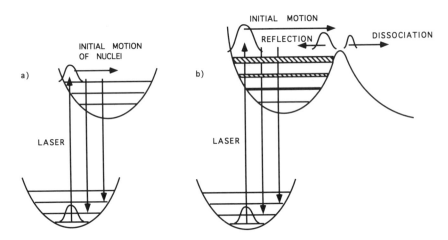

FIG. 3. Time-dependent wave-packet propagation on an excited surface. As with other figures, only a 1-D "slice" through the multidimensional PES is shown for polyatomic molecules. (a) Motion of a wave packet on a bound potential surface leads to RRS which is enhanced in modes corresponding to the direction of early motion. (b) Motion of a wave packet on the potential surface of a state that can undergo predissociation leads to RRS which need not have overtone progressions corresponding to motion along bond-breaking coordinates because reflection off of the PES barrier can change the direction of the wave-packet propagation. After reference [57].

predict intensities [60]. For these cases, the REPs have proved particularly beneficial in providing early temporal information for large species such as iodobenzene [61] and displacement from equilibrium along each normal mode coordinate in inorganic metal complexes [55]. (See Figure 4.) REPs do manifest the uncertainty broadening associated with the intermediate level. A method has been proposed for direct inversion of REPs to the time domain [62], and it shows promise for going directly from experimental data to information about the PES and transition dipole moments without modeling [63].

One interesting question is how such very fast processes can be studied with a comparatively long 10 ns laser pulse. When it is turned on, the laser radiation field places "little pieces" [36, 64] of the ground-state wave function on the excited PES As these little bits of wave-packet amplitude propagate, more new pieces are excited by the laser field. Some pieces return to vibrationally excited levels of the ground-state PES. The upward flow continues as long as the laser light is interacting with the initial ground-state wave function. The accumulation of wave-packet amplitude returning to the ground-state PES, integrated over the entire duration of the laser pulse, gives rise to the observed RS. Therefore, short time dynamics can be determined from long time measurements involving photodissociation [65].

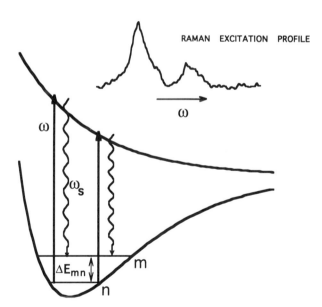

FIG. 4. Raman excitation spectrum for a photodissociating molecule. The laser and spectrometer are synchronously scanned to maintain a constant difference $\Delta E_{mn}$ between the incident and observed photon energy. Therefore, transitions to the same final state are monitored as the laser is tuned.

The paradox of high resolution RS emission spectra obtained from an ultrafast process is explained by conservation of energy [43], which dictates that the energy gained by the molecule must be equal to that given up by the inelastically scattered photon. The molecule originates from a sharp level and returns to a sharp level. The uncertainty broadening is only that associated with the combined widths of the initial and final levels, and not that of the short-lived intermediate level. In contrast, if a sufficiently short pulse laser were used, then the entire wave packet would be propagating as a unit on the excited PES. As the wave packet evolves, emission to each successively higher vibrational level (from an ensemble of wave packets) would occur at progressively later times. In principle, a fast detector could temporally resolve the delayed emission after the laser pulse. However, the transform broadening of such short pulses would cause overlap of the emission to individual vibrational levels, and the contributions of individual vibrational modes might not be readily discerned. The information about the motions of the nuclei derived from the intensities of emission to individual vibrational levels may thus be lost by the overlap of the spectral features.

## 19.2.2 Resonance Raman vs. Fluorescence

Laser excitation near or on resonance can generate fluorescence (Figure 1c) as well as Raman scattering, and is a topic of considerable interest and controversy [23, 47, 66, 67]. It is important to distinguish laser induced fluorescence (LIF) emission from RRS [68]; such differences were experimentally examined for iodine [69]. A distinction has been made by comparing the coherence width of the incident photon and that of the excited state [70]. To illustrate this, consider nonresonant Raman scattering, which appears as a frequency shifted signal. The shift is constant and so the emission frequency changes with the laser frequency. As the laser is tuned toward resonance, the picture becomes more complicated. Some emission features will continue to track the laser frequency, often with orders of magnitude increases in intensity, e.g., as for formamide [15] and nucleic acid bases [16]. These features are resonance Raman scattering. New emission features occur, which originate from specific excited levels in the molecule that have a fixed emission frequency and are associated with fluorescence. Changing the laser frequency will affect the intensity but not the frequency of this LIF emission from a particular level. Different LIF lines will come and go as the laser frequency is changed. Mixed character emission may also occur, but is usually weaker [71]. However, emission broadening from solvent induced dephasing can make it difficult to partition the contributions from RRS and fluorescence [53]. Temporal evolution of RRS and LIF processes are generally different [69]. Spontaneous Raman scattering is a coherent, inseparable "double-photon" process, with excitation and emission occurring simultaneously. In contrast, LIF involves two separable one photon processes of absorption and emission [66]. Thus, Raman scattering temporally tracks the laser pulse as long as the initial population is not depleted, whereas time dependence of fluorescence emission is governed by the lifetime of the excited state population. However, a short-lived excited state may cause LIF to temporally mimic RRS, and for continuum states such as those associated with direct dissociation, the distinction cannot be readily made by temporal means alone [47].

It is very important to correctly identify spectral features corresponding to RRS, which are often mingled with fluorescence features. Photodissociation and other reactions induced by the laser light often creates radicals and product species which undergo LIF, and such signals may be mistaken for Raman features. Only a minuscule amount of the offending reaction product may be needed to generate a LIF signal as strong as the expected RRS signal. A regular sequence of bands resulting from a fluorescence vibrational progression may resemble a Raman overtone progression. Having considered the basic differences between RRS and LIF above, a number of practical techniques may be utilized to identify RRS and LIF and also distinguish between them.

The first group of distinctions is based on frequency measurements.

1. The frequency difference between the laser and RRS frequencies remains constant. Features which do not track the changing laser frequency must then involve some dissipative process between excitation and emission, and thus can be readily distinguished from the Raman process [72]. Two or more emission scans utilizing adjacent laser frequencies will help confirm the identity of resonance Raman features.
2. Observed frequency shifts should be compared to that expected for Raman active vibrational levels of the parent species or possible reaction products. High-resolution measurements will help to assign a sequence of vibrational bands to the correct species. Diatomic fragments in particular are prone to intense LIF with long vibrational progressions, but fortunately the spectroscopy of most have been well characterized, and such information is very useful in making assignments.
3. Spectral profiles of vibrational bands are quite helpful in confirming the identity of observed features. The sharpness and shape of the Q-branch is a distinctive feature of RRS, and a fit of all of the rotational branches to a calculated Raman scattering profile, taking into account the spectrometer slit function [73], strongly supports a proper assignment, e.g., in ozone [74]. Spectral congestion will generally cause fluorescence from polyatomic species to appear to be significantly broader than the Raman Q-branch features, whereas that from individually resolved rotational levels of a diatomic species will usually be instrument limited in width.

4. If O- and S-branches can be clearly identified, these are characteristic of Raman scattering. Unfortunately, these are often hidden in the background.

A second group of distinctions is based on band intensities.

5. Signal intensity measured as a function of laser fluence ($J/cm^2$) provides useful discrimination against signals from reaction species generated by, e.g., photodissociation. RRS is linear with laser fluence until it becomes high enough to deplete the ground-state population of the parent species and saturate the signal. In contrast, photoproduct emission requires at least one photon to generate the product and usually more photons to provide excitation, so such emission usually has a higher order dependence on laser fluence [72].

6. When relative RRS intensities of transitions to different vibrational modes can be estimated from other data or theoretically, these ratios may be compared to that observed. However, intensity ratios from nonresonant RS data generally are not suitable for comparison, as resonance enhancements are usually mode selective.

7. When a long overtone progression is present, e.g., from direct dissociation, RRS commonly shows monotonic decrease in intensity with increasing vibrational levels [66], so a long nonmonotonic sequence of vibrational bands is suspect. Occasionally some Raman bands, usually the first ones in the progression, may depart from this regular pattern due to interferences from higher electronic states [50]. These interferences can be identified by using several excitation frequencies, or from a REP.

8. When REP's are obtained, considerable preresonance enhancement of RS [15, 16] is often observed well outside the absorption band which gives rise to fluorescence; the LIF intensity falls much more rapidly than that of RRS as the laser frequency is detuned from resonance.

Additional distinctions utilize time, quenching, and polarization.

9. Long lived emission features which persist after the laser light has ended may be ruled out as spontaneous Raman processes. Thus, gated detection is most beneficial, and a simple prompt/delayed gate combination has been demonstrated that simultaneously records prompt and delayed fluorescence emissions [75].

10. Fluorescence may be identified and also diminished by increased pressure and/or change of carrier gas, e.g., $N_2$, $SF_6$ or $CH_4$ in place of noble gases, addition of quenchers, sample purification, and other means [66].

11. Depolarization ratios, both linear and circular, may be quite different for LIF, RRS, and also ordinary RS [66], and such measurements are quite sensitive to intermediate state relaxation processes. Spectral profiles of vibrational bands may be sensitive to polarized detection.

12. A careful check of selection rules often reveals that they are more restrictive for RRS [52] than for an absorption–emission process. Specific spectral features that are observed but not expected to occur in a RRS process indicate some relaxation process may have occurred in the excited state and hence that the emission is LIF. Polarization sensitive detection is often beneficial for this purpose.

The above criteria are intended as a collection of experimental guidelines tailored for pulsed laser Raman systems where interfering fluorescence often comes from contaminants, reaction products, and possibly from multiphoton excitation. LIF may mimic RRS in one or more ways, and application of several of the guidelines will help assure proper identification of each spectral feature observed.

## 19.3 Product Formation and *in Situ* Monitoring

Both normal Raman and resonance Raman scattering may be used for monitoring species undergoing change (Figure 5a) such as reaction intermediates of heme proteins [76], combustion products [77], temperature- and pressure-induced phase changes of crystalline materials [78], temperature profile images of a single carbon fiber [79], conformational equilibria [80], effects due to

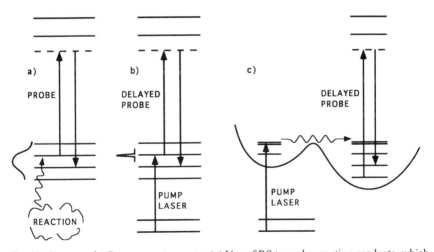

FIG. 5. Pump-probe Raman spectroscopy. (a) Use of RS to probe reaction products, which may be vibrationally excited. A nonthermal distribution of population is indicated by the small icon on the left. (b) Species produced by a pump pulse are observed by RS. A pump-probe exploration of the initial electronically excited state(s) can be monitored if the probe is prompt, and relaxation processes if it is delayed. (c) Isomerization and subsequent relaxation of a molecule in an excited state is monitored by a variable-delay probe pulse.

physisorption or chemisorption on catalytic materials [81] and surfaces [82], molecular complexes in matrix isolation [83], and weak solvent interactions [84].

For analytical Raman studies, the approach is to minimally perturb the sample with the laser Raman probe. Minimal disturbance is one of the greatest virtues of optical in situ monitoring, compared to insertion of physical probes. The DNA and aromatic amino acids may be studied inside living bacteria [85], and thermal denaturing observed *in situ* [86]. In some cases, the Raman laser plays more than one role and is involved in a controlled change of the species being studied. For example, in studying microdroplets, the laser may be used to levitate the drop, provide heat to evaporate and shrink it, and generate the Raman scatter to monitor the changing properties as it shrinks, all at the same time [87].

## 19.4 Excited States and Transient Species

RS can also be used to monitor excited states and transient species [88]. A single $\approx 10$ ns laser pulse may be used for saturation Raman spectroscopy [88], e.g., to study DNA–pyrene interactions [89] and heme protein relaxation with picosecond resolution [90]. More often, separate pulses of laser light are used to create vibrational excitation [91], electronic excitation (Figure 5b), initiate photoisomerization (Figure 5c), e.g., in diphenylbutadiene [92], or create a population of transient species (Figure 6). The preparation pulse is followed by a probe pulse [93]. One advantage of pulse laser excitation is that a larger product

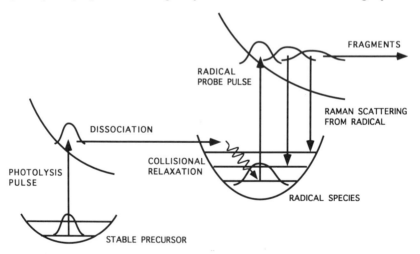

FIG. 6. Dynamical probing of radical species. Photolysis creates a high concentration of gas-phase radicals that are probed with a second pulse [94, 95]. The radical may also undergo photodissociation, and photodynamics may be explored using a delayed probe pulse applied after thermal relaxation of the radical has occurred .

concentration may be created than is sustainable in continuous mode; this is beneficial, as Raman scattering is a weak effect. The power of the pump-probe technique has been demonstrated by the first resonance Raman detection of gas-phase methyl [94] and allyl radical [95] species. The Raman scattering also provides dynamic information about the photodissociation of the radical species (Figure 6). A second advantage of pump-probe methods is that the initial pulse provides a start time for monitoring the evolution of transient species [96]. Photolytically generated transient species in solution include biphenyl radical ions in a flowing jet [97] and benzyl radicals [98]. A good review of transient species and isomerization is given in reference [6].

Tunable or multiwavelength pulsed lasers are needed for most pump-probe applications [88, 93], as the pump laser needs to be tuned to the appropriate absorption band to create the intermediate population, and resonance Raman scattering is often used to improve probe sensitivity. Pulsed, tunable visible–ultraviolet laser systems, typically with 10 ns pulse duration as discussed here, are the most widely used, such as for studying excited states of metal complexes [99], zinc porphyrins [100], and photodissociation of HbCO [101].

The time resolution of pump-probe RS currently ranges from millisecond and longer to the picosecond regime [102], such as has been used to probe the excited-state dynamics of carotenoids [103], geminate recombination of iodine atoms [104], and the *cis–trans* photoisomerization of stilbene [105]. Many more examples may be found in reference [106]. Although the laser technology is available for extremely fast pump-probe techniques and is widely used for absorption, emission, and nonlinear processes, application to Raman scattering leads to difficulties which become more pronounced with decreasing pulse width. Decreasing the pulse duration with fluence ($J/cm^2$) held constant leads to higher irradiance ($W/cm^2$) and a consequent increase in competing nonlinear processes. Femtosecond laser systems are less tunable than nanosecond systems, so the range of accessible excited electronic states is not as wide. Shorter pulses broaden the observed Raman spectral features [102, 103], causing overlap and loss of spectral information.

As previously discussed, RRS is used to probe the short time response of the nuclei to incident radiation and their rearrangement on the excited PES, including photodissociation. In such measurements, a single laser pulse is employed whose duration is generally much longer than the time scale of the event being monitored. The frequency of the scattered light is recorded, and those excited species that return to the initial PES are monitored. Raman scattering is not observed from molecules that dissociate. The acronym dissociative resonance Raman spectroscopy (DRRS) spectroscopy is applied to this process. The molecular dynamics are observed with the laser is tuned on *resonance*, causing *Raman scattering* to yield a *spectrum* from molecules which are in the process of *dissociating*. A half Fourier transform couples the time domain to the frequency domain and thereby can be used to elicit information about wave-packet propagation on the excited PES [63, 65]. The DRRS method is complementary to the pump-probe method. The former method work best for the very shortest time scales where the corresponding resonance Raman signals are the strongest. It reaches outward from the instant of initial excitation, and the sharp spectral features provide detailed information about motion of the nuclei. The latter method most easily probes products approaching completion. It reaches inward as increasingly shorter pulses are utilized; as the uncertainty broadening tends to excite much or all of the vibrational manifold, the observed features provide information about overall excited-state behavior. Both methods are vital and complementary sources of experimental information on molecular dynamics.

## 19.5 Laser Raman Spectrometer Systems

Laser Raman systems comprise a number of basic components [7]: a laser source, beam steering and focusing optics, a sample chamber, collection optics, optional polarization [107] control, and a spectrometer equipped with an optical detector (see Figure 7). Raman scattering is a linear effect, and does not benefit from the high peak power of a pulsed laser. In fact, competing multiphoton events may be a problem and the sample may be photolyzed [108]. Photoproducts have their own (misleading) Raman signatures. Often, the photon flux must be limited. For samples that are irreversibly changed by the laser light, once photolysis is complete, additional photons may well increase spurious and unwanted background or signals from photoproducts. Pulsed laser Raman systems demand particularly careful design for optimum sensitivity with emphasis on photon collection and detection. Versatile continuously tunable ultraviolet–visible [109, 110] and multiwavelength far-ultraviolet [111] pulsed laser Raman systems have been developed that are suitable for Raman studies.

---

FIG. 7. Pulsed tunable laser Raman spectrometer. A Nd : YAG laser or an excimer laser pumps a tunable dye laser or optical parametric oscillator (OPO). Stimulated Raman shifting (SRS) may be used in place of the tunable laser to provide a series of fixed wavelengths. After harmonic generation, a prism pair separates the ultraviolet beam; a symmetric double pair separator is used for REP scanning to minimize beam walk [61]. A spatial filter or an iris is used to obtain a beam with well-defined spatial characteristics, and a waveplate may be used to change the polarization before it passes into a baffled sample cell. Beam splitters send light to photodiodes to provide a trigger pulse and to monitor the absorption for RRS. Raman emission scattered at $\approx 90°$ is collected and after passing through filters and/or polarizing elements and is focused on the entrance slit of a spectrometer where it is analyzed. The radiation is directed to a photomultiplier or (intensified) CCD, which, when a photolysis laser is used, may be gated off during the pulse. (The optical plane of the Czerny–Turner spectrometer is oriented such that direction of propagation of the laser beam is normal to it; it will actually be projecting out of the plane of the page in this view.)

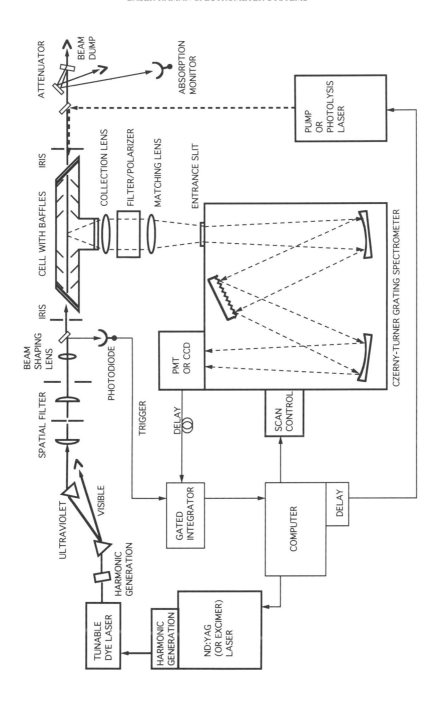

### 19.5.1 Pulsed Laser Sources

Many ultraviolet laser sources utilize excimer lasers or Q-switched Nd : YAG solid state lasers with harmonic generation [112]. The hazards associated with laser beams are well known [113, 114], and information is available on safety standards [115]. The pulse duration of these systems typically ranges from a few nanoseconds for the higher harmonics of the Nd : YAG laser up to 20 ns for an excimer laser, and a wide variety of lasers and options are commercially available. In addition to the criteria of pulse energy and repetition rate, beam quality is a crucial factor. The selection of the laser is frequently governed by the intended application. The good beam quality and narrow bandwidth of the Nd : YAG and similar solid-state lasers makes them attractive for use in Raman scattering. With harmonic generation and stimulated Raman shifting (SRS) [116], a wide variety of fixed wavelengths are available, extending deep into the vacuum ultraviolet [111, 117]. Nd : YAG lasers may be used to pump dye or Ti : sapphire lasers [113] and optical parametric oscillators (OPO), which are now undergoing rapid commercial development.

Standard excimer lasers provide a large beam with considerable divergence; the divergence is quite asymmetric and the output is not readily focused to the small sizes required in many Raman experiments. Use of unstable resonator optics [118] can significantly improve beam quality for some modest loss of pulse energy. Oscillator–amplifier combinations can also improve beam quality. Most Raman features are spectrally quite narrow compared to the $\approx 100\,\mathrm{cm}^{-1}$ bandwidth of a standard excimer laser; this will limit the resolution. Line narrowing and limited tunability can be achieved in the oscillator–amplifier systems. Excimer laser output may use SRS to provide additional wavelengths.

The maximum capabilities of pulsed Raman spectroscopy are realized when a tunable laser source is employed. Resonances may be matched accurately, and multiple wavelengths may be used to obtain Raman excitation profiles (REPs). Principles of operation are well known [119, 120], and various suitable dye laser systems [112] are available commercially and are sufficiently narrowband to provide rotational resolution. Given available dyes, computer-controlled scanning of intracavity diffraction gratings [121], and efficient nonlinear frequency conversion, laser radiation continuously tunable from the deep ultraviolet to the near infrared can be obtained. Many pulsed lasers have complex spectral [122, 123] and temporal structure which should be considered. Second harmonic generation (SHG) and mixing extends the continuous tunability to <200 nm; computer- or servo-controlled tracking is needed for continuously scanned REPs [61]. As most dye lasers have dominant vertical polarization, the SHG output will be horizontally polarized, which is transmitted with little Fresnel loss [124] by equilateral prisms. Rotation back to the vertical plane before entering the cell may be accomplished with a waveplate [107]; a pair of offset mirrors or right-angle prisms with a 90° dihedral reflection angle will form a broadband polarization changer [122]. A high repetition-rate, continuously tunable narrow-band (10 cm$^{-1}$) system based on a mode-locked Ti : sapphire has been developed for biological samples which are sensitive to ultraviolet damage [125]. Time resolution for pump-probe experiments may be increased using picosecond laser sources [126].

## 19.5.2 Laser Beam Characteristics

Beam quality considerations are important for obtaining optimum $S/N$, ratios especially where accurate intensity measurements are needed. Pulse-to-pulse variations in output energy cause intensity fluctuations in the Raman signal. However, the laser pulse amplitude can be measured by picking off a small fraction of the beam with a beam-splitter and normalizing the signal to total laser energy. It should be slightly wedged, or of sufficient thickness, that the two reflections are well separated and do not interfere. A Raman signal showing saturation due to electronic excitation or photodissociation of the sample will render this normalization less effective. Another problem is that ultraviolet light causes degradation of silicon photodetectors over time. Photodiodes are best used in the photoconductive mode to gain quick response. The applied voltage increases the charge separation inside the photodiode, decreasing the effective capacitance. Unfortunately, only a few types of ultraviolet-sensitive silicon photodiodes designed for use in the photoconductive mode are available [127]. However, visible-sensitive photodiodes may be used if the incident UV is converted to the visible using a glass window or fluorescent filter. The fluorescence converter may be viewed edge-on, as it behaves as a light pipe to guide the fluorescent emission to the photodiode. Pyroelectric detectors also provide reliable detection, but are generally less sensitive than photodiodes. In addition, when picking off light with a beam splitter (which also serves as an attenuator), it should be used near normal incidence unless the light is always $S$- or $P$-polarized. Otherwise, fluctuations in polarization of the laser light will lead to significant changes in reflectivity.

Spatial beam splitter can be a more serious problem than amplitude jitter, because it can cause the image of the experimental region to move across the entrance slit or aperture of the spectrometer, which causes the Raman signal to fluctuate and may also cause frequency errors, even if the laser amplitude is perfectly stable. This problem may be addressed by adding a total emission monitor inside the spectrometer, and normalizing the Raman signal to this [61]. Quadrant photodiodes may be used to monitor spatial jitter, and provide an error signal for correction. A spatial filter will improve beam pointing stability in the sample cell. The pinhole serves to improve beam quality (forming a $TEM_{oo}$ mode) at the expense of reduced pulse energy, and it converts spatial jitter into amplitude jitter, which is readily quantifiable. Clearly, the spatial filtering must occur before the beam pickoff [122]. Sapphire or diamond pinholes (with nonparallel substrate surfaces) are the most durable.

Laser beam propagation is described by Gaussian optics [118]; care must be taken in the optical layout, as the geometric image point of the pinhole may become separated from the Gaussian beam waist. Since light passing through the spatial filter is restricted in position, but not angle, only the geometric image point is translationally stable. The geometric image point is located by illuminating the pinhole with an extended incoherent source, its image being positioned at the geometric image point. The Gaussian beam waist may be located using a coherent source by translating a knife edge across the beam while a distant image of the beam is viewed. For a good $TEM_{oo}$ Gaussian beam, a shadow passes across the round image in the same direction as the motion of the knife edge when it is beyond the focal point, and the opposite direction before the focal point. When the knife edge is at the focal point, the image fades uniformly; a round image stays round. A negative lens may be used to increase the divergence of beam and enlarge the spot on a screen. A pencil beam in a Raman cell can be generated using long focal-length lenses, $\approx 0.5$ to 1 m. For the standard 90° collection geometry, the resulting image of this line is well matched to the spectrometer slit. The electric vector of the laser is normally polarized at right angles to direction of the collection, and it may be rotated by a waveplate [107] for depolarization studies. A pair of shorter focal-length lenses used in a telescope configuration makes the focusing adjustable. Cylindrical lenses provide a line-shaped image for solid samples and backscatter geometry; more useful area can be illuminated and a line image can be projected onto the slit of the spectrometer.

Cell designs are almost as varied as the applications of RS [128]. To effectively remove laser light scatter as the beam passes through the cell, including that originating from the laser windows, sharp-edged cone-shaped baffles are used [122]. Compared to metal sample vessels, glass cells are clean and easy to make, but are prone to fluorescence under illumination with ultraviolet laser light. Photodegradation often requires that flowing or otherwise moving samples be used. The size of Raman signals can be used to monitor the cumulative changes in a stationary sample [108]. Liquid samples can be flowed from a jet nozzle [5], and photolabile species such as hemeproteins are studied in a small flow tube [129]. Gas-dynamic focusing provides a well-defined flow geometry for gaseous samples [130], and a fuel injection valve can be used to deliver a fresh sample pulse for each laser pulse [94, 95]. Pulsed free-jet expansions also cool $CH_3I$ for improved REPs [131].

## 19.5.3 Collection Optics

The collection optics [8, 132] are one of the most important parts of the Raman system and are critical to its overall efficiency. Indeed, the relatively small solid angle over which light is collected is often the most significant loss encountered. Frequently a "large-aperture" lens is employed, but it might be expected that reflecting optics would be more effective. However, reasonable geometries seldom provide collection angles much greater than that of an $f/1$ lens. The purpose of the collection optics design is to match this low $f/\#$ to that of the spectrometer [133], usually with a little overfill to minimize adverse effects due to misalignment, spatial jitter, and lower fluence at the beam "edge" General properties of lenses may be found in references [124, 134).

The advantages of a larger collection angle may be offset by magnification of the image [135]. For example, consider a laser beam focused to a 100 micron diameter pencil, fluence of 1 J/cm$^2$ (uniform width and rectangular fluence cross-section is assumed for simplicity), passing unattenuated through the sample, and a spectrometer with $f/8$ acceptance angle and 200 micron slits using the geometry in Figure 7. Consider an ideal thin lens of 200 mm focal length and 50 mm clear aperture, and having a collection angle of $f/4$. (Even though the $f/\#$ varies inversely with the angle of collection, this approximation is widely used, as the ratio of the focal length divided by the aperture is a concept which is easy to remember.) The first lens collimates the collected light, which may then be imaged by a second lens, $f/8$ and 400 mm fl., onto the spectrometer slit. The lens pair matches the wide angle of collection to the narrower acceptance angle of the spectrometer without loss of light. The ratio of the $f/\#$'s, $4:8$, is approximately the inverse to the ratio of the cone angles, $8:4$, matching the spectrometer $f/\#$. However, this efficient angle matching comes at a price: The image size is magnified by a factor of two, enlarging it to 200 microns wide. The magnification also increases sensitivity to changes in the laser beam position.

In an attempt to improve the collection efficiency, consider substituting a 50 mm focal length lens placed 100 mm from the sample as an $f/2$ lens. The collection efficiency increases as the square of the cone angle, making this fourfold change appear very attractive. However, the magnification is increased from twofold to fourfold; the slit becomes overfilled in both height and width by factors of two, and in the first approximation no additional light enters the slit [132]. If the slit width is increased, resolution will be decreased. The laser may of course be focused more tightly to 4 J/cm$^2$, to avoid the need to widen the entrance slit. Assuming that the sample can tolerate the higher fluence and that the Raman scattering remains linear in irradiance, the scattering probability for each molecule is increased fourfold, and the collected solid angle is increased (slightly less than) fourfold, for an overall improvement of $\approx 16$-fold in signal collected per unit volume in the observed region. However, there is an eightfold shrinkage in volume of the observed cylinder, leaving only about a twofold increase in total collected light. For the other extreme, assume that the sample was already receiving the maximum acceptable fluence. Focusing the laser beam more tightly would result in a pronounced *reduction* of signal and S/N ratio.

Excessive irradiance is a routine concern for pulsed laser Raman spectroscopy. It is not difficult to generate millijoules or more of ultraviolet laser light in 10 ns pulses with a 50–200 micron beam waist, which leads to irradiance of $\geq 1$ GW/cm$^2$, sufficient to cause saturation, multiphoton processes, sample degradation, spurious emission from photoproducts, and even physical damage to a sample holder [90]. Thus, when searching for weak signals, it is often counterproductive to focus more tightly to increase the laser fluence.

An important figure of merit characteristic of collection efficiency is the product of area and acceptance solid angle termed the throughput or étendue [133, 135, 136]; the related quantity optical conductance is also used [137]. Étendue is a fixed characteristic of a particular spectrometer operating with a given resolution and wavelength. Since the collection optics exchange size and angle inversely, their étendue is constant [132]. Collection over a very large solid angle leads to an enormous étendue and provides empty magnification. Interferometric spectrometers have round entrance apertures which generally have a much larger area than the slit of a grating spectrometer of comparable size and resolution; the étendue is likewise larger (this is called the Jacquinot advantage [137]). These instruments are subject to the same constraints: Once the étendue of the particular instrument is matched, further changes in the collection optics cannot add more light. The exception is micro-Raman [138], where the image is tiny, so the largest available numerical aperture microscope objective is used, without concern that the étendue will become too large for the spectrometer. In summary, considerable care is needed to see that the optimum collection efficiency is obtained, but once this condition is achieved, it is the spectrometer, and not the collection optics, which dictates what fraction of the total Raman scattering is usable.

Additional problems are encountered with polarized Raman studies because surface reflections/transmission can introduce polarization artifacts [139–141], which are much more pronounced as the angle of incidence becomes further off-normal, such as occurs for low $f/\#$ lenses. Collection geometries at angles intermediate between those parallel or perpendicular to the laser beam propagation have been proposed [24], but the effects of polarization errors should be carefully considered before implementation. Mirrors used at 45° incidence also introduce artifacts [142]. To reduce this problem for a 180° backscatter geometry, a Lyot depolarizer is placed before the beam turning mirror in Raman optical activity (ROA) instruments [143].

Measuring absolute scattering cross sections is quite challenging, as the geometry of the collected light must be determined accurately. An integrating sphere or box [144] with highly scattering surfaces to homogenize the light is quite beneficial for accurate measurements, and a nondispersed total scatter method has been used with benzene [145]. A review presenting techniques and many molecular scattering cross sections is available [146].

Camera optics are ideal for collection of visible light. For the ultraviolet, multielement $f/1$ collection optics can be used, and the spherical aberration of an all–fused-silica lens train can be minimized with careful design. However, chromatic aberration is quite significant, and the alignment wavelength should be approximately that of the signal. Reflective optics avoid chromatic aberration [109], but introduce other problems. Spherical collectors are readily available and inexpensive, but spherical aberration blurs the image, especially if a low collection $f/\#$ is used. Ellipsoidal mirrors remove the spherical aberration, and a parabolic mirror may be used to collimate the light, which may be focused by a second mirror, or long focal-length lens, to match the input $f/\#$ of the spectrometer. The sample partially obstructs the collected light; in resonance Raman, target molecules absorb in the spectral region of the incident light, and reabsorption of the Raman light must be considered [41, 147]. Off-axis reflectors [7] and Cassegrain-type collectors may mitigate this problem [148]. For backscatter geometries [111], the collection optics must allow entry of the laser beam, such as by a hole in an off-axis ellipsoid or a flat mirror, or by a small laser reflector placed in front of the collection optics. Polarization may be altered in these off-axis geometries [143].

## 19.6 Spectrometers

The RS light emerging from the collection optics must be separated from that resulting from spurious background processes and elastic scattering, and spectrally analyzed. This is accomplished using a variety of spectrometers, and key aspects of their choice and operation are considered. Dispersive spectrometers [73, 132] separate the various wavelengths in space, usually using a diffraction grating. Two major types are in general use. For the Czerny–Turner monochromator, light passes through an entrance slit and falls onto a curved mirror; the collimated reflected beam proceeds to a plane diffraction grating. The diffracted light is reflected by a second adjacent mirror to form an image of the entrance slit at the exit plane. The optical path through the instrument is usually in the shape of the letter "M," with the grating as close to the plane of the entrance and exit slits as is practical (Figure 7). The aberrations introduced by the first off-axis "collimator" (usually spherical) mirror are largely compensated by the second "camera" mirror [73, 133]. The exit plane may contain a slit, usually the same size as the entrance slit, to pass one wavelength at a time; the monochromator grating is rotated, or "scanned," to provide a spectrum. Multichannel detectors located in the exit plane allow collection of many different wavelengths simultaneously. Another popular spectrometer type uses a concave grating which self-focusing; this combines all three optical elements into one, and so only a single optical component is needed between the entrance and exit slits [133]. For applications requiring high dispersion and high resolution, the

Czerny–Turner instrument is more suitable. Use of toroidal mirrors can provide an exceptionally flat and stigmatic field, which is particularly useful for imaging [149]. Frequency calibration is usually done with spectral lamps or standard Raman scatterers [7], while intensity calibration uses broadband lamps [150, 151], and species with well-characterized RS intensity [7]. Continuous intensity calibration has been incorporated into REPs by simultaneously monitoring total emission [61].

Both ruled and holographic gratings are used in spectrometers. Holographic gratings tend to have a broad efficiency curve as well as reduced stray light and the absence of "grating ghosts," whereas ruled gratings can provide higher peak efficiency at the blaze wavelength [152]. Ruled gratings may also be used in higher orders to provide increased dispersion. This approach is particularly useful in the deep ultraviolet, where first-order gratings are not usually designed to be used at a high angle of incidence. Holographic gratings are generally not very efficient when used above first order. All gratings have a strong polarization dependence on wavelength, but this is considerably diminished with increasing order number [153]. Care is needed to avoid detecting other orders in a grating spectrometer. Higher orders are usually easily rejected by a colored short-wavelength "cutoff" filter. In the ultraviolet, blocking a lower order of visible light is not as easy, especially in the 200 nm range, because the selection of filters is very limited. Use of a "solar blind" [95] detector is also an excellent approach to suppress visible light.

A single-grating instrument with a ruled grating will suppress scattered light by about four orders of magnitude when the spectrometer is tuned well away from the laser line, and the holographic grating will improve this to about five orders. For gas samples and a well-baffled cell, this rejection level is adequate. For condensed-phase samples, especially solids, the diffuse elastic laser scattering can be quite intense. Double monochromators [128], which are essentially two monochromators in series, are often used. Triple monochromators have also been employed, with the first two stages being used in a subtractive configuration [7, 93, 154]. The light is dispersed onto a rectangular aperture which passes the entire desired spectral region, with one edge of the aperture positioned to block the scattered laser light. The second stage recombines this light, minus the intense scattered light, into its exit slit, which in turn serves as the entrance slit for the third stage.

A two-stage system is generally not recommended for use with a multichannel detector (MCD), since the exit slit of the first stage provides a high degree of stray light rejection by passing only a narrow range of wavelengths [6], and it may have an insufficiently flat field [132]. However, a double monochromator has been used successfully with a CCD-type MCD [155], in a scanning mode [156], and an intensified photodiode array [148]. Use of first-stage dispersion which is less than the second stage has proved successful in additive and subtractive double monochromators [93]. A crossed dispersed echelle spectrometer with a charged coupled device (CCD) detector provides a very wide spectral range [157], but the small slit height [158] needed to prevent overlap of different orders reduces the amount of light entering the spectrometer.

The many optical elements of a multistage grating spectrometer can cause significant loss of light. Filters can provide an efficient means to block the scattered laser light for fixed excitation wavelengths, allowing a single-stage spectrometer to be used. Colored glass filters are effective if the frequency shift is sufficiently large [154], although filter fluorescence may add to the background [159]. Dielectric band-pass and wideband interference filters have a sharper cutoff, but usually have spectral structure on the band-pass region which will need calibration for intensity measurements. Dielectric filters can be angle-tuned with about a 10% shift to shorter wavelengths at a 45° angle, but with reduced transmission and sharpness of the cutoff. Interference filters often have a colored glass blocking material on one side, and a marked increase in fluorescence has been observed when the colored side is turned toward the light source to be blocked [159].

Holographic notch filters provide visible laser line rejection with minimal loss of the Raman signal [160]. They can reject up to six orders of magnitude of stray laser light, allowing use of a single-stage spectrometer. As the transmission frequency is angle-sensitive, they should be located in a collimated beam. A pair of volume holographic gratings used in a subtractive dispersion configuration provide a high-throughput tunable filter [161]. Multi-reflection "chevron" filters, constructed from band-pass filters, transmit the laser wavelength (to an absorber), yet reflect over 80% of the desired Raman light, providing a rejection exceeding $10^8$ [162]. This approach is suitable for use in the ultraviolet. Atomic vapor filters provide remarkably sharp notch filters for the atomic resonance lines [163]. Chemical filters have proved useful in the ultraviolet [86, 164]. A crystalline colloidal Bragg diffraction filter has also been used [165], although some effort is needed to construct it. Tunable acousto-optic and liquid crystal filters [166] or spatial filters [167] may be employed when multiple-wavelength excitations are needed. Fourier transform techniques in data processing may also be used to remove fluorescence features which are spectrally broader than the Raman features [168].

## 19.6.1 Fourier-Transform Instruments

The use of nondispersive spectrometers has increased rapidly. Fourier-transform instruments [169, 170] have come into widespread use in the past decade, although an early analysis suggested that this approach would not be practical because of the lack of suitable laser line filters [171]. F-T instruments are mainly used in the near infrared with 1064 nm radiation from the cw

Nd : YAG laser. Near-infrared excitation minimizes fluorescence interference. The high throughput (Jacquinot advantage) and the multiplex advantage of simultaneously detecting all wavelengths (Fellgett advantage) [133, 170] with infrared detectors has been very beneficial for analytical applications and, e.g., for temperature-jump and photolysis [172]. A number of factors must be considered before applying FT spectrometers with pulsed lasers. Most FT instruments are designed for continuous scanning and may not be suitable for pulsed signals. Step-scan and synchronized versions have been developed [173, 174] that suggest time resolution could reach 5 ps [175], but with a commensurate increase in complexity. Alignment and positioning of the traveling mirror needs to be accurate to a fraction of a wavelength of light, and is therefore more critical for short wavelengths, such as the deep ultraviolet.

There is a "multiplex disadvantage" [126, 176] with the FT system, in that essentially all wavelengths are recorded simultaneously by the single-channel detector. Intense scattered light will have a comparatively high level of shot noise associated with it, and this increases the noise level for *all* of the spectral elements. Very high levels of rejection are needed for the laser light, and effective filters are vital. A premonochromator has been used as a tunable filter [177]. Fluorescence may occur from photolysis products, especially diatomic species which give numerous spectrally narrow but intense emission features. Choosing an FT spectrometer for a dynamic system must be done very carefully [173]. For systems that are detector-noise–limited, such as with most infrared detectors, FT systems have the advantage. For a quantum-noise–limited detector such as a photomultiplier, the advantage of an FT instrument over a monochromator is not so great, as the more resolution elements are added, the greater the overall noise level that is added to each channel [178]. In quantum-noise–limited detection, a spectrometer with a multichannel detector is preferable from the $S/N$ viewpoint. Source fluctuations (i.e., from the laser) aggravate the multiplex disadvantage [133]. The rapid development of excellent infrared multichannel detectors may soon provide serious competition to FT Raman systems [169, 179], including the scanning multichannel method [180] for an extended spectral range.

Other analyzers include the Hadamard transform spectrometer, which works best for a system that is detector-noise–limited, and hence is most attractive for the near infrared [181]. The scanning Fabry–Perot interferometer (FPI) [182], while excellent for the highest-resolution applications, or for the wide acceptance angle of atmospheric probes in compact satellite payloads, is a rather specialized instrument. Another approach is to use a filter in place of the monochromator. To select the Raman shift, the laser rather than the filter is tuned; this allows for direct imaging, which is preferred over multiplexing [183]. This approach may not provide accurate intensity information for RRS, however, as the scattering cross section may vary considerably with wavelength. Short scans may provide a useful means to obtain high-resolution data for a band shape, or for separating closely spaced bands if a narrow band-pass filter is used. A novel method, resonance ionization detection (RID), also eliminates the need for a spectrometer [184]. The laser is tuned, and Raman scatter at the selected wavelength is absorbed by an atomic vapor as the first step in a multiphoton ionization technique. The ions can be detected with high efficiency.

## 19.6.2 Photomultiplier Detectors

Many Raman experiments make use of photomultipliers [185] which have a photocathode inside a transparent evacuated glass bulb from which an incident photon ejects an electron, typically with 20% quantum efficiency near the peak of the spectral sensitivity. Photomultipliers are available in a wide variety of sizes and types (end-on, side-on), with sensitivity that may or may not be uniform across the input face. The entrance window material usually determines the short-wavelength cutoff, and the photocathode material [127, 186] determines the long-wavelength cutoff and the range of spectral sensitivity. A bialkali photocathode has high quantum efficiency throughout the ultraviolet and much of the visible; it is excellent for many pulsed laser resonance Raman applications. Cesium telluride cuts off at about 320 nm; hence it is called "solar blind" and is good for suppressing stray visible light [85], but different photomultipliers have considerable variation in the degree of suppression of visible light. Red-sensitive materials such as the multialkali and gallium arsenide photocathodes have lower work functions; hence, thermally emitted electrons cause a considerably larger "dark current." Red-sensitive detectors are often cooled to reduce the dark current. Such red detectors should not be exposed to bright fluorescent light, as this temporarily increases the dark current and may permanently degrade the red sensitivity. Ordinary glass and UV glass windows begin to absorb <320 nm and <250 nm respectively, so synthetic fused silica should be used for the deep ultraviolet. $MgF_2$ windows may be used to provide good transmission to about 120 nm.

After leaving the photocathode, the ejected photoelectron is multiplied by a series of dynodes, with a typical pulse of $10^6$ electrons falling on the anode. The gain may be quickly gated off [187] to avoid saturating the detector during an intense photolysis pulse. Single-photoelectron sensitivity is readily achieved [188]. Careful photomultiplier selection and circuitry design is needed to maintain a fast, linear response [186]. A $10 \times$ preamplifier incorporated into the housing is useful for boosting weak signals and provides impedance matching to a 50 ohm coaxial cable [122]. A gated integrator is normally used to capture the signal produced during the laser pulse, thereby discriminating against signals associated with long-lived fluorescence, or other non-Raman background emission [78, 95]. In pulse experiments, time delays associated with electron transit times in the photomultiplier must be considered. Transit times vary considerably, depending on the voltage applied across the dynde chain, and the corresponding change in delay must be considered.

For high-speed response, microchannel plate (MCP) photomultipliers are used. A single plate has a gain of about three orders of magnitude, and additional plates may be included for $100 \times$ more gain each. The transit time, and especially the transit time spread, is very small, and in the time-correlated photon counting (TCPC) mode, these detectors can provide time resolutions in the tens of picoseconds. When picosecond laser sources are used for fast time response and/or temporally based fluorescence suppression, the MCP-photomultiplier is an outstanding detector, offering a combination of high speed and sensitivity. However, spectrometer design must be considered; the unequal path lengths for different rays in a large grating monochromator can stretch pulses by hundreds of picoseconds [189]. A subtractive double monochromator may be preferred. Even conventional photomultipliers have achieved 0.4 ns resolution in TCPC mode [190]. Streak cameras provide the fastest response of all and now are available with photon counting sensitivity [102].

### 19.6.3 Multichannel Detection (MCD)

Single-channel detection wastes much of the available light from a complex Raman spectrum. When many different frequencies are to be monitored, measuring each simultaneously will be more efficient. It is essential to distinguish between multichannel and multiplex detection. Multiplex detection allows many spectral frequencies to simultaneously fall on the same detector. Thus, as the range of frequencies is expanded, the total amount of light falling on the detector increases, and the associated noise also increases. After the transform is performed to recover the spectrum, contributions to the noise on each feature come from all of the incident frequencies. A few strong features can show an outstanding signal-to-noise ratio, but weak a feature in the midst of some very strong ones will suffer from increased shot noise. For multichannel detection, however, each spectral channel is essentially independent, and the shot noise level for each spectral frequency is independent of the amount of light falling on other channels. In practice, excessive light at one frequency can cause "blooming" or spillover into adjacent channels, and raise the background.

For maximum utilization of the Raman-scattered light, multichannel detection is appealing. The first widely used detection system for Raman scattering was a multichannel device—the photographic plate. Although vidicon camera tubes with a silicon intensified target (SIT) [191] restarted multichannel detection, silicon photodiode arrays [148, 189] were the first major system to gain widespread application. A typical array consists of a line of several hundred photodiode elements, typically 25 micrometers wide and only 2.5 mm high. This is less than the height of the image of the entrance slit formed at the exit plane, leading to reduced coupling efficiency. The dispersed spectrum from the spectrometer (it is essential that the spectrometer have a flat field in the exit plane) falls on the array, with each adjacent pixel collecting a different wavelength.

Although silicon photodiodes have excellent sensitivity at the red end of the spectrum, the quantum efficiency is poor in the ultraviolet. In addition, there is leakage current and readout noise when the array is scanned [154]. Such arrays lack the sensitivity of photomultipliers, so the multichannel gain is significantly offset by a considerably lower signal-to-noise ratio per channel. Unlike photomultipliers, however, the photodiode can store the charge pair created by each incident photon, so that the spectrum may be effectively integrated on the array for minutes to hours for very weak signals. This reduces the number of readouts needed for a given total data collection time, and hence lessens the readout noise. Thermoelectric cooling to as low as $-30°C$ reduces the leakage current, and it is essential for extended collection times to avoid saturating the charge capacity of the pixels. Readout times are on the order of milliseconds, so fast time resolution is not available. Residual charges remaining after readout may contaminate subsequent spectra.

An image intensifier [192] is often used in front of a diode array. A semitransparent photocathode produces photoelectrons and is similar to that of a photomultiplier. Amplification is most commonly done with a multichannel plate. Photoelectrons are multiplied in individual $\approx 10$-micron diameter parallel channels and strike a phosphor screen. The resulting amplified image is transferred to the array by an optical-fiber coupler. Double MCP intensifiers have a gain of $> 10^5$ and generate such large optical signals from the phosphor that individual photoelectrons may be observed, as with a photomultiplier. Gated intensifiers with time resolution on the order of a few nanoseconds are available, so the time resolution lost by using a photodiode array can be recovered by gating the intensifier on and off as needed [189].

The charge-coupled device (CCD) is rapidly supplanting the diode array detector for multichannel detection [132]. $S/N$ ratio comparisons with several other types of spectrometers are given in reference [178]. A CCD array has $\geq 10^6$ pixels, each about 20 micrometers square. They may be read out individually, but normally the pixels are "binned" together into slit-shaped groups. Readout times are reduced by binning, and less electrical noise is added. Since the height of the array is comparable to the slit height, the overfilling in this dimension is not a serious problem. Low-noise CCDs are designed for scientific and astronomical [193] applications, and cooling to $-110°C$ greatly reduces the dark charge accumulation [7]. Coupling a CCD to a cross-dispersal echelle spectrometer [158, 194] takes advantage of the 2-D format and considerably extends the spectral range, although the entrance slit height must be small. The noise level of CCDs designed for spectroscopic use is so low that intensification is not needed, although identification and removal of "cosmic ray spikes" may be needed for the most sensitive applications. The quantum efficiency in the red and near infrared can be above 40%, and may approach 80% for the "backside thinned" devices [132]. In the ultraviolet, much of the light is lost as it is absorbed entering the detector, and scintillator-type coatings are used to convert the short wavelengths to visible wavelengths. This reduces overall quantum efficiency to about 20%–30% [132]. The drawback of limited range imposed by the finite size of the detector surface is overcome by scanning multichannel detection schemes [179] or an echelle spectrometer. An image intensifier is necessary for fast gated detection; for other applications, however, a careful analysis is useful in determining whether or not it will provide an improvement in the $S/N$ ratio [192].

## 19.7 Conclusion

Raman spectroscopy can be applied to study a wide variety of systems undergoing change, and many more examples may be found in recent books and reviews [195–199]. The applications are astonishingly diverse, including femtosecond dynamics of photodissociation, determining nuclear displacements upon excitation of complex molecules, and studying isomerization, radicals, combustion, and even living bacteria *in situ*. The ingenious apparatus which has been developed for this work is almost as diverse as the applications. One of the more unusual features of Raman studies is that the time scale of the phenomena studied may be very much faster than the time scale of the experiment. Rapid motions of nuclei on a PES, which are the essence of chemical reactions, are captured by the information contained in Raman scattering.

## Acknowledgment

Acknowledgment is made to Peter B. Kelly and Bruce R. Johnson for reviewing the manuscript, to Chung-Yi Kung for assistance in preparing Figures 2 and 4, and for support from the National Science Foundation (Grant #CHE-92-20278) and the Robert A. Welch Research Foundation.

## References

1. Raman, C. V., and Krishnan, K. S. (1928). *Nature* **121**, 501.
2. Landsberg, G., and Mandelstam, I. (1928). *Naturewissenschaften* **16**, 557.
3. Long, D. A. *et al.* (1989). *Chem. Br.* **25**, 589–622.
4. Austin, J. C., Jordon, T., and Spiro, T. G. (1993). In *Biomolecular Spectroscopy, Part A*, R. J. H. Clark and R. E. Hester (eds.), Vol. 20, pp. 55–127. John Wiley & Sons, Ltd., New York.

5. Mathies, R. (1979). In *Chemical and Biological Applications of Lasers*, C. B. Moore (ed.), Vol. 4, pp. 55–99. Academic Press, New York.

6. Hamaguchi, H. (1987). In *Vibrational Spectra and Structure: A Series of Advances*, J. R. Durig (ed.), Vol. 16, pp. 227–309. Elsevier, Amsterdam.

7. Ferraro, J. R., and Nakamoto, K. (1994). *Introductory Raman Spectroscopy*. Academic Press, Boston.

8. Long, D. A. (1977). *Raman Spectroscopy*. McGraw-Hill, New York.

9. Szymanski, H. A. (1970). *Raman Spectroscopy: Theory and Practice*. Plenum, New York.

10. Herzberg, G. (1991). *Infrared and Raman Spectra of Polyatomic Molecules*, Molecular Spectra and Molecular Structure, Vol. 2. Krieger Publishing Co., Malabar, Florida.

11. Wilson, E. B., Jr., Decius, J. C., and Cross, P. C. (1980). *Molecular Vibrations: The Theory of Infrared and Raman Vibrational Spectra*. Dover, New York.

12. Myers, A. B., and Mathies, R. A. (1987). In *Biological Applications of Raman Spectroscopy: Resonance Raman Spectra of Polyenes and Aromatics*, T. G. Spiro (ed.), Vol. 2, pp. 1–58. John Wiley & Sons, New York.

13. Tang, J., and Albrecht, A. C. (1970). In *Raman Spectroscopy: Theory and Practice*, H. A. Szymanski (ed.), Vol. 2, pp. 33–68. Plenum, New York.

14. Clark, R. J. H., and Dines, T. J. (1982). In *Advances in Infrared and Raman Spectroscopy*, R. J. H. Clark and R. E. Hester (eds.), Vol. 9, pp. 282–360. Heyden, London.

15. Hildebrandt, P., Tsuboi, M., and Spiro, T. G. (1990). *J. Phys. Chem.* **94**, 2274.

16. Tsuboi, M., Nishimure, Y., Hirakawa, A., and Peticolas, W. L. (1987). In *Biological Applications of Raman Spectroscopy: Resonance Raman Spectra of Polyenes and Aromatics*, T. G. Spiro (ed.), Vol. 2, pp. 109–179. John Wiley & Sons, New York.

17. Sension, R. J. *et al.* (1992). *J. Chem. Phys.* **96**, 2617.

18. Okamoto, H., Sekimoto, Y., and Tasumi, M. (1994). *Spectrochim. Acta* **50A**, 1467.

19. Clark, R. J. H., and Dines, T. J. (1986). *Angew. Chem. Int. Ed. Engl.* **25**, 131–158.

20. Kiefer, W. (1974). *Appl. Spectrosc.* **28**, 115–131.

21. Clucas, W. A., Armstrong, R. S., and Nugent, K. W. (1992). In *Thirteenth International Conference on Raman Spectroscopy*, W. Kiefer, M. Cardona, G. Schaack, F. W. Schneider, and H. W. Schrötter (eds.), Vol. 13, pp. 92–93. John Wiley & Sons, Chichester.

22. Hamaguchi, H. (1985). In *Advances in Infrared and Raman Spectroscopy*, R. J. H. Clark and R. E. Hester (eds.), Vol. 12, pp. 273–310. John Wiley & Sons, Chichester.

23. Johnson, B. B., and Peticolas, W. L. (1976). In *Ann. Rev. Phys. Chem.*, Vol. 27, pp. 465–491. Annual Reviews, Inc., Palo Alto, California.

24. Mortensen, O. S., and Hassing, S. (1980). In *Advances in Infrared and Raman Spectroscopy*, R. J. H. Clark and R. E. Hester (eds.), Vol. 6, p. 1. Wiley, New York.

25. Page, J. B. (1991). In *Light Scattering in Solids VI: Topics in Applied Physics*, M. Cardona and G. Güntherodt (eds.), Vol. 68, pp. 17–72. Springer-Verlag, Berlin.

26. Ribeiro, M. C. C., Barreto, W. J., and Santos, P. S. (1994). *J. Raman Spectrosc.* **25**, 923.

27. Champion, P. M., and Albrecht, A. C. (1982). *Ann. Rev. Phys. Chem.* **33**, 353–376.

28. Hizhnyakov, V., and Tehver, I. (1988). *J. Raman Spectrosc.* **19**, 383.

29. Page, J. B., and Tonks, D. L. (1981). *J. Chem. Phys.* **75**, 5694.

30. Blazej, D. C., and Peticolas, W. L. (1980). *J. Chem. Phys.* **72**, 3134.

31. Stallard, B. R., Champion, P. M., Callis, P. R., and Albrecht, A. C. (1984). *J. Chem. Phys.* **80**, 70.
32. Albrecht, A. C., Clark, R. J. H., Oprescu, D., Owens, S. J. R., and Svendsen, C. (1994). *J. Chem. Phys.* **101**, 1890.
33. Cable, J. R., and Albrecht, A. C. (1986). *J. Chem. Phys.* **84**, 4745.
34. Joo, T., and Albrecht, A. C. (1993). *J. Phys. Chem.* **97**, 1262–1264.
35. Heller, E. J., Sundberg, R. L., and Tannor, D. (1982). *J. Phys. Chem.* **86**, 1822–1833.
36. Heller, E. J. (1981). *Acc. Chem. Res.* **14**, 368–375.
37. Williams, S. O., and Imre, D. G. (1988). *J. Phys. Chem.* **92**, 3363.
38. Chaseman, D., Tannor, D. J., and Imre, D. G. (1988). *J. Chem. Phys.* **89**, 6667.
39. Leforestier, C. *et al.* (1991). *J. Comput. Phys.* **94**, 59–80.
40. Hartke, B., Kiefer, W., Kolba, E., Manz, J., and Strempel, J. (1992). *J. Chem. Phys.* **96**, 5636.
41. Morikis, D., Li, P., Bangcharoenpaurpong, O., Sage, J. T., and Champion, P. M. (1991). *J. Phys. Chem.* **95**, 3391–3398.
42. Sundberg, R. L., and Heller, E. J. (1982). *Chem. Phys. Lett.* **93**, 586.
43. Imre D., Kinsey, J. L., Sinha, A., and Krenos, J. (1984). *J. Phys. Chem.* **88**, 3956.
44. von Dirke, M., Heumann, B., Schinke, R., Sension, R. J., and Hudson, B. S. (1993). *J. Chem. Phys.* **99**, 1050.
45. Sension, R. J., and Hudson, B. S. (1989). *J. Chem. Phys.* **90**, 1377.
46. Markel, F., and Myers, A. B. (1993). *J. Chem. Phys.* **98**, 21.
47. Shapiro, M. (1993). *Chem. Phys. Lett.* **212**, 444.
48. Ci, X., and Myers, A. B. (1992). *J. Chem. Phys.* **96**, 6433.
49. Reber, C., and Zink, J. I. (1992). *J. Phys. Chem.* **96**, 571.
50. Galica, G. E., Johnson, B. R., Kinsey, J. L., and Hale, M. O. (1991). *J. Phys. Chem.* **95**, 7994.
51. Lao, K. Q., Person, M. D., Xayariboun, P., and Butler, L. J. (1990). *J. Chem. Phys.* **92**, 823.
52. Wang, P. G., and Ziegler, L. D. (1993). *J. Phys. Chem.* **97**, 3139.
53. Johnson, A., and Myers, A. B. (1995). *J. Chem. Phys.* **102**, 3519.
54. Hirakawa, A. Y., and Tsuboi, M. (1983). In *Vibrational Spectra and Structure: A Series of Advances*, J. R. Durig (ed.), Vol. 12, pp. 145–204. Elsevier, Amsterdam.
55. Zink, J. I., and Shin, K.-S. K. (1991). In *Adv. Photochem.*, D. H. Volman, G. S. Hammond, and D. C. Neckers (eds.), Vol. 16, pp. 119–214. Wiley-Interscience, New York.
56. Kash, P. W., Waschensky, G. C. G., Butler, L. J., and Franci, M. M. (1993). *J. Chem. Phys.* **99**, 4479.
57. Schinke, R. (1993). *Photodissociation Dynamics: Spectroscopy and Fragmentation of Small Polyatomic Molecules*. Cambridge University Press, Cambridge, U.K.
58. Chung, Y. C., and Ziegler, L. D. (1988). *J. Chem. Phys.* **89**, 4692.
59. Campbell, D. J., and Ziegler, L. D. (1993). *Chem. Phys. Lett.* **201**, 159.
60. Lee, S.-Y., and Mathies, R. A. (1988). *Chem. Phys. Lett.* **151**, 9.
61. Kung, C.-Y., Chang, B.-Y., Kittrell, C., Johnson, B. R., and Kinsey, J. L. (1993). *J. Phys. Chem.* **97**, 2228.
62. Remacle, F., and Levine, R. D. (1993). *J. Chem. Phys.* **99**, 4908.
63. Johnson, B. R., and Kinsey, J. L. (1993). *J. Chem. Phys.* **99**, 7267.
64. Williams, S. O.,, and Imre, D. G. (1988). *J. Phys. Chem.* **92**, 6636.
65. Johnson, B. R., and Kinsey, J. L. (1995). In *Femtosecond Chemistry*, J. Manz and L. Wöste (eds.), p. 353. VCH, Weinhein.

66. Behringer, J. (1975). *Mol. Spectrosc. Chem. Soc., London [Specialist Periodical Reports]* **2**, 100–172; **3**, 163–280.
67. Kono, H., Nomura, Y., and Fujimura, Y. (1991). In *Adv. Chem. Phys.*, I. Prigogine and S. A. Rice (eds.), Vol. 80, p. 403. Wiley, New York.
68. Lee, D., and Albrecht, A. C. (1985). In *Advances in Infrared and Raman Spectroscopy*, R. J. H. Clark and R. E. Hester (eds.), Vol. 12, pp. 179–213. John Wiley & Sons, Ltd., Chichester.
69. Rousseau, D. L., and Williams, P. F. (1976). *J. Chem. Phys.* **64**, 3519.
70. Friedman, J. M., and Hochstrasser, R. M. (1974). *Chem. Phys.* **6**, 155–165.
71. Ziegler, L. D. (1994). *Acc. Chem. Res.* **27**, 1–8.
72. Stevens, R. E., Kittrell, C., and Kinsey, J. L. (1995). *J. Phys. Chem.* **99**, 11067.
73. Stewart, J. E. (1970). *Infrared Spectroscopy: Experimental Methods and Techniques.* Marcel Dekker, New York.
74. Chang, B.-Y., *et al.* (1994). *J. Chem. Phys.* **101**, 1914.
75. Kittrell, C., Le Floch, A. C., and Garetz, B. A. (1993). *J. Phys. Chem.* **97**, 2221.
76. Kitagawa, T., and Ogura, T. (1993). In *Biomolecular Spectroscopy, Part B*, R. J. H. Clark and R. E. Hester (eds.), Vol. 21, p. 139. John Wiley & Sons, Chichester.
77. Murphy, W. F. (1991). In *Analytical Raman Spectroscopy*, J. G. Grasselli and B. J. Bulkin (eds.), Vol. 114, pp. 425–451. John Wiley & Sons, Inc., New York.
78. Sharma, S. K. (1989). In *Raman Spectroscopy: Sixty Years On*, H. D. Bist, J. R. Durig, and J. F. Sullivan (eds.), Vol. 17B, pp. 513–568. Elsevier, Amsterdam.
79. Veirs, D. K., Ager, J. III, Loucks, E. T., and Rosenblatt, G. M. (1990). *Appl. Opt.* **29**, 4969.
80. Durig, J. R., and Little, T. S. (1989). In *Recent Trends in Raman Spectroscopy*, S. B. Banerjee and S. S. Jha (eds.), pp. 93–112. World Scientific, Singapore.
81. Bartlett, J. R., and Cooney, R. P. (1987). In *Spectroscopy of Inorganic-Based Materials*, R. J. H. Clark and R. E. Hester (eds.), Vol. 14, pp. 187–283. John Wiley & Sons, New York.
82. Campion, A. (1987). In *Vibrational Spectroscopy of Molecules on Surfaces*, J. T. Yates, Jr. and T. E. Madey (eds.), pp. 345–415. Plenum Press, New York.
83. Redington, R. L. (1983). In *Vibrational Spectra and Structure: A Series of Advances*, J. R. Durig (ed.), Vol. 12, pp. 323–395. Elsevier, Amsterdam.
84. Laane, J. (1983). In *Vibrational Spectra and Structure: A Series of Advances*, J. R. Durig (ed.), Vol. 12, pp. 405–467. Elsevier, Amsterdam.
85. Dalterio, R. A., Nelson, W. H., Britt, D., and Sperry, J. F. (1987). *Appl. Spectrosc.* **41**, 417.
86. Baek, M., Nelson, W. H., Britt, D., and Sperry, J. F. (1988). *Appl. Spectrosc.* **42**, 1312.
87. Schaschek, K., Popp, J., and Kiefer, W. (1993). *J. Raman Spectrosc.* **24**, 69.
88. Asher, S. A. (1988). *Ann. Rev. Phys. Chem.* **39**, 537–588.
89. Cho, N., and Asher, A. (1993). *J. Am. Chem. Soc.* **115**, 6349.
90. Li, P., Sage, J. T., and Champion, P. M. (1992). *J. Chem. Phys.* **97**, 3214.
91. Ryabov, E. A., and Letokhov, V. S. (1989). In *Recent Trends in Raman Spectroscopy*, S. B. Banerjee and S. S. Jha (eds.), pp. 28–39. World Scientific, Singapore.
92. Hester, R. E. (1989). In *Raman Spectroscopy: Sixty Years On*, H. D. Bist, J. R. Durig, and J. F. Sullivan (eds.), Vol. 17A, pp. 351–362. Elsevier, Amsterdam.
93. Atkinson, G. H. (1982). In *Advances in Infrared and Raman Spectroscopy*, R. J. H. Clark and R. E. Hester (eds.), Vol. 9, pp. 1–62. Heyden, London.
94. Westre, S. G., Gansberg, T. E., Kelly, P. B., and Ziegler, L. D. (1992). *J. Phys. Chem* **96**, 3610.

95. Liu, X., Getty, J. D., and Kelly, P. B. (1993). *J. Chem. Phys.* **99**, 1522.
96. Peticolas, W. L. (1989). In *Raman Spectroscopy: Sixty Years On,* H. D. Bist, J. R. Durig, and J. F. Sullivan (eds.), Vol. 17A, pp. 467–484. Elsevier, Amsterdam.
97. Sasaki, Y., and Hamaguchi, H. (1994). *Spectrochim. Acta* **50A**, 1475.
98. Langkilde, F. W. *et al.* (1994). *J. Chem. Phys.* **100**, 3503.
99. Mabrouk, P. A., and Wrighton, M. S. (1986). *Inorg. Chem.* **25**, 526.
100. Kumble, R., Hu, S., Loppnow, G. R., Vitols, S. E., and Spiro, T. G. (1993). *J. Phys. Chem.* **97**, 10521.
101. Rodgers, K. R., and Spiro, T. G. (1994). *Science* **265**, 1697.
102. Hamaguchi, H., and Gustafson, T. L. (1994). *Ann. Rev. Phys. Chem.* **45**, 593–622.
103. Koyama, Y., and Mukai, Y. (1993). In *Biomolecular Spectroscopy, Part B,* R. J. H. Clark and R. E. Hester (eds.), Vol. 21, p. 49. John Wiley & Sons, Chichester.
104. Xu, X., Yu, S.-C., Lingle, R., Jr., Zhu, H., and Hopkins, J. B. (1991). *J. Chem. Phys.* **95**, 2445.
105. Phillips, D. L., Rodier, J.-M., and Myers, A. B. (1993). *Chem. Phys.* **175**, 1.
106. Lau, A., Siebert, F., and Werncke, W., eds. (1993). *Time-Resolved Vibrational Spectroscopy VI: Springer Proceedings in Physics,* Vol. 74. Springer Verlag, Berlin.
107. Kliger, D. S., Lewis, J. W., and Randall, C. E. (1990). *Polarized Light in Optics and Spectroscopy.* Academic Press, Boston.
108. Killough, P. M., De Vito, V. L., and Asher, S. A. (1991). *Appl. Spectrosc.* **45**, 1067.
109. Asher, S. A., Johnson, C. R., and Murtaugh, J. (1983). *Rev. Sci. Instrum.* **54**, 1657.
110. Asher, S. A. (1993). *Anal. Chem.* **65**, 59A and 201A.
111. Hudson, B., and Sension, R. J. (1989). In *Raman Spectroscopy: Sixty Years On,* H. D. Bist, J. R. Durig, and J. F. Sullivan (eds.), Vol. 17A, pp. 363–390. Elsevier, Amsterdam.
112. Murray, J. R. (1987). In *Laser Spectroscopy and Its Applications,* L. J. Radziemski, R. W. Solarz, and J. A. Paisner (eds.), pp. 91–174. Marcel Dekker, Inc., New York.
113. Littman, M. (1995). In *Electromagnetic Radiation,* F. B. Dunning and R. G. Hulet (eds.), manuscript in preparation. Academic Press, San Diego.
114. Sliney, D., and Wolbarsht, M. (1980). *Safety with Lasers and Other Optical Sources: A Comprehensive Handbook.* Plenum Press, New York.
115. Winburn, D. C. (1990). *Practical Laser Safety,* A. L. Kling (ed.), Occupational Safety and Health. Marcel Dekker, New York (1990).
116. White, J. C. (1992). In *Tunable Lasers: Topics in Applied Physics,* L. F. Mollenauer, J. C. White, and C. R. Pollock (eds.), Vol. 59, pp. 115–207.
117. Hudson, B., *et al.* (1986). In *Advances in Laser Spectroscopy,* B. A. Garretz and J. R. Lombardi (eds.), Vol. 3, pp. 1–32. John Wiley & Sons, New York.
118. Siegman, A. E. (1986). In *Lasers,* pp. 581–660. University Science Books, Mill Valley, California.
119. Duarte, F. J., and Hillman, L. W., eds. (1990). *Dye Laser Principles.* Academic Press, Boston.
120. Schäefer, F. P., ed. (1990). *Dye Lasers: Topics in Applied Physics,* Vol. 1. Springer-Verlag, Berlin.
121. Duarte, F. J., ed. (1991). *High-Power Dye Lasers: Springer Series in Optical Sciences,* Vol. 65. Springer-Verlag, Berlin.
122. Kittrell, C. (1995). In *Molecular Dynamics and Spectroscopy by Stimulated Emission Pumping,* H.-L. Dai and R. W. Field (eds.), Vol. 4, pp. 109–147. World Scientific, Singapore.

123. Kajava, T. T., Lauranto, H. M., and Salomaa, R. R. E. (1992). *Appl. Opt.* **31**, 6987.
124. Hecht, E. (1987). *Optics.* Addison-Wesley, Reading, Massachusetts.
125. Doig, S. J., and Prendergast, F. G. (1995). *Appl. Spectrosc.* **49**, 247.
126. Diem, M. (1993). In *Introduction to Modern Vibrational Spectroscopy,* M. Diem (ed.), pp. 109–174. Wiley-Interscience, New York.
127. Hamamatsu Corp. (1994). *Photodiode and Photomultiplier Catalogues.* Photodiode Model #1722-02. Bridgewater, New Jersey.
128. Barańska, H., and Łabudzińska, A. (1987). In *Laser Raman Spectroscopy: Analytical Applications,* H. Barańskia, A. Łabudzińska, and J. Terpiński (eds.), pp. 47–78. Ellis Horwood, Chichester.
129. Ogura, T., and Kitagawa. (1988). *Rev. Sci. Instrum.* **59**, 1316.
130. Robinson, J. C., Fink, M., and Mihill, A. (1992). *Rev. Sci. Instrum.* **63**, 3280.
131. Wang, P. G., Zhang, Y. P., Ruggles, C. J., and Ziegler, L. D. (1990). *J. Chem. Phys.* **92**, 2806.
132. McCreery, R. L. (1994). In *Charge-Transfer Devices in Spectroscopy,* J. V. Sweedler, K. L. Ratzlaff, and M. B. Denton (eds.), pp. 227–279. VCH, New York.
133. Thorne, A. P. (1988). *Spectrophysics.* Chapman and Hall, London.
134. Kingslake, R. (1978). *Lens Design Fundamentals.* Academic Press, San Diego.
135. Fryling, M., Frank, C. J., and McCreery, R. L. (1993). *Appl. Spectrosc.* **47**, 1965.
136. Bland-Hawthorn, J., and Cecil, G. (1996). In *Atomic, Molecular, and Optical Physics: Atoms and Molecules,* F. B. Dunning and R. G. Hulet (eds.), Vol. 29B, pp. 000–000. Academic Press, San Diego, California.
137. Schrader, B. (1994). In *Fourier Transform Raman Spectroscopy,* D. B. Chase and J. F. Rabolt (eds.), pp. 49–71. Academic Press, San Diego, California.
138. Pallister, D. M., and Morris, M. D. (1994). *Appl. Spectrosc.* **48**, 1277.
139. Barron, L. D., and Vrbancich, J. (1985). In *Advances in Infrared and Raman Spectroscopy,* R. J. H. Clark and R. E. Hester (eds.), Vol. 12, pp. 215–272. John Wiley & Sons, Ltd., Chichester.
140. Dawson, R. (1972). *Spectrochem. Acta* **28A**, 715.
141. Teboul, V., Godef, J. L., and Le Duff, Y. (1992). *Appl. Spectrosc.* **46**, 476.
142. Nafie, L. A., and Che, D. (1994). *Adv. Chem. Phys.* **85**, 105–149.
143. Hecht, L., and Barron, L. D. (1994). *J. Raman Spectrosc.* **25**, 443–451.
144. Li, B., and Myers, A. B. (1990). *J. Phys. Chem.* **94**, 4051.
145. Schomacker, K. T., Delaney, J. K., and Champion, P. M. (1986). *J. Chem. Phys.* **85**, 4240.
146. Schrötter, H. W., and Klöckner, H. W. (1979). In *Raman Spectroscopy of Gases and Liquids: Topics in Current Physics,* A. Weber (ed.), Vol. 11, pp. 123–166. Springer Verlag, Berlin.
147. Strekas, T. C., Adams, D. H., Packer, A., and Spiro, T. G. (1974). *Appl. Spectrosc.* **28**, 324.
148. Kaminaka, S., and Kitagawa, T. (1992). *Appl. Spectrosc.* **46**, 1804.
149. Niemczyk, T. M., and Gobeli, G. W. (1990). In *Optical Spectroscopic Instrumentation and Techniques for the 1990s,* B. J. McNamara (ed.), Vol. 1318, p. 33. SPIE, Bellingham, Washington.
150. Ouillon, R., and Adam, S. (1982). *J. Raman Spectrosc.* **12**, 281.
151. Purcell, F. J., Kaminski, R., and Russavage, E. (1980). *Appl. Spectrosc.* **34**, 323.
152. Loewen, E. G. (1983). In *Applied Optics and Optical Engineering,* R. R. Shannon and J. C. Wyant (eds.), Vol. 9, pp. 33–71. Academic Press, New York.
153. Maystre, D., Neviére, M., and Petit, R. (1980). In *Electromagnetic Theory of Gratings,* R. Petit (ed.), Vol. 22, pp. 160–225. Springer Verlag, Berlin.

154. Deffontaine, A., Bridoux, M., and Delhaye, M. (1984). *Rev. Phys. Appl.* **19**, 415.
155. Deckert, V., *et al.* (1955). *Appl. Spectrosc.* **49**, 253.
156. Engert, C., Michelis, T., and Kiefer, W. (1991). *Appl. Spectrosc.* **45**, 1333.
157. Scheeline, A., Bye, C. A., Miller, D. L., Rynders, S. W., and Owen, R. C., Jr. (1991). *Appl. Spectrosc.* **45**, 334.
158. Pelletier, M. J. (1990). *Appl. Spectrosc.* **44**, 1699.
159. Bristow, M. P. F. (1979). *Appl. Opt.* **18**, 952.
160. Schoen, C. L., Sharma, S. K., Helsley, C. E., and Owen, H. (1993). *Appl. Spectrosc.* **47**, 305.
161. Pallister, D. M., Govil, A., Morris, M. D., and Colburn, W. S. (1994). *Appl. Spectrosc.* **48**, 1015.
162. Puppels, G. J., de Grauw, C. G., te Plate, M. B. J., and Greve, J. (1994). *Appl. Spectrosc.* **48**, 1399.
163. Pelletier, M. J. (1993). *Appl. Spectrosc.* **47**, 69.
164. Chou, P.-T., Studer, S. L., and Martinez, M. L. (1991). *Appl. Spectrosc.* **45**, 513.
165. Flangh, P. L., O'Donnell, S. E., and Asher, S. A. (1984). *Appl. Spectrosc.* **38**, 847.
166. Morris, H. R., Hoyt, C. C., and Treado, P. J. (1994). *Appl. Spectrosc.* **48**, 857.
167. Trulson, M. O., Lueck, H. B., and Freidrich, D. M. (1994). *Appl. Spectrosc.* **48**, 720.
168. Mann, C. K., and Vickers, T. J. (1987). *Appl. Spectrosc.* **41**, 427.
169. Chase, D. B., and Rabolt, J. F. (1994). In *Fourier Transform Raman Spectroscopy*, D. B. Chase and J. F. Rabolt (eds.), pp. 1–48. Academic Press, San Diego.
170. Hendra, P., Jones, C., and Warnes, G. (1991). *Fourier Transform Raman Spectroscopy: Instrumentation and Chemical Applications*, M. Masson and J. F. Tyson (eds.), Ellis Horwood Series in Analytical Chemistry. Ellis Horwood, New York.
171. Schildkraut, E. R., and Hirschfeld, T. B. (1973). In *Laser Raman Gas Diagnostics*, M. Lapp and C. M. Penny (eds.), pp. 379–388. Plenum Press, New York.
172. Cutler, D. J., and Petty, C. J. (1991). *Spectrochim. Acta* **47A**, 1159.
173. Hartland, G. V., Xie, W., and Dai, H.-L. (1992). *Rev. Sci. Instrum.* **63**, 3261.
174. Palmer, R. A. (1993). *Spectroscopy* **8**, 26.
175. Jas, G. S., Wan, C., and Johnson, C. K. (1994). *Spectrochim. Acta* **50A**, 1825.
176. Williams, R. (1989). *Appl. Spectrosc. Reviews* **25**, 63–79.
177. Stubley, E. A., and Horlick, G. (1985). *Appl. Spectrosc.* **39**, 811.
178. Meaburn, J. (1976). *Detection and Spectrometry of Faint Light*. Vol. 56, Astrophysics and Space Science Library. D. Reidel Publishing, Boston.
179. Deckert, V., and Kiefer, W. (1992). *Appl. Spectrosc.* **46**, 322.
180. Panitz, J.-C., Zimmerman, F., Fischer, F., Häfner, W., and Wokaun, A. (1994). *Appl. Spectrosc.* **48**, 454.
181. Hammaker, R. M., *et al.* (1990). In *Raman and Luminescence Spectroscopies in Technology II*, F. Adar and J. E. Griffiths (eds.), Vol. 1336, pp. 124–134. SPIE, Bellingham, Washington.
182. Vaughan, J. M. (1989). *The Fabry–Perot Interferometer*. The Adam Hilger Series on Optics and Optoelectronics. Adam Hilger, Bristol.
183. Puppels, G. J., Grond, M., and Greve, J. (1993). *Appl. Spectrosc.* **47**, 1256.
184. Smith, B. W., Omenetto, N., and Winefordner, J. D. (1990). *Spectrochim. Acta: Future Trends in Spectroscopy*: 45A Supplement, 101–111.
185. Young, A. T. (1974). In *Astrophysics, Part A: Methods of Experimental Physics*, N. Carleton (ed.), Vol. 12A, pp. 1–94. Academic Press, New York.
186. Burle Industries Inc. (1980). *Photomultiplier Handbook*. Burle Technologies, Inc., Lancaster, Pennsylvania.

187. Benard, D. J., and Boehmer, E. (1994). *Appl. Spectrosc.* **48,** 1293.
188. Wright, A. G. (1982). In *Biomedical Applications of Laser Light Scattering,* D. B. Sattelle, W. I. Lee, and B. R. Ware (eds.), pp. 409–419. Elsevier Biomedical Press, Amsterdam.
189. Kamogowa, K., Fujii, T., and Kitagawa, T. (1988). *Appl. Spectrosc.* **42,** 248.
190. Howard, J., Everall, N. J., Jackson, R. W., and Hutchinson, K. (1986). *J. Phys. E: Sci. Instrum.* **19,** 934.
191. Freeman, J. J., Heaviside, J., Hendra, P. J., Prior, J., and Reid, E. S. (1981). *Appl. Spectrosc.* **35,** 196.
192. Talmi, Y. (1994). In *Charge Transfer Devices in Spectroscopy,* J. V. Sweedler, K. L. Ratzlaff, and M. B. Denton (eds.), pp. 133–196. VCH, New York.
193. Cooper, J. B., Aust, J., Stellman, C., Chike, K., and Myrick, M. L. (1994). *Spectrochim. Acta* **50A,** 567.
194. Chase, B. (1994). *Appl. Spectrosc.* **48,** 14A.
195. Bist, H. D., Durig, J. R., and Sullivan, J. F., eds. (1989). *Raman Spectroscopy: Sixty Years On,* Vol. 17A & B. Elsevier, Amsterdam.
196. Clark, R. J. H., and Hester, R. E. eds. (1989). *Time Resolved Spectroscopy: Advances in Spectroscopy,* Vol. 18. John Wiley & Sons, Chichester.
197. Yu, N.-T., and Li, X.-Y. eds. (1994). *Fourteenth International Conference on Raman Spectroscopy,* Vol. 14. John Wiley & Sons, Chichester.
198. Morresi, A., Mariani, L., Distefano, M. R., and Giorgini, M. G. (1955). *J. Raman Spectrosc.* **26,** 179–216.
199. Johnson, B. R., Kittrell, C., Kelly, P. B., and Kinsey, J. L. (1996). *J. Phys. Chem.* submitted for publication.

# Index